Comparative vertebrate endocrinology

*For Hans Heller and Harry Waring
who introduced me to
comparative endocrinology*

Comparative vertebrate endocrinology

P. J. BENTLEY

Mount Sinai School of Medicine of
The City University of
New York

CAMBRIDGE UNIVERSITY PRESS
CAMBRIDGE
LONDON · NEW YORK · MELBOURNE

5 9 6
B 477c

Published by the Syndics of the Cambridge University Press
The Pitt Building, Trumpington Street, Cambridge CB2 IRP
Bentley House, 200 Euston Road, London NWI 2DB
American Branch: 32 East 57th Street, New York, N.Y. 10022
296 Beaconsfield Parade, Middle Park, Melbourne 3206, Australia

© Cambridge University Press 1976

Library of Congress catalogue card number: 75–10235

hard covers ISBN: 0521 20726 6
paperback ISBN: 0521 09935 8

First published 1976

Composed in Great Britain by
William Clowes & Sons Limited, London, Colchester and Beccles
Printed in the United States of America by Vail-Ballou Press, Inc.,
Binghamton, New York

Contents

[v]

Preface

This book has been written primarily for use as a textbook by undergraduate, as well as graduate, students. It is hoped that it may serve as a basis for course work in Comparative Endocrinology and also as an auxiliary text to aid the teaching of Comparative Animal Physiology. In order to gain most from this book, the reader should have a basic knowledge of zoology and animal physiology. I have, nevertheless, attempted to put the endocrinology that is described into a broader biological framework by relating it to the animal's physiology, ecology and evolutionary background. This is one of the reasons why I have departed from the more usual format of previous textbooks in this area which generally deal with each endocrine gland in succession, chapter by chapter. Instead, I have attempted to describe certain broad and basic biological processes the functioning of which is often coordinated by the secretion from several endocrine glands.

No attempt has been made to describe invertebrate endocrinology as the rapid growth of this area really justifies a separate textbook. The book by K. G. Highnam and L. Hill (*Comparative Endocrinology of the Invertebrates*, Elsevier: Amsterdam, 1970) deals admirably with this subject.

It has not been possible in a book of this nature to give a complete list of original references. There are far too many of these and many of the earlier observations are already a part of the 'classical literature'. Instead, I have attempted to refer the reader to more recent papers and reviews that contain references to the material described and can act as useful 'starting points' for those students who wish to study the subject further. In order to keep abreast of developments in the various subject areas described, the current literature should be consulted. The principal journals where papers on these subjects are published are: *General and Comparative Endocrinology*, *Journal of Endocrinology*, *Endocrinology* and *Comparative Biochemistry and Physiology*. There are, in addition, many papers that appear in the standard physiological journals, especially *Journal of Physiology* and *American Journal of Physiology*.

P.J.B.

Mount Sinai School of Medicine of
The City University of New York
September 1974

Some commonly used abbreviations in endocrinology

ACTH	adrenocorticotrophic hormone
ADH	antidiuretic hormone
AMP	adenosine 3',5'-monophosphate
AVP	arginine-vasopressin
CBG	cortisol-binding globulin
COMT	catechol-*O*-methyl-transferase
CRH	corticotrophin-releasing hormone (= CRF)
CT	calcitonin
FSH	follicle-stimulating hormone
GH	growth hormone
Gn-RH	gonadotrophin-releasing hormone (= LH/FSH-RH)
HCG	human chorionic gonadotrophin
HCS	human chorionic somatomammotrophin (= HPL)
HIOMT	hydroxyindole-*O*-methyl-transferase
HPL	human placental lactogen
HTF	heterothyrotrophic factor
ICSH	interstitial cell-stimulating hormone (= LH)
-IF	– inhibiting-factor
-IH	– inhibiting-hormone
LH	luteinizing hormone (= ICSH)
LTH	luteotrophic hormone (= prolactin)
LVP	lysine-vasopressin
MAO	monoamine oxidase
MI	melanophore index
MSH	melanocyte- (or melanophore-) stimulating hormone
MRH	melanocyte-stimulating hormone-releasing hormone
PNMT	phenylethanolamine-*N*-methyl-transferase
P-	prolactin (prefix as in P-RH)
PTH	parathormone
-RF	– releasing factor
-RH	– releasing hormone
-R-IH	– release-inhibiting hormone
SHBG	sex hormone-binding globulin
T_3	tri-iodothyronine

T_4	tetra-iodothyronine (= thyroxine)
TBG	thyroid hormone-binding globulin
TRH	thyrotrophin-releasing hormone
TSH	thyroid-stimulating hormone (= thyrotrophin)

1. Introduction

This book describes a method of transferring information within vertebrates. Such communication is necessary in order to coordinate physiological processes with each other and to the happenings in the external environment. Even unicellular organisms synchronize their various internal life processes. In such small creatures, however, local accumulations of metabolites may exert a direct control on biochemical reactions, while external stimuli have relatively widespread effects so that specialized pathways for communication may not be as necessary. Thus when the distances involved are short, physical processes such as conduction, convection and diffusion may be adequate for the integration of the physiological processes. Nevertheless, even unicellular organisms possess specific coordinating systems such as that seen in the protozoan *Tetrahymena* (Blum, 1967) which possesses adrenaline. This hormone has similar metabolic actions in this protozoan to those which it has in vertebrates.

The problems of communication and coordination are greater in multicellular than in unicellular organisms. There are several reasons for this, especially their larger size. As the linear distances between the different parts of an animal increase, simple physical communications become relatively slower and less precise, and so not as effective. In multicellular organisms, the cells are usually specialized and perform different functions which, in combination, are essential for the animal's life. Thus some tissues may be concerned with the formation of reproductive germ cells, several others with the preparation of suitable nutritive materials and yet others with building morphological structures. The ultimate successful completion of these processes will be determined by the effectiveness of the communication between the tissues themselves and the external environment.

The transfer of information in animals

There are three principal ways by which cells in multicellular organisms can communicate with each other. Firstly, when they are in close juxtaposition, and are only separated by narrow fluid-filled spaces, direct electrical and chemical interactions can occur. Cells also maintain some structural connections with each other across which they may also communicate. Secondly, contact between more remote cells can be maintained along tracts of nerve cells which are merely tissues that are specialized for such exchanges of information. Thirdly, chemicals may be released, for example from the endocrine glands, into the blood which carries them to special sites that are physicochemically programmed to react and respond to them.

The endocrine glands are tissues that, unlike exocrine glands, have no ducts but release their secretions, called 'hormones', directly into the blood passing through them. It is with the diversity of such hormonally controlled processes that we are going to be principally concerned in this book. It should, however, always be recalled that the endocrine gland represents only a single facet of the animal's communication network, and that nerves are also important. Endocrinologists and neurophysiologists have often only concentrated on their own special fields of study to the relative exclusion of the rest of the animal's physiology. This is unfortunate as the complete animal is an academically and esthetically pleasing thing to see and contemplate, and any single facet taken from the whole becomes less interesting and is physiologically nonsensical. The relations of nerves and endocrines, however, can also be considered from a more direct standpoint as it is apparent that the functions of each are related to each other. They are often mutually interdependent and may even act together to control a single process. Nerve cells thus can respond to hormones in a manner that influences behavior and endocrine glands often receive information and directions from the brain. Both hormones and nerves can act together to control the melanophores in certain fishes. Some hormones, including adrenaline, vasopressin and oxytocin, are even made by nerve cells.

Neural versus humoral coordination

It is uncertain which came first; nerves or hormones. Why do animals have both? It may help us to understand hormones if we compare their respective properties and roles in the body.

The neural transfer of information occurs along distinct morphological pathways made up of chains of nerve cells with their long axons. Transmission along these avenues is fast (up to about 100 m/sec) and is directed precisely to specific sites in the body. Neural transmission involves a series of electrical events interrupted at intervals by a local release of chemicals (transmitters), and is concluded by the release of these, principally acetylcholine or a catecholamine (such as noradrenaline) close to the effector tissue. Such a transmitter is then rapidly destroyed near its site of action. Further stimulation will be dependent on subsequent neural transmission.

The effect is thus rapid in onset, short in duration and can be localized with considerable accuracy.

The hormones, on the other hand, are released into the blood which carries them towards their effector site(s). In most instances this is outside the cardiovascular system so that the hormone must also cross capillaries and diffuse through the intercellular spaces to the site of its action. Not surprisingly, hormonal responses are slower than those mediated by nerves. Hormones are dispersed very widely in the body and so come into contact with a great variety of cells with which an interaction, in most instances, would not be fruitful. The problem of ensuring that hormones only act at specific sites is largely solved by the multiplicity in their chemical structures. (There are over 40 different known hormones in a mammal.) Complementing such variations are parallel differences in the chemical structures of the sites where they interact ('receptors') with their target (or 'effector') cells.

A hormone can exert widespread effects by interacting with different effector tissues (for instance estrogens act on the uterus, mammary glands, liver, brain, etc.). The characteristics of the receptors in each may differ just as the response will vary. A hormone thus may act very specifically at each of several sites in the body and yet, at the same time, exert many different actions.

Perhaps the most physiologically significant difference between neural and humoral communication is in the duration of the actions of the transmitters involved. Because their transmitters are rapidly destroyed, nerves must be repetitively stimulated if their effects are to be prolonged. While hormones also have a finite period of survival, the duration of their effects varies from less than a minute to several days. Some hormones, once released into the circulation, survive in it for many hours. When some reach their receptor sites, the initiated response may be of a persistent nature that is not readily terminated. Thus if an endocrine gland is removed, it may be several days before physiological signs of its absence became apparent. Hormones are thus sometimes described as exerting their effects slowly but persistently, in contrast to the more rapid and transient actions of nerves. There are, however, exceptions to such a generalization.

What is comparative endocrinology?

Comparative endocrinology concerns the study of the endocrine glands in different species of animals, both vertebrates and invertebrates. Its aims are analogous to the older and more classical disciplines of comparative anatomy and comparative physiology. The prime academic objective is to reconstruct evolutionary pathways by the study of extant species. Fig. 1.1 shows the phylogenetic relationships of the vertebrates and this emphasizes

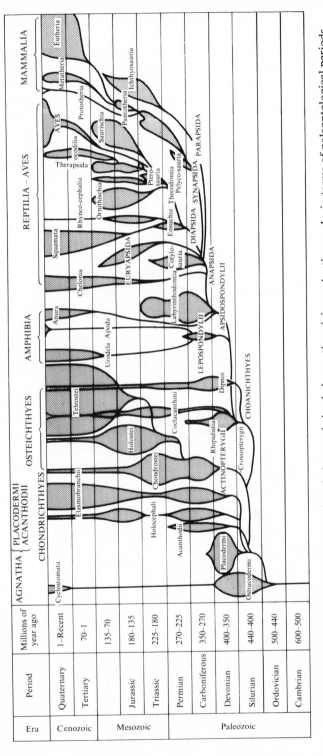

Fig. 1.1. A classification of vertebrates in relation to their phylogenetic origins and a time scale, in terms of paleontological periods. (From Torey, 1971.)

the extant groups which may be particularly interesting in such studies. The mere examination of the endocrine system of some bizarre and exotic vertebrate does not alone constitute 'Comparative Endocrinology' (it may be 'Animal Endocrinology') unless the data can be considered in relation to that in other, phyletically related species. Such information can be used to help confirm, complete and even extend our knowledge of the phylogenetic relationships between vertebrates and to follow the evolution of endocrine mechanisms. The lungfishes (Dipnoi) may afford us an example. These fishes have long been considered, on the basis of morphological information, to be close to the original line of evolution connecting the bony fishes (Osteichthyes) and the Amphibia. As we shall see later, homologous vertebrate hormones often exhibit considerable differences in their chemical structure. Many such differences are apparent between the hormones in fishes and tetrapods. The structure of several hormones present in lungfishes, however, show a greater similarity to those in tetrapods than those in other fishes. For instance, a neurohypophysial hormone called mesotocin is present in amphibians, reptiles and birds but in bony fishes the homologous hormone is isotocin (which differs from mesotocin by a single amino acid substitution) with the exception of the lungfishes, which have mesotocin. It has also been found that the growth hormone and prolactin present in lungfishes are more like those in tetrapods than in other fishes.

Apart from contributing to the over-all phyletic study of vertebrates, the comparative endocrinologist aspires to reconstruct the lines of evolution within the endocrine system itself. This can be done by examining and comparing in different species, the morphology of the endocrine tissues, the structures and activities, both immunological and pharmacological, of their secreted hormones and their different physiological roles.

The uses of comparative endocrinology

The classic, or academic aims, of comparative endocrinology have been described above. The provision of such intellectual satisfaction is not, however, sufficient justification for all! There are, indeed, a number of other contributions that such studies can make to biology, and some examples of these are given below.

The process of reproduction in vertebrates is dependent on the endocrine secretions and an understanding of this relationship can provide information that may be usefully applied when, for esthetic or economic reasons, we may wish to increase, or decrease, the fecundity of a species. This type of study thus constitutes a contribution to the field of 'biological control' (Bern, 1972).

Knowledge of the endocrine system in man has largely been made possible by experiments on other animals. This has principally involved mammals like rats, rabbits, and monkeys but also some more exotic and bizarre creatures. Quantitative measurements of gonadotrophins and melanocyte-stimulating hormone (MSH) were originally made (and sometimes are still) using the responses of the clawed toad (*Xenopus laevis*), while prolactin levels can be measured by its effects on the pigeon's crop-sac or on the behavior of a newt. Oxytocin is assayed by utilizing its ability to decrease the blood pressure of chickens, and the rate of water movement across the toad's urinary bladder can be used to distinguish between two, chemically different, mammalian antidiuretic hormones (ADHs).

The responsiveness of a toad's urinary bladder to ADH and aldosterone is used to study the 'mechanism of action' of these hormones on membrane permeability. Such preparations provide useful 'models' of hormonal effects on the mammalian kidney.

The relationship of the structure of a molecule, to its biological activity, is a field of considerable interest to biologists. The diversity, or polymorphism, in the structure of vertebrate hormones, together with their disparate effects on different tissues and in various species, offers a natural 'laboratory' for such studies. Nature has had a long time and wide opportunities to experiment with the effects of changes in molecular structures on the activities of such excitants. At present, this is most clearly seen among the neurohypophysial hormones of which there are at least nine known chemical variants among the vertebrates. These hormones are peptides containing eight amino acids and often only differ from one another by a substitution at a single chemical locus. They are very reactive molecules and can exert actions at many different sites ranging from the uterus and mammary gland to blood vessels, the kidney, and the amphibian skin and urinary bladder. Analogous effector tissues in different phyletic groups exhibit different abilities to respond to each such hormone, be it a natural one or a variant made in the chemist's laboratory. There are available, and in use, more than 20 different effector-preparations that can be used to study the effects of changes in chemical structure among these hormones on its biological effectiveness. Natural variants of hormones, in which the biological activity has been altered in some way, may be of potential use to man. For instance, calcitonin (a hormone concerned with the regulation of calcium in the body) from the salmon ultimobranchial bodies is far more potent in man than the natural hormone he possesses.

The diversity of vertebrates as a background for endocrine variation

There are some 42 000 extant species of vertebrate animals. The vertebrates originated some 400 million years ago as creatures who apparently lived in the sea or, possibly, in fresh water. They subsequently evolved and occupied almost every conceivable habitat in the oceans, in fresh-water rivers and lakes, and on the land. Their abodes range from the cold polar regions to hot equatorial ones, from deserts to swamps, from high mountains to the ocean deeps. The considerable morphological and physiological diversity of vertebrates mirrors their success in this multitude of environmental conditions. It is thus not surprising to find that the endocrine system exhibits inter-specific differences that reflect adaptations to such different environments. Nevertheless, it is also somewhat unexpected to find that considerable similarities are still apparent in the endocrine systems of species as distantly related as the hagfish (Cyclostomata) and man.

The endocrine glands of vertebrates have special roles to play in the regulation of many types of physiological processes which include reproduction, osmoregulation, intermediary and mineral metabolism, and growth and development (Table 1.1). The nature of the responses to hormones differ considerably but can be classified into several major groups including their actions on membrane permeability, muscular contraction, the transformation of substrates involved in intermediary metabolism and growth, and a controlling (or trophic) action on other endocrine glands (Fig. 1.2).

Many, though not all, of the endocrine glands are essential for life and the reproduction and survival of the species. In other instances, however, their immediate importance for survival is not clear. Animals cannot reproduce if the endocrine function of their gonads is compromised and death soon follows complete destruction of the adrenal cortex. Life may be shortened if the Islets of Langerhans fail to produce sufficient insulin and normal growth, development and maturation of the young will not occur if the secretion of pituitary growth hormone or thyroid hormone is inadequate. On the other hand, antidiuretic hormone from the neuro-hypophysis, is not essential for life though in its absence very large volumes of urine are secreted by the kidney. In man this is an annoying condition as prolonged sleep is not possible and even during the waking hours it can lead to social difficulties but it is not fatal. If drinking water were in limited supply, however, dehydration could be a potential problem and absence of this hormone may then affect survival. It should also be remembered that while too little of a hormone can constitute a problem, too much may also result in physiological difficulties. Hormone imbalances can result from genetic abnormalities, the presence of tumors, and accidental disruption of the events controlling secretion of the hormone. A few examples of such endocrine dysfunction and their effects are summarized in Table 1.2.

Endocrine glands, or tissues, have been identified among all of the vertebrates. Those common to the major groups (from the Cyclostomata to the Mammalia) are the pituitary, thyroid, endocrine pancreas, adrenal

TABLE 1.1. *The secretions of the endocrine glands*

Gland	Hormones	Target tissues
Pituitary		
Adenohypophysis		
Pars distalis	Follicle stimulating hormone, FSH	Ovary and testis
	Luteinizing hormone, LH (also called interstitial-cell stimulating hormone, ICSH)	Ovary and testis
	Thyrotrophic hormone, TSH	Thyroid
	Corticotrophic hormone, ACTH (adrenocorticotrophic hormone)	Adrenocortical tissue
	Growth hormone, GH (somatotrophic hormone)	Liver forms somatomedins which alter tissue metabolism (liver, muscle, adipose tissue)
	Prolactin (luteotrophic hormone, LTH)	Mammary glands, fish gills, tadpole metamorphosis, corpus luteum, kidney, skin, etc.
	Lipotrophin	Adipose tissue
Pars intermedia	Melanocyte stimulating hormone, MSH	Melanocytes, pigmentation and color change
Neurohypophysis		
Pars nervosa	Vasopressin, ADH, vasotocin	Kidney, amphibian skin and urinary bladder
	Oxytocin	Mammary gland, uterus
Hypothalamus	Pituitrophins; FSH/LH-RH, P-IH, MSH-R-IH, CRH, TSH-RH, GH-RH, etc.[1]	Release of hormones by the adenohypophysis

Thyroid gland	Thyroxine (T$_4$) Tri-iodothyronine (T$_3$)	Tissue metabolism and differentiation; calorigenic (mammals), morphogenetic (amphibians)
Parathyroid glands	Parathormone, PTH	Bone, kidney and (?) gut
Ultimobranchial bodies ('C' cells in mammalian thyroid)	Calcitonin, CT (also called thyrocalcitonin)	Bone and (?) kidney
Adrenal glands		
Cortex (interrenals in sharks and rays)	Cortisol, corticosterone, cortisone, 1α-hydroxycorticosterone	Tissue metabolism (liver, muscle), proteins to amino acids, gluconeogenesis Intestine (teleosts)
	Aldosterone	Na and K in kidney, sweat and salivary glands, gut, amphibian skin and bladder, fish gills
Medulla (chromaffin tissue)	Noradrenaline (norepinephrine) Adrenaline (epinephrine)	Tissue metabolism (liver, muscle, adipose tissue), glycogenolysis, mobilization fatty acids, calorigenic, constriction and relaxation of smooth muscle
Islets of Langerhans		
Alpha-cells	Glucagon	Liver (glycogenolysis), adipose tissue (fatty acid release)
Beta-cells	Insulin	Liver, muscle and adipose tissue (amino acids to protein, glucose to fat and glycogen)
Gonads		
Ovary		
Graafian follicle	Estrogens (estradiol)	Female sex organs and characters, mammary glands, brain
Corpus luteum and interstitial tissue	Progestins (progesterone)	Uterus and mammary glands

TABLE 1.1 – *continued*

Gland	Hormones	Target tissues
Testis		
Interstitial tissue (Leydig cells)	Androgens (testosterone)	Male sex organs and characters, sperm maturation, brain
Sertoli cells (?)	Androgens	Sperm maturation
Placenta (pregnant eutherian mammals)	Estrogen (estriol), progesterone	Uterus, mammary glands, fetus
	Chorionic gonadotrophin, HCG	Corpus luteum
	Placental lactogen, HPL (somatomammotrophin, HCS)	Mammary glands
Gut		
Stomach (pyloric mucosa)	Gastrin	Stimulates secretion of gastric juice
Intestine (mucosa)	Enterogastrone	Inhibits secretion of gastric juices
	Secretin	Stimulates secretion of pancreatic juices from exocrine pancreas and hormones from endocrine pancreas
	Cholecystokinin–pancreozymin	Enzyme secretion from pancreas and hormones from endocrine pancreas
	Enteroglucagon	As for glucagon
Kidney		
Tubular cells	1,25-dihydroxycholecalciferol (1,25-$(OH)_2$-vitamin D_3)	Intestine, bone
Juxtaglomerular cells	Renin	Plasma $\alpha2$-globulin-angiotensinogen→ angiotensin (targets: adrenal cortex, vascular smooth muscle)
Putative endocrine glands		
Pineal gland	Melatonin	Hypothalamus (inhibits release MSH and gonadotrophins), melanocytes (larval anurans, cyclostomes ?)

Corpuscles of Stannius (some bony fishes)	'Hypocalcin'	Calcium metabolism, osmoregulation
Urophysis (some fishes)		Osmoregulation, smooth muscle contractions (?)
Thymus	Thymic hormone(s)	Immunological maturation via induction of immunological competence of lymphocytes, etc.

[1] For explanation see list of abbreviations.

TABLE 1.2 *Some effects of endocrine dysfunction in man*

Gland	Secretory activity	Abnormality	Principal effects
Pituitary			
Adenohypophysis (growth hormone)	←	Giantism, acromegaly	Excessive growth
	→	Dwarfism	Retarded growth
Neurohypophysis	→	Diabetes insipidus ADH	Excessive loss of water in urine
	←	Schwartz–Bartter syndrome ADH	Low plasma Na
Thyroid gland	←	Graves disease	High metabolic rate and nerve cell activity
	→	Myxedema	Low metabolic rate
	→	Cretinism	Inadequate development and growth
Parathyroids	←	Hyperparathyroidism	Hypercalcemia, polyuria, reduced bone calcium
	→	Hypoparathyroidism	Muscle tetany
Islets of Langerhans			
Beta-cells	→	Diabetes mellitus	Hypoglycemia, muscle wasting
Adrenal cortex	→	Addisons disease	Renal Na loss and K retention (low plasma Na, high plasma K), low blood pressure, muscle weakness
	←	Cushings syndrome	High blood pressure, obesity, retarded growth
Adrenal medulla	←	Phaeochromocytoma	Hyperglycemia, high blood pressure
Ovaries	→		Sterility, failure to develop or maintain secondary sex characters
Testis	→		As above

← Increased activity.
→ Decreased activity.

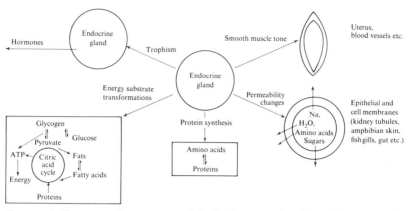

Fig. 1.2. A diagrammatic summary of the basic types of actions of hormones in vertebrates.

chromaffin and cortical tissues, and gonads. The parathyroid glands have been found in the tetrapods but not in the fishes. The ultimobranchial bodies or their homologues, the thyroid 'C' cells, have been identified in all groups except the cyclostomes. Such tissues secrete more than 40 different hormones in a mammal. If we include all the naturally occurring analogues of these hormones that occur among the vertebrates, we can account for at least twice this number of hormones and there are undoubtedly many more.

Apart from these glands, other tissues that may have an endocrine function (putative endocrine glands) have also been identified in a number of vertebrates. A tissue is considered to have an endocrine function if it releases a product into the circulation that has an excitatory or inhibitory effect on some distant effector gland organ or tissue. The precise status of the pineal body as an endocrine gland is still equivocal though many consider that it has such a function especially as one of its products, melatonin, has been identified in the blood. Two other tissues, that are present in some fishes, the urophysis and corpuscles of Stannius also have a putative endocrine status.

It is conceivable, indeed likely, that other endocrine glands exist among the vertebrates. Even within the last 10 years a new endocrine tissue, the ultimobranchial bodies (the thyroid 'C' cells in mammals), has been identified. This tissue is concerned with the regulation of calcium levels in some vertebrates. Despite the busy, even frantic, activity of endocrinologists the hormonal role of this gland had previously not been confirmed.

Despite their anatomical and embryological homologies the endocrine glands of vertebrates display considerable diversity in their morphological arrangements, the chemical nature of their secretions and even their physiological role in the body. It is principally about these differences that we will be concerned in the succeeding chapters.

Conclusions

Physiological processes are coordinated with the aid of both nerves and hormones. Each of these mechanisms has special characteristics that may be suited to the needs of the particular process involved and they often both operate together. During the course of geological time vertebrates have evolved and acquired morphological features and physiological processes that have permitted them to adapt to changing environments and to occupy a variety of ecological habitats. Such biologically important changes are accompanied by the neural and humoral processes necessary for their coordination. Contemporary species of vertebrates exhibit considerable structural and functional diversity that can be related to the nature of the life that they lead and to their ancestry. They are classified into systematic groups that are also thought to reflect their evolution. Thus a comparison of the endocrine function of contemporary species of vertebrates is not only of importance in fully understanding how they live today, in a particular environmental situation, but also may tell us how such hormonally mediated processes evolved.

2. Comparative morphology of the endocrine tissues

Endocrine glands and tissues display a diversity in their gross morphological and histological patterns. This is particularly apparent when comparing species from phyletically distant groups. In some instances the physiological significance of these differences has been recognized but in most this is not so and may be related to the initial pattern of embryonic growth. If, however, one intuitively suspects a close relationship between structure and function, then the lack of a known correlation may merely reflect our ignorance.

The endocrines may display several different types of morphological variation. Their position in the body may not be the same. This variation can be of a minor nature, such as is seen with the ultimobranchial bodies which can be situated near the heart or the thyroid gland. In some fish, however, thyroid tissue may vary in position from the branchial region to the kidney. Endocrine cells may show varying degrees of association and be scattered in small segments, or 'islets', or be closely associated as a compact gland enclosed in a capsule. Such aggregation of an endocrine tissue is commonly seen as one ascends the evolutionary (or the phyletic) scale. In addition, different endocrine tissues may display diverse associations with each other, as for instance the conglomeration of chromaffin and interrenal (or adrenocortical) tissue in the adrenal gland. Their relationship to the neural and vascular tissues can be very important. Pituitary tissues thus cannot function properly if they are transplanted to other parts of the body (ectopic transplant) or if the small blood vessels between the gland and the brain are cut. The major blood vessels not only carry hormones away from endocrine tissues but also supply them with nutrients and controlling stimuli. The pattern of the vasculature within the gland can also be important for its correct functioning.

The types of cells that make up an endocrine gland are, not surprisingly, similar in homologous glands among the vertebrates. Such similarities as reflected by their microscopic anatomy (size, shape, the presence of inclusions, granules, etc.) and their reactions with dyes (tinctorial relationships) serve to aid in their identification. More recently, antibodies to specific hormones have been used to identify the cells where they are formed. These antibodies may be labelled with radioactive materials or

[15]

fluorescent dyes so that the precise locus where they react can be seen. The histological appearance of endocrine cells may change somewhat at different times depending on their secretory state. This characteristic can be used to predict their activity and physiological role. Inactive thyroid cells thus have a flattened, instead of columnar appearance which is typical of their active state, while neurohypophysial tissue that is depleted of its hormone has little stainable (with Gomori chrome–alum hematoxylin) neurosecretory material.

The pituitary gland (see Fig. 2.1)

The pituitary is a conglomerate of tissues and cells that reflect the ten major hormones it secretes. These hormones help regulate the activities of the thyroid, adrenal cortex and gonads, and contribute to the control of various other physiological activities including water and salt metabolism, growth, lactation, parturition and the pigmentation of the skin. A comparative account of the anatomy of this gland has recently been provided by Holmes and Ball (1974). Embryologically, the pituitary arises as a result of a downgrowth of tissue (the infundibulum) from the brain and an upgrowth (the hypophysis) from the roof of the mouth. Enclosed within these tissues is a piece of mesoderm that forms a net of blood vessels sometimes called the 'mantle plexus'. The pituitary lies in close apposition to the hypothalamus at the base of the brain. In mammals it is usually enclosed in a small, bony chamber, the sella turcica, from which it is connected by a stalk of nervous tissue to the brain, just behind the optic chiasma. The hypophysis partly differentiates into the adenohypophysis that secretes seven or eight hormones which, so the histologists tell us, are formed by a similar number of distinctive types of cells. These are most descriptively labelled by the name of the hormone they secrete followed by the suffix *troph*. We thus have thyrotrophs, gonadotrophs, somatotrophs and so on. An alternative terminology utilizes the Greek alphabet: α-cells = somatotrophs, β-cells = gonadotrophs and so on. The adenohypophysis can be divided on a gross morphological basis into three or four sections: the pars tuberalis, the pars distalis (sometimes with a rostral and caudal section) and the pars intermedia. The latter gives rise to the melanocyte-stimulating hormone (MSH) while the rest of the hormones come from the pars distalis. The pars tuberalis lies between the pars distalis and the brain (in the region of the median eminence) and is associated with the blood vessels that connect the two.

The neural, or infundibular, tissue forms the neurohypophysis which basically lies caudally to the adenohypophysis; hence the terms anterior and posterior lobes of the pituitary. The neurohypophysis is connected to

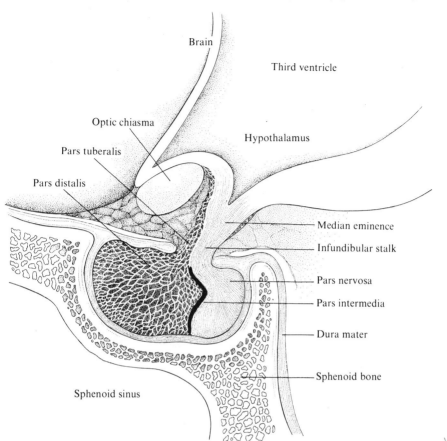

Fig. 2.1. The pituitary gland of man. It lies in a bony chamber at the base of the brain to which it is connected by a stalk. The pars intermedia is quite small in man but may be much larger in other species. (From R. Guillemin and R. Burgus, *The Hormones of the Hypothalamus.* Copyright © 1972 by Scientific American, Inc. All rights reserved.)

the brain by the infundibular stalk. The two hormones (ADH and oxytocin in mammals) it secretes are formed in nerve cells (by a process called neuro-secretion) which originate in the supraoptic and paraventricular nuclei in the brain of amniotes or the preoptic nucleus of amphibians and fishes. The axonal tract running from the bodies of these nerve cells, in the nuclei, to the periphery, where they are stored and released, is called the supra-opticohypophysial tract. The neurohypophysis thus consists of the distal parts of nerve cells interspersed with glial cells and pituicytes. The function of the latter is unknown.

'Primitive type' as in some reptiles

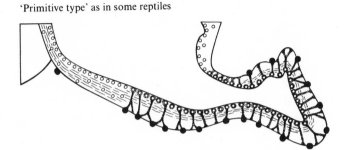

Birds and reptiles

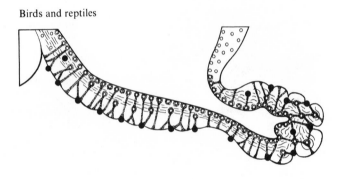

Mammals

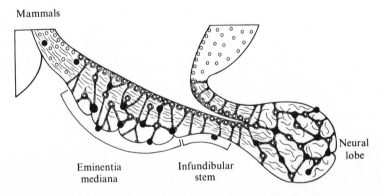

Eminentia
mediana

Infundibular
stem

Neural
lobe

Fig. 2.2. Histological differentiation of the amniote neurohypophysis.
The 'primitive form' is seen in reptiles such as the Rhynchocephalia, Chelonia
and some Lacertilia. Solid black lines are the blood vessels, nerve fibers are
thinner lines. (From Wingstrand, 1951.)

Three regions of the neurohypophysis can be distinguished. The rostral *median eminence* is part of the wall of the hypothalamus and lies in close conjunction with the adenohypophysis to which it is usually connected by a system of portal blood vessels which originate from the mantle plexus. The median eminence is contiguous with the *infundibular stalk* which connects it to the most prominent part of the neurohypophysis, the *pars nervosa* (or *neural lobe*). The latter is much more highly developed in terrestrial tetrapods than in the fishes. The phyletic development of the amniote neurohypophysis is shown in Fig. 2.2.

Comparative morphology of the pituitary

The diverse morphology of the vertebrate pituitary provides us with some information (albeit equivocal) about the nature of the evolutionary changes that may have taken place in this gland. Attempts have been made to choose or construct, the pituitary that is considered most typical of each major phyletic group. Considerable differences from a 'median gland' may nevertheless exist among various species within each systematic group.

The structure of the pituitaries of fishes, from the Cyclostomata to the Dipnoi are shown in Fig. 2.3. The cyclostomes have a simple type of pituitary in which the different regions are only loosely associated with each other. The parts of the adenohypophysis in these phyletic prototypes are often termed the pro-, meso- and meta-adenohypophysis. They are thought to correspond respectively to the cephalic part of the pars distalis (or possibly the pars tuberalis), the caudal pars distalis and the pars intermedia of other vertebrates. The close proximity of the adenohypophysis to the brain may not be functionally essential in these lowly fishes. Considerable intraspecific variation occurs and ectopic transplants of the adenohypophysis to others parts of the body do not appear to compromise its function, at least in hagfishes (Myxinoidea) (Fernholm, 1972).

The actinopterygian fishes possess a pituitary in which there is a close association between the various component tissues. The homologies of these tissues to those in tetrapods have on occasion been difficult to recognize but they undoubtedly exist. The neurohypophysis is not a very discrete tissue in fishes (there is no distinct neural lobe) and shows considerable admixture with the pars intermedia into which it sends finger-like projections and shares a common blood supply (see Fig. 2.3). Portal blood vessels connecting the median eminence and adenohypophysis have been described in all groups of actinopterygians except the teleosts. Considerable variation has been observed among the latter in which the blood supply to the adenohypophysis passes initially through the neurohypophysis. No clear portal system, as seen in other actinopterygians, is apparent in

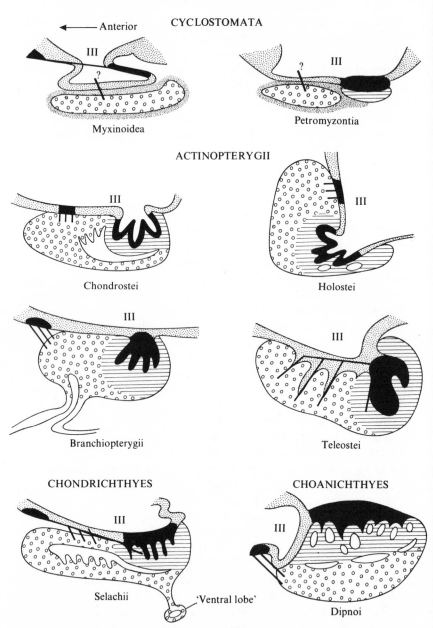

Fig. 2.3. The pituitary glands of fishes. Diagrammatic representation from midsagittal section.

Small dots = nervous tissue; black = neurohypophysial tissue; large open dots = pars distalis; horizontal lines = pars intermedia.

teleosts. At least five distinct types of cells have been identified in the fish pituitary as shown in Fig. 2.4 which is that of a teleost, the eel *Anguilla anguilla*. These cells are present in separate zones in contrast to the tetrapods and lungfishes (Dipnoi) where they are intermingled with each other.

The chondrichthyean fish (sharks and rays) have a pituitary which on superficial examination looks rather different from that of other fishes. It displays, however, a similar basic structure, though gross differences, such as a rather large pars intermedia, are often apparent. A characteristic and distinct lobe of the adenohypophysis lies below the pars distalis, this is called the '*ventral lobe*', which has been shown to be the site of formation of a gonadotrophin. Like most actinopterygians, chondrichthyeans have a distinct portal blood system connecting the hypothalamus and adenohypophysis but this special blood supply does not extend to the ventral lobe.

The pituitary of lungfishes shows more similarities to that of tetrapods than to those of other fishes. This is especially interesting in view of the special phyletic relationship that is usually considered to exist between

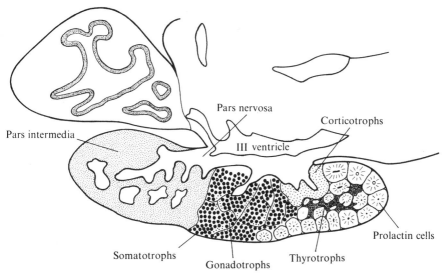

Fig. 2.4. Diagram from midsagittal section of the pituitary of a teleost fish (the eel *Anguilla anguilla*).

The cells (trophs) that produce the hormones in the pars distalis can be seen to lie generally in distinct zones. (From Olivereau, 1967.)

Thick black lines = blood vessels which carry neurosecretory products to the adenohypophysis, or in Myxinoidea to the neurohypophysis.

III = third ventricle. (From Ball and Baker, 1969, modified slightly according to Holmes and Ball, 1974.)

lungfishes and tetrapods. The gross similarities between the amphibian and lungfish pituitaries can be seen in Fig. 2.5. The different types of cells in the adenohypophysis are intermingled in lungfishes (not separated as in other fishes), just as in the tetrapods. The neurohypophysis of the lungfishes also displays the beginnings of the differentiation of a neural lobe.

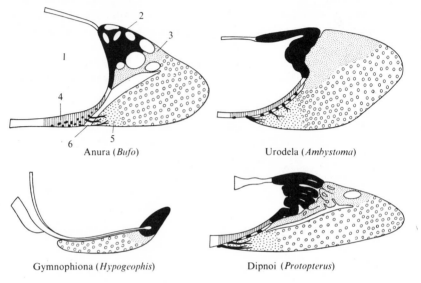

Anura (*Bufo*)   Urodela (*Ambystoma*)

Gymnophiona (*Hypogeophis*)   Dipnoi (*Protopterus*)

Fig. 2.5. The pituitaries of the three main groups of the Amphibia compared to that of a dipnoan (lungfish).

1 = Saccus infundibuli; 2 = neural lobe; 3 = pars intermedia; 4 = median eminence; 5 = 'zona tuberalis'; 6 = portal blood vessels. (From Wingstrand, 1966. Originally published by the University of California Press; reprinted by permission of The Regents of the University of California.)

The neural lobe, which is a characteristic of the tetrapods, is formed by the enlargement, in a posterior direction, of the neurohypophysis. Wingstrand (1966) has suggested that this change may be related to a terrestrial manner of life in which the secreted hormones had a special significance. This theory is consistent with measurements showing a much greater amount of stored hormonal material in the neurohypophysis of tetrapods than in that of fishes (Follett, 1963).

The basic morphological pattern of the tetrapod pituitary is well exemplified in the reptiles (Fig. 2.6). It can, however, be seen that even here differences between the major systematic groups occur. Such variations usually reflect the relative degree of development of the neurohypophysis and the presence, reduction or absence of the pars tuberalis. The reptilian adenohypophysis has distinct cephalic and caudal zones.

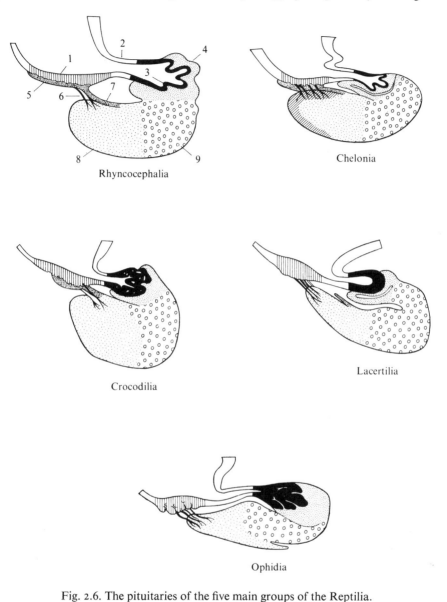

Fig. 2.6. The pituitaries of the five main groups of the Reptilia.
1 = Median eminence; 2 = infundibular stem; 3 = neural lobe (pars nervosa);
4 = pars intermedia; 5 = pars tuberalis; 6 = portal blood vessels; 7 = pars
tuberalis interna; 8 = cephalic lobe of pars distalis; 9 = caudal lobe of pars
distalis. (From Wingstrand, 1966. Originally published by the University of
California Press; reprinted by permission of The Regents of the University of
California.)

The pituitaries from more than 100 species of birds have been examined and, as in the reptiles, the pars distalis has two distinct regions. It is interesting that a pars intermedia has not been identified among the birds. The hormone typically secreted by this tissue, MSH, has nevertheless been identified in the pituitary of the domestic chicken (Shapiro *et al.*, 1972). This absence of a pars intermedia is not unique to birds; it is not present in elephants or whales either.

Among the mammals, considerable morphological differences exist in the intimate arrangement of the tissues within the pituitary (see Hanstrom, 1966). The detailed embryonic development of the pars distalis differs from that of other amniotes. The mammalian pars distalis arises mainly from the aboral lobe of Rathkes pouch, which in birds and reptiles forms the caudal section of the pars distalis. The oral lobe, which forms the cephalic part of the pars distalis in the last two groups, fails to develop in mammals (Wingstrand, 1966). In addition to being absent in whales and elephants, the pars intermedia is very much reduced in adult primates, including man. The simplest type of pituitary is seen in the echidna, *Tachyglossus aculeatus* (an egg-laying monotreme) and some rodents and insectivores. It is interesting that the echidna shows a pattern that is considered to be like a 'primitive' mammal. The echidna's pituitary is, however, typically mammalian in its embryonic origins. It has nevertheless some features, including a prominent porto-tuberal tract between the median eminence and pars distalis, which are seen more often in birds and some reptiles.

The endocrine glands of the pharynx: thyroid, parathyroids and ultimobranchial bodies

Apart from the adenohypophysis, which has its origins in the roof of the mouth, three (or four if one includes the thymus) other endocrine glands arise from the pharyngeal tissues: the thyroid gland from the floor of the pharynx, the parathyroids from the II, III and IV gill pouches and the ultimobranchial bodies from the last, VI, pair of these.

The thyroid

Thyroid tissue is present in all vertebrates though its gross morphological arrangement varies somewhat. Its hormones have ubiquitous effects on tissue metabolism, differentiation and maturation. The 'thyroid-unit' is a follicle in which a group of epithelial cells surround a central cavity which is filled with a glycoprotein secretion called thyroglobulin (Fig. 2.7). The encompassing cells have a columnar appearance when they are most active and a flattened one when they are least active. Thyroid follicles have a

(a)

(b)

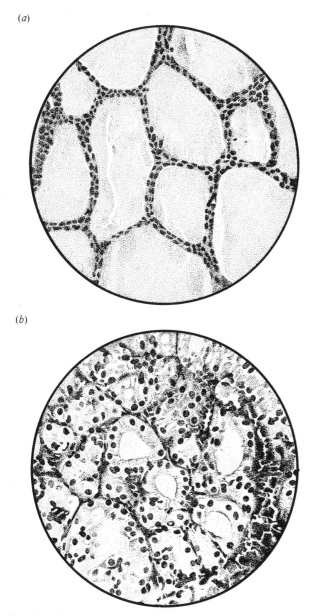

Fig. 2.7. The thyroid gland of the laboratory rat showing the follicles surrounded by epithelial cells.

(a) The inactive condition where the cells are flattened and the follicles are distended with 'colloid' which contains the thyroglobulin. (b) The active condition where the epithelial cells are columnar and little colloid is present.

remarkable ability to trap inorganic iodide which can be stored and transposed into hormones which are, in turn, stored in the follicle cavity. It is probably the only endocrine gland that stores its products outside of the cells.

In man, the thyroid gland is situated in the region of the neck and it has a generally comparable position in other vertebrates. In cyclostomes and most teleost fishes the thyroid follicles lie scattered along the blood vessels under the pharynx. Occasionally they may be found further afield, even in the kidneys. In chondrichthyean fish (sharks and rays) and some teleosts, like the Bermuda parrot fish and tuna, the follicles are aggregated into a distinct glandular mass. This pattern persists in higher vertebrates; there are two such aggregates in amphibians, birds and many reptiles. In lizards these two lobes are joined, a situation that is also usually characteristic of mammals.

TABLE 2.1. *The thyroid in the phylum chordata.* (From Rall, Robbins and Lewallen, 1964)

Subphylum	Class	Species[1]	Thyroid gland	Thyroid-like[2] activity
Hemichordata[3]		*Glossobalanus minutus*	−	−
Protochordata				
Urochordata	Ascideacea	*Ciona intestinalis*	−	+
(Tunicate)	(sea-squirt)	*Clavelina lepadiformis*	−	+
	Larvacea		−	+
	Thaliacea	*Salpa maxima*	−	+
Cephalochordata	Amphioxi	*Branchiostoma lanceolatum* (amphioxus)	−	+
Vertebrata	Agnatha	(Lamprey)		
	(cyclostoma)	ammocoete larva	−	+
		adult	+	+
		(Hagfish)	+	+
	Chondrichthyes	(Shark)	+	+
	(elasmobranch)			
		(Skate)	+	+
	Osteichthyes		+	+
	(teleost)		+	+
	Amphibia		+	+
	Reptilia		+	+
	Aves		+	+
	Mammalia		+	+

[1] Not a complete list. Common names are given in parentheses.
[2] That is, synthesis of iodothyronines.
[3] Not usually classified as Chordata at present time, but included for reference in the light of earlier discussions.

The thyroid appears to have the longest phylogenetic history of any endocrine gland (see Table 2.1). It is not only present in vertebrates; tissues that may be homologous, though not having the characteristic follicular units, have been identified in protochordates, including *Amphioxus* (Cephalochordata) and various ascidians (sea-squirts, Urochordata). The development of the thyroid in lampreys can be followed during the metamorphosis of its ammocoete larva. This beast collects small particles of food by filtering water that passes, with the help of ciliary action, through its pharynx. This process is aided by a ventral outgrowth from the floor of the mouth called the endostyle or subpharyngeal gland. An analogous tissue also exists in *Amphioxus* and ascidians. It secretes a sticky mucus that traps the food particles before they can pass out across the gills. This action has been likened to that of a 'moving flypaper'. Embryologically, the endostyle of the lamprey ammocoete larva differentiates to form the adult thyroid. This has given rise to speculation as to whether or not the endostyle in the ammocoete and in protochordates has some thyroid function.

The endostyle does not contain thyroid-like follicles. It has, however, been shown (see Barrington, 1962), like the thyroid, to be able to accumulate selectively and concentrate radioactive iodide. This has not only been demonstrated in the lamprey ammocoete but also in *Amphioxus* and several ascidians. The iodine is bound in organic form with tyrosine and organo-iodine compounds (*iodothyronines*), including, possibly, small amounts of thyroxine (for a summary see Table 2.1).

Iodine readily reacts with proteins containing the amino acid tyrosine. Indeed, in one extensive investigation when cows were being fed experimental diets containing thyroid compounds to improve their milk yields, these were made by incubating proteins, such as casein, with iodine at an appropriate pH and temperature. Thus, the spontaneous formation of organo-iodine compounds in nature would not be surprising. Indeed among the ascidians, the outer tunic or coat contains scleroproteins that combine with iodine; possibly even more readily than the tissues associated with the endostyle. Iodinated tyrosines have also been isolated in many other non-vertebrates, including coelenterates. Barrington (1962) has conjectured about the possibility that the spontaneous occurrence and availability of such compounds in nature may have led to their use as hormones. Subsequently their formation may have become more localized in special tissues.

Parathyroid glands and ultimobranchial bodies

These glands secrete hormones that contribute to the control of calcium in the body fluids. Embryologically, there are initially three pairs of parathyroids in tetrapods but those from the II pair of branchial pouches usually disappear; two or sometimes only one pair persist. When one pair persist they are usually derived from the III gill pouch. Two pairs of parathyroids are usually present in amphibians. Among reptiles this number varies. One pair is present in the Crocodilia, two pairs in the Chelonia and Ophidia, and one to three pairs in the Lacertilia. Birds and mammals have one or two pairs. The number found does not seem to follow any phyletic pattern.

A pair of ultimobranchial bodies is present in all the vertebrates from the birds to the chondrichthyean fish. They are apparently absent in cyclostomes. In mammals this tissue is embedded in the thyroid gland, where it makes up the parafollicular or 'C' cells.

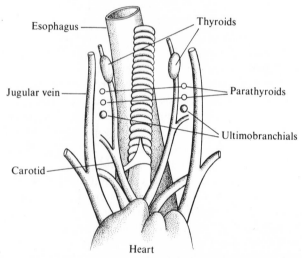

Fig. 2.8. The position of the thyroid gland, the parathyroids and the ultimo-branchial bodies in the domestic fowl, *Gallus domesticus*. (From Copp, Cockcroft and Keuk, 1967*b*. Reproduced by permission of the National Research Council of Canada.)

The association of the thyroid, parathyroids and ultimobranchial tissues may be somewhat complex. The last two tissues are histologically different from the thyroid and they lack the typical follicular structure of that gland. Their positions in the domestic fowl are shown in Fig. 2.8. The morphological distribution of these three glands has often made it difficult to dissociate the effects of the latter two, which both elaborate secretions having opposite effects on blood calcium concentrations. The parathyroids, especially in mammals, are usually closely associated with the thyroid though not to the same extent as the ultimobranchial tissues. Removal of the mammalian thyroid, including that in man, is often associated with low blood calcium levels and an associated muscle tetany. This is the result of removal of or damage to the parathyroid gland; an observation that furnished an important clue as to its possible endocrine significance. The concomitant absence of the 'C' cells was not initially apparent and only became so after examination of the effects of thyroid extracts on plasma calcium levels.

In order to elucidate the respective roles of the 'C' cells and the parathyroids in mammals, morphological variations between species have been usefully exploited. In dogs, which are common experimental animals, there are two pairs of parathyroids, one embedded deeply in the thyroid. Rats, however, only have a single pair of parathyroids which are at the surface of the thyroid and so can be destroyed with a cautery. In neither species is it possible to isolate the blood supply of the parathyroids from that of the thyroid, which contains the 'C' cells, so that many crucial endocrine experiments cannot be performed on them. Sheep and goats have two pairs of parathyroids and one of these is situated near the thymus, where it has a separate blood supply from that of the thyroid. These animals have played an important role in elucidating the respective roles of the endocrines in calcium metabolism (Hirsch and Munson, 1969). In addition, the pig has also proved to be useful and in this species the thyroid has no attached parathyroid tissue so that one can deal with the 'C' cells in relative isolation from the former.

In non-mammals the ultimobranchial bodies are usually separated from the thyroid. Nevertheless in birds, for example the domestic fowl, they may contain parathyroid tissue (Copp, 1972). In pigeons, calcitonin is not only found in the ultimobranchials but also in the parathyroids and thyroid.

The admixture of these three distinct endocrines may have some fundamental significance but this is unknown. It has been suggested that the differentiation of the 'C' cells is aided by their association with the thyroid tissue. It is also possible that, as has been observed in the adrenals (see later), some functional symbiosis may occur. Much of the variation, however, would appear to be the result of embryological complications. Although this has certainly helped to hide their effective roles from the endocrinologist, it has nevertheless provided him with some fascinating intellectual exercises during which he has utilized for his own purposes much of the interspecific variation that initially served to deceive him.

The adrenals

The adrenal glands are so named from their position adjacent to the kidneys. In mammals, they are a composite gland made up of two distinct tissues arranged in two zones, an outer cortex surrounding an inner medulla. The cortex is mesodermal in its origins and secretes several steroid hormones involved in the regulation of intermediary and mineral metabolism. It is called the interrenal or adrenocortical tissue. The medulla, on the other hand, is neural tissue homologous to that of the sympathetic ganglia.

(a)   (b)

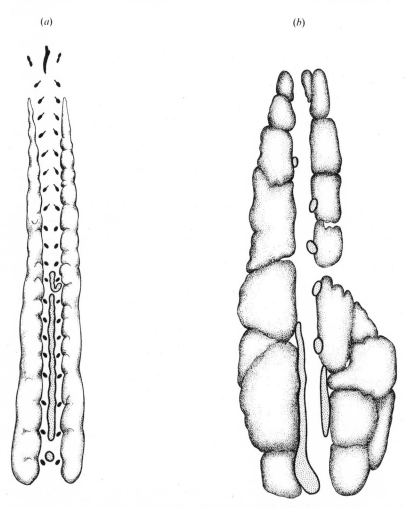

Fig. 2.9. Adrenal tissues in the Chondrichthyes.
(a) The smooth dogfish (*Mustelus canis*). Double row of black dots = chromaffin tissue lying between the two kidneys. Interrenal (adrenocortical) tissue = stippled. This lies in several pieces between the kidneys. (b) The skate (*Raja laevis*). The broken U-shaped interrenal is shown lying between the kidneys. (From Hartman and Brownell, 1949.)

Because it stains dark brown with chromic acid it is called chromaffin tissue. It secretes the catecholamine hormones adrenaline and noradrenaline (also called epinephrine and norepinephrine). These have several roles, including the mobilization of fats and carbohydrates as well as influencing the tone of many blood vessels. While these two endocrine tissues are closely associated in mammals, they are separated in many fishes. In the cyclostome fishes the chromaffin tissues are widely dispersed in small islets along certain blood vessels. Putative adrenocortical tissues have been identified embedded in the posterior cardinal veins. This tissue has, however, not been conclusively shown to secrete steroid hormones (Weisbart and Idler, 1970). In teleost fishes adrenocortical tissues lie along the posterior cardinal veins in the anterior part of the kidney (the head-kidney) where they may or may not be associated with the chromaffin tissue. In sharks and rays (Chondrichthyes) the adrenocortical tissue forms a more compact glandular mass lying between the kidneys; hence the name interrenals. In the dogfish (Fig. 2.9*a*) islets of chromaffin tissue lie along the inner borders of the kidneys while the interrenal forms a fairly complete mass between them. The adrenocortical tissue of the skate, on the other hand, forms several lobules (Fig. 2.9*b*).

In the Amphibia, chromaffin and adrenocortical tissues are usually associated with each other, lying in islets on the ventral surface of the kidney (Fig. 2.10). Considerable differences can be seen between various species. In urodeles, they are in scattered groups; in *Siren* (Fig. 2.10*a*) they lie in rows between the kidney, while in *Necturus* and *Amphiuma* (Fig. 2.10*b, c*) they are on its surface. Anurans, like the leopard frog (*Rana pipiens*), have contiguous strips of adrenal tissue (Fig. 2.10*d*). It is interesting that in the African lungfish (*Protopterus*), adrenocortical tissues lie in islets along the post-cardinal veins and on the ventral surface of the kidney, a pattern similar to that seen in urodeles (Janssens *et al.*, 1965).

In anamniotes, the adrenocortical tissues and the mesonephric kidney have a common embryological origin so that their close association is not unexpected. Amniotes, however, have a metanephric kidney (the mesonephros is not seen in adults) so that the kidneys and adrenals are less intimately connected. The adrenals, more predictably, form separate compact masses of tissue lying near the kidneys. Considerable variations nevertheless still exist as seen among the different major groups of the reptiles (Fig. 2.11). The chromaffin tissues of reptiles are more closely intermingled with the adrenocortical tissues than they are in amphibians. This admixture of the two tissues is even more apparent in birds (Fig. 2.12) where the adrenals may be fused to form a single gland. Mammals have paired adrenal glands (Fig. 2.13) with a distinct cortex and medulla. It is interesting that this is not as well defined in the echidna *Tachyglossus*

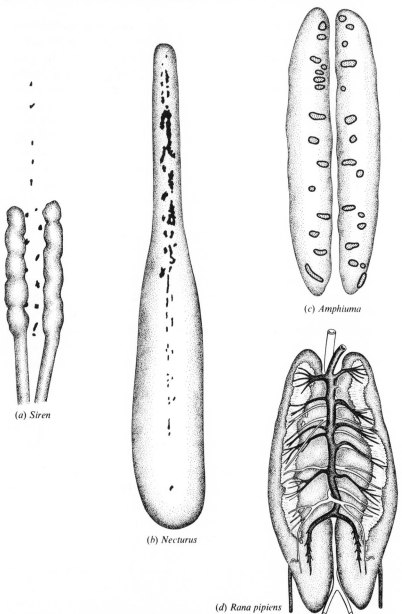

(a) Siren

(b) Necturus

(c) Amphiuma

(d) Rana pipiens

Fig. 2.10. Adrenal tissues in the Amphibia.
Urodela: (a) *Siren*; (b) *Necturus*; (c) *Amphiuma*. The adrenal tissue is shown as the dark area lying on the ventral surface or between the kidneys.
Anura: (d) *Rana pipiens*. The adrenal tissues lie in two strips (light color) along the outer ventral border of each kidney.
(From Hartman and Brownell, 1949.)

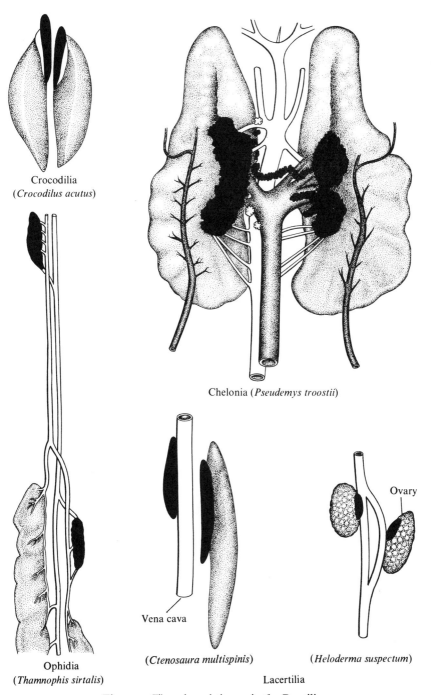

Crocodilia
(*Crocodilus acutus*)

Chelonia (*Pseudemys troostii*)

Ophidia
(*Thamnophis sirtalis*)

Vena cava

(*Ctenosaura multispinis*)

Lacertilia

Ovary

(*Heloderma suspectum*)

Fig. 2.11. The adrenal tissues in the Reptilia.
The adrenals are shown in black in relationship to the kidney(s) (shaded).
(From Hartman and Brownell, 1949.)

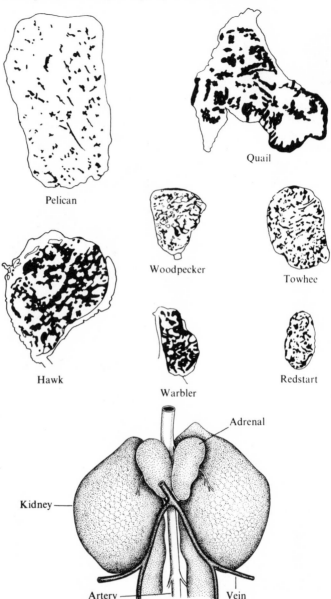

Fig. 2.12. The adrenals in birds.
Top. Cross-section of the adrenal glands from various species showing the distribution of the chromaffin tissue (black) and the adrenocortical tissue (white).
Bottom. The adrenals of the herring gull (*Larus argentatus*).
(From Hartman and Brownell, 1949.)

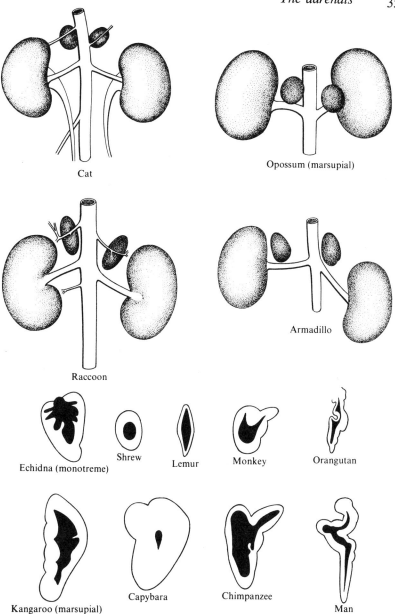

Fig. 2.13. The adrenals in mammals.

Top. The position of the adrenals in relation to the kidneys in a variety of species.

Bottom. Cross-section of the adrenals from various species showing the chromaffin tissue (black) and adrenocortical tissue (white).

(From Hartman and Brownell, 1949.)

aculeatus (an egg-laying monotreme), whose adrenals are considered to be similar to those of reptiles (Wright, Chester Jones and Phillips, 1957).

The relative amounts of adrenocortical and chromaffin tissues vary in different species. As shown in Table 2.2, there are similar amounts of both

TABLE 2.2. *Weights of the adrenal medulla and cortex in various species.* (From Hartman and Brownell, 1949)

Animal	Weight	Medullae (g)	Cortices (g)	Proportion
Fowl	2000	0.1	0.1	1:1
Dog	15000	0.25	1.25	1:5
Cat	3000	0.02	0.35	1:17.5
Rat	200	0.002	0.04	1:20
Rabbit	3000	0.01	0.4	1:40
Guinea-pig	500	0.008	0.5	1:62.5

in the domestic fowl; in the dog, adrenocortical tissue is five times more predominant, while in the guinea-pig there is more than 60 times as much adrenocortical as chromaffin tissue. Adrenocortical tissues may also show considerable variability depending on the season, diet and physiological condition of the animal. It is well known that in many reptiles and amphibians the adrenocortical tissue regresses in winter and proliferates in summer. This can also be related to breeding and has been observed in teleost fish (see for instance Robertson and Wexler, 1959; Chan and Phillips, 1971; Lofts, Phillips and Tam, 1971). Birds from marine habitats, where a lot of salt is present in the diet, have larger adrenals than those species where fresh water is freely available. Glaucous-winged gulls (*Larus glaucescens*) reared with only salt solutions to drink have much larger adrenals than those given fresh water (Holmes, Butler and Phillips, 1961).

The mammalian adrenal cortex is histologically composed of three types of cells situated in three layers or zones. These are the outer *zona glomerulosa* (round cells, rich in mitochondria and poor in lipids), an intermediate *zona fasciculata* (columnar cells, rich in lipids) and a smaller inner *zona reticularis* (flattened cells poor in lipids) (see Fig. 2.14). The zona glomerulosa is not apparent in all mammals; such as some mice, lemurs and monkeys. It appears that the three zones are each principally (though possibly not exclusively) involved in the formation of different hormones; the zona glomerulosa forms aldosterone, the zona fasciculata cortisol and corticosterone, while the zona reticularis can secrete androgenic steroids. This last zone hypertrophies in certain conditions associated with an excess production of these hormones in man. Aldosterone assists regulation of

(a)

Capsule ———

Zona glomerulosa ———

Zona fasciculata ———

Zona reticularis ———

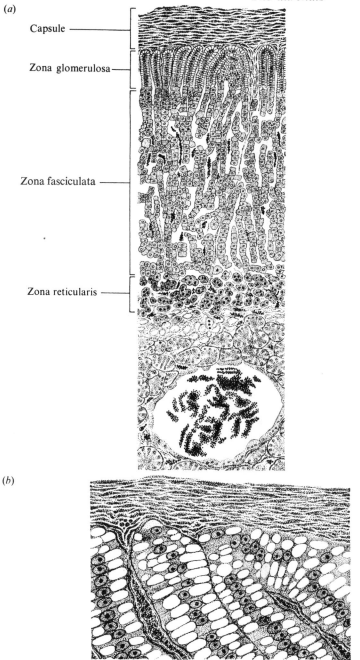

(b)

Fig. 2.14. (a) Histological section of the adrenal of a mammal, the racoon, showing the zonation of the adrenal cortex. (b) Enlargement of the capsular glomerular zone. (From Hartman and Brownell, 1949.)

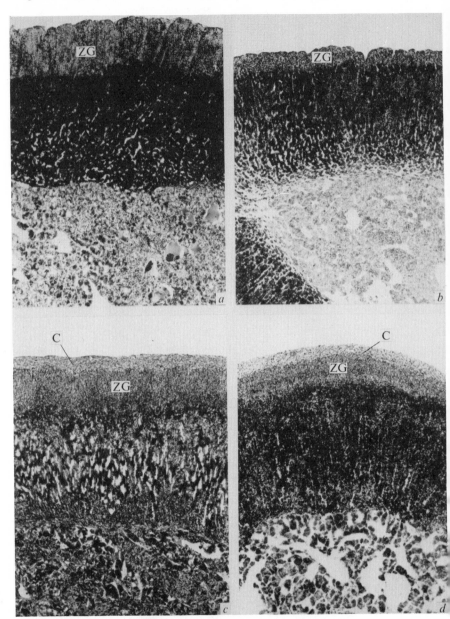

Fig. 2.15. The adrenal glands, in section, from two marsupials from sodium-deficient and sodium-replete areas.
Macropus giganteus (kangaroo): (*a*) Sodium-deficient; (*b*) sodium-replete.
Vombatus hirsutus (wombat): (*c*) Sodium-deficient; (*d*) sodium-replete.

sodium metabolism in mammals. It is therefore not unexpected to find that the zona glomerulosa can undergo considerable hypertrophy in mammals, such as rabbits and kangaroos, that live in areas where the salt content of the diet is low (Fig. 2.15) (Blair-West *et al.*, 1968).

Intraspecific differences in the size of the adrenals, the volumes of the cortex and medulla, and the cells of the cortex may be determined genetically. Shire (1970) has observed such differences in several strains of mice. In one strain, 'CBA', both the cortex and medulla are larger than in another, 'strain A'. The volume of the medulla in CBA mice is 0.35 mm^3 per 25 g body weight, while it is only 0.18 mm^3 in strain A. Similarly the cortex has a volume of 1.5 mm^3 in the CBA strain and 0.82 in the other mice. The differences have been shown to reflect genetic variation at one or two gene loci for the cortex and at at least two such loci in the case of the medulla. The CBA mice have a well-developed zona glomerulosa but attempts to correlate this with more production of aldosterone (as compared to a less well-endowed strain) have not been successful (Stewart *et al.*, 1972).

Among the vertebrates, there appears to be an evolutionary trend towards a more intimate association of the adrenocortical and the chromaffin tissue. This tendency may partially reflect their embryogenesis; such as the tissue aggregation that follows the loss of the mesonephros. One is tempted to ask if the relationship of the two endocrine tissues has any functional significance. The adrenocortical tissue can certainly function in the absence of the chromaffin tissue as seen *in vitro* in the laboratory. There is, however, some question as to the chromaffin cell's ability to produce optimal amounts of catecholamines. The mammalian adrenal medulla contains an enzyme, phenylethanolamine-N-methyl-transferase or PNMT, that, by methylating noradrenaline, converts it to adrenaline. The formation of this enzyme is induced by steroid hormones from the adrenal cortex (Pohorecky and Wurtman, 1971). The concentrations of the steroids must be high; far higher than normally present in the systemic circulation. This is achieved by the direct transfer of the steroids to the medulla through a local portal blood system. PNMT activity has also been identified in frogs but the enzyme is different from that in mammals and cannot be induced by corticosteroid hormones. It seems likely that such a physiological relationship between the adrenal cortex and the medulla may have arisen following their morphological juxtaposition.

Note that the zona glomerulosa is wider in sodium-deficient animals. In the wombat a thicker capsule lies at the outer border of this cell layer. C = capsule, ZG = zona glomerulosa. (From Blair-West *et al.*, 1968, and J. R. Blair-West, personal communication.)

The endocrine hormones of the gut

Several hormones are formed, and released from the posterior part of the foregut and its derivative glands. The most notable of these are insulin and glucagon from the pancreas; gastrin from the stomach; and secretin, cholecystokinin–pancreozymin, entero-glucagon and enterogastrone from the duodenum and upper parts of the jejunum. These hormones integrate and control the processes that result from feeding; which include the secretion of digestive enzymes and the concentrations of the absorbed nutrients in the blood.

Gastrin is present in the pyloric gland area of the stomach where, with the aid of immunofluorescent antibodies, it has been identified in special endocrine cells that are called *G-cells* situated in the deeper regions of the pyloric glands. Secretin has been identified, using similar techniques, in numerous epithelial cells, the *S-cells*, lining the crypts and villi of the duodenum and jejunum. At the time of writing, the site of formation of cholecystokinin–pancreozymin has not, apparently, been identified but it is considered to be only a matter of time before this is done. The other intestinal hormones; entero-glucagon and entero-gastrone, have not been specifically isolated and chemically purified, and although they are thought, on physiological grounds, to be formed in the upper regions of the intestine their precise sites of formation are uncertain. Indeed, although immuno-logical evidence for the existence of entero-glucagon has been obtained, entero-gastrone has remained a rather elusive 'hormone'.

Insulin and glucagon, as well as small amounts of gastrin, are found in special tissues situated in the pancreas called (after their discoverer) *Islets of Langerhans* (see Falkmer and Patent, 1972). Several types of cells have been identified in these tissues. The hormone-secreting cells contain granules and are distinguished from each other by their different appear-ances following exposure to histological fixatives and dyes. They are named in several ways: *A-cells* (also called α- or α_2-) which form glucagon, *B-cells* (β-) forming insulin and *D-cells* (δ or α_1-) which probably contain gastrin. B-cells are present in all vertebrates while A-cells are absent in cyclostome fishes and some urodele amphibians. The status of the D-cells is not clear and they may represent a stage in the development of the A-cells. As insulin-like hormones have also been identified in a number of inverte-brates the presence of comparable types of cells has also been sought in them but with equivocal results.

The morphological disposition of hormone-secreting cells of the pancreas differs considerably among vertebrates (for summary see Table 2.3). They are most often associated with the exocrine pancreas. This gland is formed from one or more diverticula of the gut and contains exocrine acinar cells

TABLE 2.3. *Evolution of the endocrine pancreas in vertebrates.* (From Pictet and Rutter, 1972)

Pisces	Agnatha	Myxinoidea (Hagfish):	Endocrine pancreas is a separate organ derived from the biliary duct; exocrine pancreas partially incorporated into the liver.
		Petromyzontia (Lamprey):	Endocrine pancreas is a separate organ derived from the duodenum.
	Chondrichthyes	Elasmobranchii (Shark, Torpedo):	Endocrine pancreas is formed of cells located around the medium- and small-sized pancreatic ducts; exocrine pancreas completely separated from the liver.
		Holocephali (Chimaera):	Endocrine cells are integrated in the exocrine tissue; they are accumulated in clusters, which are not vascularized.
	Osteichthyes	Crossopterygii Coelocanthini (*Latimeria*):	Endocrine cells located around the ducts as in Elasmobranchii. First evidence of islet formation: some B-cells are organized around the capillaries between the acini; exocrine pancreas separated from the liver.
		Teleostei: (95% of the living fishes, the most evolved fishes):	Endocrine pancreas either forms a separate organ (principal islets or Brockmann bodies) (goosefish, toadfish, etc.) or is integrated to exocrine pancreas.
		Dipnoi:	Forms a separate organ, 'principal islets', as in teleosts.
Amphibia	Urodela		Endocrine cells scattered in the exocrine tissue or gathered in islets. Some species have no A-cells.
	Anura		Endocrine cells originate from the pancreatic ductule during larval stage and accumulate in islets, which persist during and after metamorphosis.
Reptilia			Endocrine cells form islets. Sometimes they accumulate around duct as an external coat (in some snakes).
Aves Mammalia			Endocrine pancreas mostly found as vascularized islets free of exocrine tissue.

and their ducts. The endocrine secretory cells are formed from the ducts and in several fishes they lie around them. The cyclostome fishes lack such a discrete pancreas and the B-cells are found in the submucosal part of the anterior region of the intestine or around the bile duct.

In mammals the endocrine cells lie in small groups (or islets) that are about 100 to 200 μm in diameter. They are quite distinct from the exocrine tissue. A man has about 2 million of these while the guinea-pig has 15 to 40 thousand. These islets may be an admixture of the three types (A-, B- and D-) of cells or principally composed of a single type; as seen in rabbits and rats. B-cells are about five times more common than A-cells in mammals. Cyclostome fishes only have B-type cells. In the hagfishes (Myxinoidea), they are situated in a distinct gland, around the bile duct near its entrance to the intestine (Fig. 2.16a). Lampreys (Petromyzontia) also have a well-defined region of B-cells that extends along the submucosa of the duodenum. The B-cells form follicles (*Follicles of Langerhans*, Barrington, 1942) enclosing a cavity containing material which may represent the storage site for the insulin.

The Chondrichthyes also have different distribution patterns of A- and B-cells. In the Selachii (sharks and rays), the cells are usually situated near or around the ducts of the exocrine pancreatic tissue. In the Holocephali (chimaeroid fishes), this tissue extends more amongst the exocrine cells though it retains its association with the ducts. (see Fig. 2.16b).

There is also considerable diversity in the arrangement of the A- and B-cells among bony fishes (Osteichthyes). Some teleosts (toadfish and goosefish) show an aggregation of these cells into two pea-sized glands called '*principal islets*' or *Brockmann bodies* (Fig. 2.16c). Others, such as the eel (Fig. 2.16c), have islets of tissue scattered among the exocrine tissue as in mammals. In lungfishes (Dipnoi), the tissue is congregated into teleostean-like principal islets, while the Coelacanth (Crossopterygii) has a Selachii-type arrangement where the tissue is associated with the ducts of the acinar cells.

Some of the urodele Amphibia lack A-cells but in a few species, like the axolotl, the tissue forms distinct islets. The anurans, on the other hand, have assumed the tetrapod-like pattern which persists in the reptiles and birds. The latter, however, exhibit some distinctive characteristics not found in other vertebrates. There is a much higher proportion of A-cells which probably accounts for the large amount of glucagon that can be extracted from their pancreas. Three types of islets are present: dark islets containing mainly A-cells, light islets with B-cells and mixed islets with both types.

What is the significance of all this morphological variation in the distribution of A- and B-cells? 'Why are the Islets of Langerhans?' (Henderson,

(*a*)   CYCLOSTOME (*Myxine*)

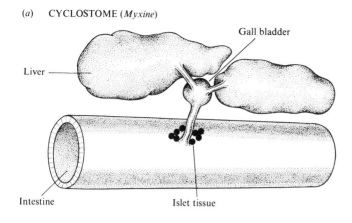

(*b*)   CHONDRICHTHYES

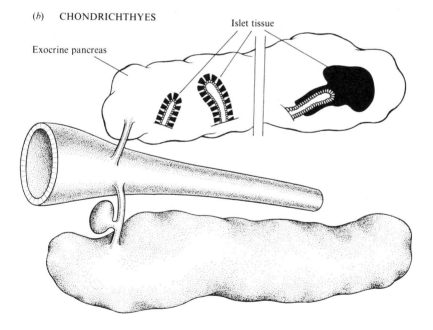

Fig. 2.16. The various types of pancreas in fishes.
(*a*) Cyclostome-type (*Myxine*): ring-like arrangement around the bile duct.
(*b*) Chondrichthyean-types. Left = many elasmobranchs; right = Holocephali.

(c)   ACTINOPTERYGII (Type i)

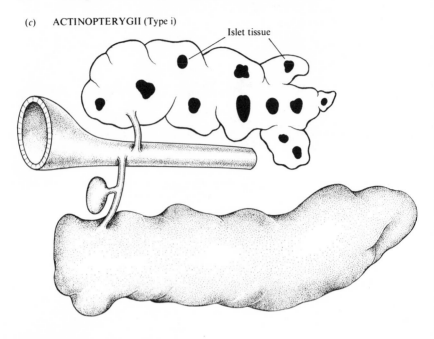

ACTINOPTERYGII (Type ii)

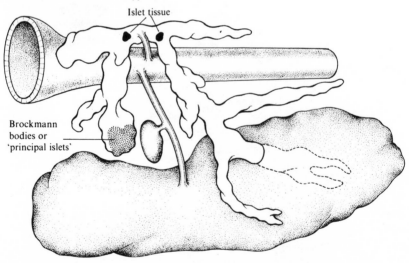

(c) Actinopterygian-types. (i) *Anguilla*; this is present in a few teleosts and is similar to that in tetrapods. (ii) The more general teleost with Brockmann bodies or 'principal islets'. (From Epple, 1969.)

1969). Why, in so many different vertebrates, is the endocrine tissue dispersed so widely amongst the exocrine tissue? While in some fishes the endocrine tissues may tend to retain the site of their embryonic origins from the acinar duct epithelium, there seems to be an evolutionary tendency towards aggregation of the cells. This aggregation sometimes results in relatively large pieces of tissue, the 'principal islets' of the teleosts and dipnoans. Among tetrapods, the small islet arrangement is universal. Henderson (1969) has conjectured that this may reflect a dependence of the exocrine acinar cells on the endocrine tissues that secrete insulin, glucagon and gastrin. The small islets increase the surface area of the endocrine tissues and promote their contact with the exocrine tissue so that high local levels of hormones can be maintained. It has indeed been noticed that, in diabetes mellitus (a lack of insulin) in man, the exocrine pancreas tends to atrophy and become invaded with excess fat and fibrous tissue. The admixture of the A-cells and the B-cells also may have functional significance and there is evidence to suggest that glucagon may have an insulinogenic effect on the A-cells.

The gonads

While the ovaries and testes are not essential for the survival of the individual, they are for the propagation of the species. They have a dual but related function; the production of ova and sperm as well as several hormones that are concerned with the development of the germ cells and the fertilized egg.

The gonads are formed from the dorsal coelomic epithelium. The adrenocortical (or interrenal) tissue has a similar origin and both secrete related steroid hormones. The primordial gonadal tissue consists of a cortex and a medulla. The latter differentiates into the testis, the former the ovary. The steroidogenic cells that are present in the gonads (as well as the adrenals) have a distinctive structure. They contain a very well developed endoplasmic reticulum, mitochondria that have tubular cristae, and usually lipids that histochemically behave like cholesterol. A notable characteristic that helps histological identification of steroidogenic cells in the gonads is the presence of the enzyme 3β-HSDH (Δ^5-3β-hydroxysteroid dehydrogenase). It is responsible for the conversion of certain precursors of the steroid hormones into progesterone and androstenedione which may subsequently be converted to other hormones.

The gonads are usually paired structures lying in the body cavity near the kidneys. The testes of most mammals (except a few such as whales, the elephant and guinea-pig), however, are suspended outside the abdominal cavity in the scrotal sac. Some species have only a single gonad which usually reflects the degeneration of the other or possibly their fusion. The cyclostome fishes only have a single testis and ovary in the median line, while in nearly all birds only the left ovary reaches full development. A single ovary is also sometimes present in teleost and chondrichthyean fishes.

Considerable diversity exists in the relationship of the gonads to their excretory ducts. These are lacking altogether in cyclostomes where the eggs and sperm are released into the body cavity and thence through pores to the exterior. Some fishes, especially teleosts, have gonaducts that are merely extensions of the ovaries. In most vertebrates including many bony fishes, sharks and rays, and tetrapods, the germ cells pass through a homologous series of ducts derived from the Wolffian duct in the male and the Mullerian duct in the female.

Testis

The testis shows a considerable degree of uniformity among the vertebrates (see Dodd, 1960; Lofts, 1968; Lofts and Bern, 1972). The formation of sperm occurs in two principal ways. In amniotes, it differentiates from the germinal epithelium situated in the wall of the seminiferous tubules. Anamniotes, on the other hand, do not have such continuous tubules but those which they have may be branched and divided into lobules. The sperm in these instances is differentiated into groups contained within small envelopes called 'cysts' (cystic spermatogenesis). Two principal types of cells are concerned with the secretion of the male sex hormones. (1) in tetrapods (except for urodele amphibians) groups of cells lying between the seminiferous tubules and called *interstitial* or *Leydig cells* (Fig. 2.17a) are the site of formation of testosterone. In urodele amphibians and fishes, the walls of the testicular lobules contain cells called '*boundary cells*' that apparently have a similar role (Fig. 2.17b). Embryologically, the interstitial and boundary cells are thought to be homologous. Some fishes, including lampreys, possess tetrapod-like interstitial cells. (2) Associated with the basement membrane of the seminiferous tubules are the *Sertoli cells*. Their function has for a long time been controversial and it was suggested that they were concerned with the nourishment of the sperm (hence they are also called 'sustentacular cells'). It now seems clear that they are the site of production of sex hormones that are probably involved in the growth and maturation of the sperm. These cells have been identified in nearly all vertebrates that have been examined. In those species that have a cystic spermatogenesis, they lie juxtaposed to the heads of the sperm in the spermatic cysts.

The testes undergo considerable structural changes associated with the periodic, or cyclical, breeding behavior. These changes are reflected in the size and lipid content of the interstitial and Sertoli cells.

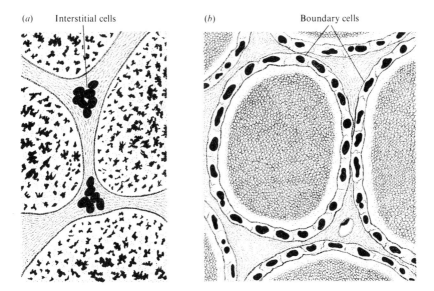

Fig. 2.17. The testis.
(*a*) Sections from a reptile, the viper, showing the lipid-filled interstitial (Leydig) cells. (*b*) A teleost fish; the pike, showing the boundary cells in the wall of the testicular lobule. These are the homologues of the interstitial cells which are absent in most of the fishes. (From Gorbman and Bern, 1962, based on data from B. Lofts and A. J. Marshall.)

The ovaries

The ovaries contain several distinct types of cells and tissues (see Lofts and Bern, 1972; Dodd, 1960) (Fig. 2.18). (1) The *germinal epithelium* that envelops the ovaries and from which the developing eggs and their surrounding sheet of granulosa cells are formed. (2) The latter is called the *theca interna* which, in turn, is surrounded by a fibrous *theca externa*. The complete structure, ovum and its surrounding membranes, is called the *Graafian follicle*. (3) Following the maturation and expulsion of the ova the granulosa cells may form the *lutein cells* that compose the *corpus luteum*. Also present in the ovary are structures formed as the result of the degeneration (atresia) of unovulated follicles, the *corpora atretica*. (4) Lying in between all these structures is the *interstitial tissue* which anatomically, and probably functionally, is analogous to the tissue of the same name in the testis.

Estrogenic hormones are formed by the cells of the theca interna and possibly the granulosa cells of the developing follicles. Progesterone arises from the lutein cells of the corpus luteum and, at least in the rat, the interstitial tissue. The latter also produces androgenic hormones.

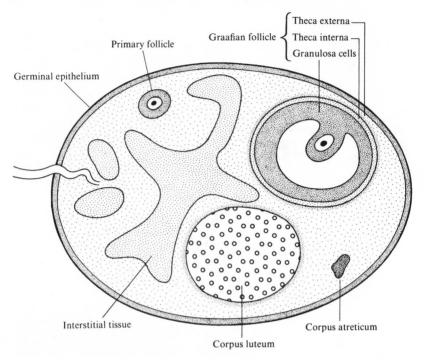

Fig. 2.18. The ovary. Diagrammatic representation of the various tissues present. (From Lofts and Bern, 1972.)

The basic structure and endocrine role of the ovaries persists throughout the vertebrates. There are some points worthy of comment. The corpus luteum is not present in all species, especially oviparous ones. It is absent in birds which nevertheless have progesterone in their ovaries and blood. Corpora lutea appear sporadically among other vertebrate groups. They are present in ovoviviparous and viviparous amphibians, chondrichthyeans and some teleosts. A corpus luteum also appears following ovulation in *Myxine* (hagfish) while all reptiles (including oviparous ones) develop a corpus luteum after ovulating. Indirect, histological, and more direct chemical evidence indicates that it produces progesterone in all groups. It is possible that the corpora atretica also produce hormones but many consider this to be doubtful.

Putative endocrine glands

The pineal body

Some argue that, at least in mammals and probably also in birds, the pineal body deserves the title of gland and should not be classified as a putative endocrine. It forms melatonin which exerts an antigonadotrophic action in some vertebrates and pales the skin in larval amphibians and cyclostomes.

The pineal (see Ariëns Kappers, 1965, 1970; Oksche, 1965; Wurtman, Axelrod and Kelly, 1968) originates as a sac-like evagination from the dorsal part of the brain (the diencephalon) (Fig. 2.19*a*, *b*). It lies beneath the cranium in the mid-line position and is also called, especially in mammals, the *epiphysis cerebri*. Embryologically, its hollow stem maintains contact with the III ventricle and this pattern persists in many non-mammals.

The pineal body is present in most vertebrates. It is, however, absent in hagfish (myxinoid cyclostomes), crocodiles, at least two species of chondrichthyeans (*Torpedo ocellata* and *T. marmorata*) and several mammals, including whales. The pineal is very small in the elephant and rhinoceros.

In man, the pineal is a relatively simple knob of tissue but in other vertebrates it is more complex. It appears to have undergone considerable changes in its structure (and no doubt function) during its evolution. Noteworthy is the differentiation of an associated tissue that comes to lie nearer to the roof of the skull. This is variously called the *parapineal* (in fishes), *frontal-* (in amphibians) and *parietal-* (in reptiles) *organ*. In many species, especially lizards, it penetrates through the brain case and lies just beneath the skin. The parapineal is connected to the roof of the brain, or the pineal body proper by nerve cells. In fishes, amphibians and certain reptiles (not the snakes), the parapineal and even the pineal body contain photoreceptor cells. In lizards and the tuatara (Rhyncocephalia), the pineal takes the form of a well-differentiated 'third eye' that even contains struc-

(*a*)

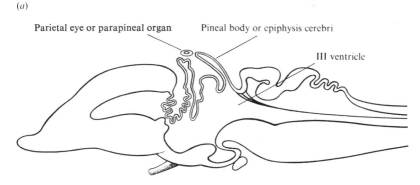

Fig. 2.19. (*a*) The pineal body in relation to the brain (in section) of a lower vertebrate. (From Bargmann, 1943.)

(b)

LAMPREY

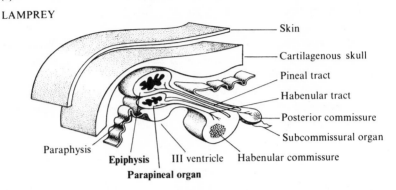

- Skin
- Cartilagenous skull
- Pineal tract
- Habenular tract
- Posterior commissure
- Subcommissural organ

Paraphysis

Epiphysis   III ventricle   Habenular commissure

Parapineal organ

TELEOST FISH

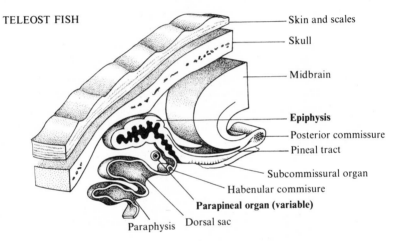

- Skin and scales
- Skull
- Midbrain
- **Epiphysis**
- Posterior commissure
- Pineal tract
- Subcommissural organ
- Habenular commisure
- **Parapineal organ (variable)**

Paraphysis   Dorsal sac

ALBINO RAT

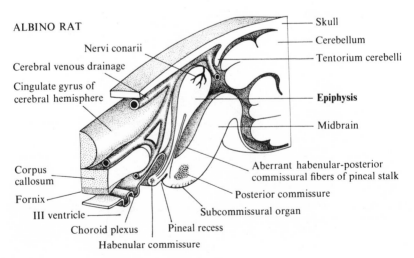

Nervi conarii

Cerebral venous drainage

Cingulate gyrus of
cerebral hemisphere

- Skull
- Cerebellum
- Tentorium cerebelli
- **Epiphysis**
- Midbrain
- Aberrant habenular-posterior
  commissural fibers of pineal stalk
- Posterior commissure
- Subcommissural organ

Corpus
callosum

Fornix

III ventricle

Choroid plexus   Pineal recess

Habenular commissure

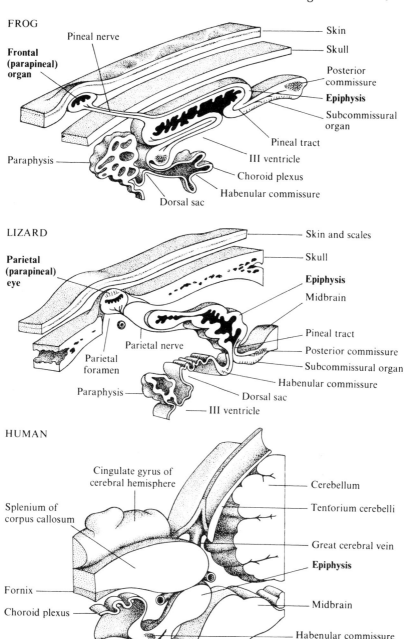

(b) The pineal of various vertebrates in relation to the dorsal diencephalic roof region. (From Wurtman *et al.*, 1968.)

tures analogous to the cornea and lens of lateral eyes. In such vertebrates the pineal has a sensory function.

The pineal of mammals lacks functioning photosensory cells but these have been modified to form secretory cells (*pinealocytes*) that seem to have an endocrine function. The presence of photosensory cells in the bird pineal is equivocal but pinealocytes are present. Whether or not the pineal of other vertebrates secretes hormone-like products is uncertain. In the simplest interpretation of pineal evolution, it may be viewed as evolving from a sensory structure in fish to an endocrine gland in birds and mammals.

Embryologically, the pineal body is a hollow sac-like structure and this persists in most vertebrates. In mammals, it is filled with parenchymous cells (the pinealocytes) which are derived from the photoreceptor cells seen in lower vertebrates. Some birds (many Passeriformes) exhibit the primitive

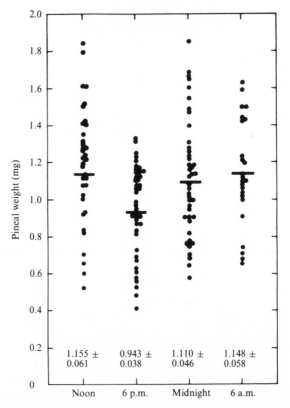

Fig. 2.20. Rhythmical changes in the weight of the rat's pineal throughout the day. These changes are associated with changes in various biochemical products that are present in the gland. (From Axelrod, Wurtman and Snyder, 1965.)

hollow condition while others, like the domestic fowl, have a solid pineal containing pinealocytes. In mammals and birds, the pineal organ has an autonomic sympathetic innervation arising from the superior cervical ganglion, the destruction of which (in rats anyway) has profound effects. In other vertebrates, the pineal seems to maintain direct neural contact with the brain but an autonomic innervation is also evident in fish, amphibians and reptiles. The pineal has been shown to form several substances that have potent biological effects. These include noradrenaline (in nerve cells) and serotonin and melatonin (in the pinealocytes). Melatonin has been identified in the circulation and may represent the pineal's endocrine product. Melatonin and serotonin have also been identified in the bird and amphibian pineal. One of the enzymes responsible for its formation, HIOMT (hydroxyindole-*O*-methyl-transferase), has been found in the pineal of fish and all the tetrapod groups as well as in the lateral eyes of fish, amphibians, reptiles and birds (but not in mammals).

The pineal of rats undergoes a number of changes associated with the exposure of the animals to light. It increases in weight in rats kept in the dark and decreases on exposure to light. Weight changes can be seen to follow a circadian rhythm over a normal 24-hour period (Fig. 2.20). As we shall see later, this change in weight may reflect its role (?) as a 'neuro-endocrine transducer' in the control of reproductive cycles.

The pineal body has often been described as a 'vestigial' organ. Despite the implied slight about its usefulness, it has shown considerable phylo-genetic persistence during which it appears to have evolved from a mainly sensory organ to an endocrine one.

Corpuscles of Stannius and the renal juxtaglomerular cells

These cells are both associated with the kidneys and may possibly produce some secretions that are similar to each other. Hence for convenience they will be dealt with together.

The corpuscles of Stannius

These tissues are so named after their discoverer (Stannius, 1839). They appear to form a hormone that decreases blood calcium levels and may also have osmotic effects in fishes. They are distinct well-vascularized masses of tissue that are present near the kidney of some, but not all, fishes. Their number varies considerably, 40 to 50 are present in the bowfin, *Amia calva* (Holostei), while in the Teleostei the number varies from two to six. They are absent in some teleosts, including the Salmonidae, as well as the Chondrostei (sturgeons) (Krishnamurthy and Bern, 1969). In the corpuscles, one, or sometimes two, types of cells can be distinguished that are usually arranged in rows to form lobules around a central core. These cells contain granules, the number and appearance of which can be seen to vary conspicuously with changes in the breeding cycle, the life cycle and the osmotic environment. This has led to the suggestion that they may have an endocrine role. Indeed, as will be related later, extracts of these tissues have been shown to contain materials that in some instances may alter sodium and calcium metabolism as well as exhibit a vasopressor effect similar to that of a kidney hormone called renin.

The corpuscles of Stannius are usually derived from the pronephric duct (in some instances they are formed from the mesonephric duct). The presence of the renin-like material in the tissue extracts may be a reflection of this origin.

The Holostei have the largest number of corpuscles of Stannius and it has been suggested that the reduction among the Teleostei, culminating in their absence in the Salmonidae, may reflect an evolutionary trend. In one study (Krishnamurthy and Bern, 1969) 28 species of teleosts from 16

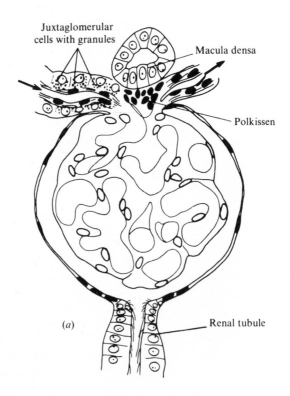

Juxtaglomerular cells with granules

Macula densa

Polkissen

(*a*)

Renal tubule

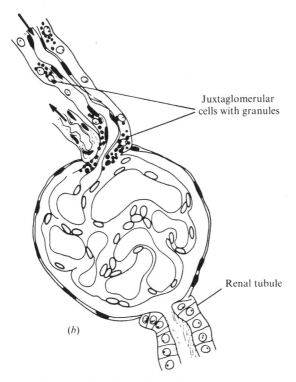

Juxtaglomerular
cells with granules

Renal tubule

(*b*)

Fig. 2.21. The juxtaglomerular apparatus.
(*a*) The laboratory rat. (*b*) The bullfrog (*Rana catesbeiana*) showing the absence
of a macula densa and polkissen. (From Sokabe *et al.* (1969). *Texas Reports Biol.
Med.* **27**, 3.)

different families were examined but no conclusions as to any evolutionary
relationships in their morphology could be drawn.

The juxtaglomerular apparatus

The kidney is the site of formation of a protein called *renin*, considered by
some to be an enzyme and by others a hormone, which can be released into
the circulation. This secretion initiates the formation of a peptide, called
angiotensin, in the plasma, which contributes to the regulation of sodium
retention in the body and possibly also increases blood pressure. Renin is
widely considered to be formed by cells situated near the renal glomerulus
at a site called the juxtaglomerular apparatus. In mammals it consists of
juxtaglomerular cells (on the afferent glomerular arteriole) and the *macula
densa*, which is a thickening of the distal renal tubule in the region where it

abuts onto a glomerular area called the *polkissen* (or extraglomerular mesangium) (Fig. 2.21*a*). In non-mammals the situation is less complex (Fig. 2.21*b*) as the macula densa and polkissen are absent (the former may be present in birds). Many species, however, still have the juxtaglomerular cells (see Fig. 2.22). They contain 'granules' that can be stained histo-

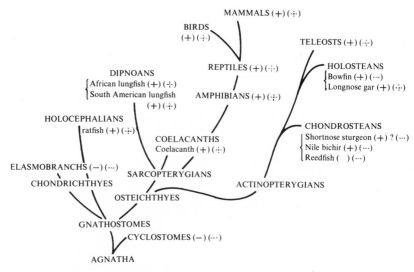

Fig. 2.22. Phylogenetic distribution of juxtaglomerular granules (stained by Bowie's method) and renin activity in the kidneys of vertebrates: (+) or (−), presence or absence of renin; (·÷·) or (···), presence or absence of granules. It can be seen that while the two are always associated with each other in tetrapods and most fishes this is not so in all holosteans or chondrosteans. (From Nishimura, Ogawa and Sawyer, 1973, modified by H. Nishimura.)

logically in a distinctive way (Bowies method). Such juxtaglomerular granules occur in arterioles often distant from glomeruli in teleost, dipnoan and coleacanth fishes and even in aglomerular fishes but in most vertebrates they occur in the afferent glomerular arteriole. The latter type of cells are present in birds, reptiles, amphibians and teleost fishes but neither these cells nor Bowie's granules have been identified in the elasmobranchs or cyclostomes (see Ogawa *et al.*, 1972). These observations parallel the identification of renin in the kidneys of these groups. An inconsistency exists in that juxtaglomerular cells with granules stainable by the Bowie method have not been found in many of the Chondrostei or Holostei even though there is evidence to suggest the presence of renin in such fish. Renin has also been tentatively identified in the corpuscles of Stannius in teleost fishes (Chester Jones *et al.*, 1966; Sokabe *et al.*, 1970) but it is not clear if Bowie's granules are also present in these tissues (compare for instance Sokabe *et al.*, 1970; Krishnamurthy and Bern, 1969).

The macula densa may be concerned with the regulation of the release of renin in mammals (and birds?). Renin and angiotensin contribute to the regulation of sodium levels in the bodies of mammals. It has been suggested that local changes in the sodium permeability of the macula densa may regulate the release of renin from the juxtaglomerular cells (Vander, 1967). Sodium-depletion increases the release of renin and this response has been shown to be dependent on a renal 'vascular receptor' (Gotshall *et al.*, 1973). In this respect it is noteworthy that a reduction in renal blood flow, such as results from hemorrhage or stimulation of the renal nerves, also promotes renin release in mammals but this response can be blocked by a β-adrenergic blocking drug (propranolol) that acts distally to the renal vascular muscle (Coote *et al.*, 1972). Thus several steps appear to be involved in the release of renin but more precise evidence as to the role (if any) of the macula densa is lacking. It is interesting that non-mammals may also utilize the renin–angiotensin system to aid sodium-regulation in the body and as these animals lack a macula densa it presumably does not have an essential role in this process in all vertebrates. It remains possible, however, that it could mediate some effect on the release of renin that is special to the mammals.

The urophysis

Tucked away beneath the vertebrae in the tail of teleost fishes is a lump of tissue that has been called the urophysis and which may influence their osmoregulation and assist smooth muscle contraction in the urinogenital tract. This 'gland' was first described in 1813 and has since been identified in some 400 different species of teleost fishes. Despite its widespread distribution and systematic persistence, its physiological role is still not understood but has captured considerable attention from several endocrinologists (see Fridberg and Bern, 1968).

The urophysis, like the neurohypophysis, is composed of neural tissue, whose cell bodies are situated in the posterior part of the spinal cord. Axons of these cells pass outside the spinal column ventrally where they make contact with blood vessels (that lead through the kidneys) to form a neuro-hemal junction. Such an arrangement is admirably suited to the discharge of endocrine secretions into the circulation. Whether this occurs or not is, however, equivocal. These nerve cells contain granules and appear to be typical neurosecretory cells such as those seen in the neurohypophysis, though their tinctorial characteristics differ. Extracts of this tissue show several biological activities; they can alter the permeability of some membranes to water and sodium, contract certain smooth muscle preparations, increase the blood pressure of eels and lower the blood pressure of rats.

A distinct neurohemal urophysis has only been identified in teleost fish. The chondrichthyeans, however, also possess neurosecretory-type cells in the caudal part of the vertebral column. These cells are giant neurons, about 20 times the size of an ordinary motor neuron and, in *Raja batis*, extend along the last 55 vertebrae. They are called *Dahlgren cells* after their discoverer and send their axons out of the ventral part of the spinal

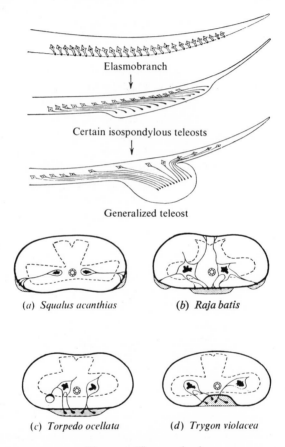

(*a*) *Squalus acanthias*   (*b*) *Raja batis*

(*c*) *Torpedo ocellata*   (*d*) *Trygon violacea*

Fig. 2.23. The urophysis.

Top. Proposed evolution of the teleost urophysis from elasmobranchs which have neurosecretory Dahlgren cells. Longitudinal section through the tail.

Bottom. (*a*) to (*d*). Different configurations of the Dahlgren cells among the elasmobranchs.

Transverse sections through the spinal cord. The vascular beds are shaded and the menix is represented by a heavy line. (From Fridberg and Bern, 1968.)

cord to make contact with blood vessels there. The tissue is thus more widespread along the spinal cord. Comparable tissues appear to be present in holostean fishes but not in cyclostomes.

The types of such structures present in chondrichthyeans and teleosts are shown in Fig. 2.23. The diffuse distribution of tissue seen in the sharks and rays may represent a primitive pattern which has subsequently evolved among teleosts to form a discrete aggregation. The widespread distribution and systematic persistence make one suspect that the urophysis serves a physiological role. Its histological and cytological appearance, similarity to the neurohypophysis and its neural and vascular connections suggest an endocrine gland. Conclusive evidence is, however, not yet available.

The thymus

The thymus, like the parathyroids and ultimobranchial bodies, is derived, embryologically, from the gill pouches and so is situated in the region of the neck or upper thorax. In fishes all the gill pouches may be involved but in amniotes usually only numbers III and IV. The thymus is present in all groups of vertebrates, from cyclostomes to mammals, but is more prominent in young and larval forms than adults where, in some species, it may even disappear. The thymus tissue may exist in varying degrees of agglomeration, from a single bilobed organ, as in mammals, to paired structures, as in frogs, or in several dispersed nodes, as in the domestic fowl. The ultimobranchial bodies and even parathyroids may on occasions be associated with the thymus.

The endocrine function of the thymus is not an established one. A number of biologically active materials have, however, been extracted from it and these (several proteins and a steroid) have been called 'thymic hormones' (see Luckey, 1973). During early life the thymus produces lymphocytes that are involved in the development of immunity. It has been proposed that the so-called thymic hormone(s) may initiate development of immune competence (ability of cells to react to the presence of foreign substances or antigens) in the lymphocytes and bone marrow. Such 'hormones' thus could be considered to be involved in embryonic differentiation.

One of these 'hormones' (see Bach *et al.*, 1972; Goldstein *et al.*, 1974) has been isolated from the thymus and is a polypeptide with a molecular weight of about 1000. It has been called thymosin and has been identified in the blood of the mouse and man. The circulating concentrations decline with age and after thymectomy. In man the thymosin concentration in the blood starts to decrease at 30 to 40 years of age and it disappears after about 50 years. There is considerable medical interest in these discoveries as the blood levels have also been shown to be low in diseases where immune responses are deficient and it fails to decline with age in people suffering from myasthenia gravis. The latter disease (a failure of skeletal muscles to be able to contract adequately) can often be ameliorated by thymectomy. Another intriguing possibility is that the decline in the activity of the thymus contributes to the process of aging and an increased susceptibility to infectious diseases. It seems likely that this interesting tissue will gain the status of a true endocrine gland.

Conclusions

It can be seen that while the endocrine glands display considerable inter-specific differences in their morphology many of these variations can be placed into categories that correspond to major systematic groups of vertebrates. It would thus appear that the endocrines have evolved in a relatively orderly manner that may be influenced by broad structural considerations, such as the animal's shape, size and pattern of embryonic differentiation, as well as its particular hormonal requirements. Such evolutionary changes are not, however, confined to the glands' morphology for, as we shall see in the next chapter, considerable variation also occurs in the chemical structures of the hormones themselves.

3. The chemical structure and polymorphism, and evolution of hormones

In a mammal the endocrine glands secrete more than 40 distinct hormones. In addition, different species may form many hormones that, while being structurally analogous, nevertheless display chemical differences. Such natural variants are usually characteristic of a single species and represent a polymorphism of the excitant's molecular structure. This change has a genetic basis. For example, it may only be the substitution of a single amino acid in the molecule of a peptide hormone or it may be much more extensive. The biological effects of such differences can be considerable or negligible.

Vertebrate hormones belong to two principal classes of chemical compounds. Some are made from cholesterol. These are the steroid hormones from the adrenal cortex and the gonads. The others are made up of amino acids and range in complexity from those, like adrenaline, that are derived from a single tyrosine molecule, to others like the pituitary growth hormone that contain about 190 such units. The molecular weights can vary from about 200 to 30 000.

What properties do these molecules have that make them suitable to be hormones? What characteristics may be important for their utilization as such? Armed with considerable hindsight about endocrine physiology some answers can be offered. The basic requirements will not be the same for all hormones but will depend on what they do. The steroid hormones are poorly soluble in water but readily soluble in lipids. This will facilitate their penetration into the cell and fixation at intracellular sites. Such lipid solubility will also be important if a hormone is to penetrate the blood–brain barrier. Transport in the blood is essential for a hormone to fulfil its physiological role so that, if they are hydrophobic molecules, they must either be effective at very low concentrations or be attachable to protein components which carry them to their sites of action. This binding is especially prominent among the steroid and thyroid hormones. An ability to interact with other biological molecules is also important for 'triggering' the excitant effects of hormones. They must be capable of interacting with a receptor molecule in, or on, the effector tissue. Such an interaction must not be of a strong covalent nature but must involve chemical forces whereby an equilibrium of a reversible nature occurs. Above all, a hormone must have a high degree of specificity towards its target receptor site. Necessarily this is a property

[61]

of both structures. The manner by which it is accomplished is still largely conjectural. Hormones have complex three-dimensional structures which contain various components that may be electrically charged, hydrophilic or hydrophobic, acidic or basic, and so on. Such properties together may constitute a 'key' to which the receptor acts as a complementary 'lock'. Obviously, large hormone molecules offer the possibilities of more complex 'keys' and greater specificity. They are also more liable to genetically mediated structural changes. The latter are very common events in nature. The contribution of size to specificity of effect is, however, not at all clear. Indeed, it is difficult to comprehend why such gigantic molecules as the pituitary adenohypophysial hormones are necessary to mediate their effects. There is no evidence to indicate that evolution towards smaller, more compact hormone molecules has occurred, such as would perhaps be expected if they initially contained much superfluous material. Indeed some hormone molecules, for instance parathormone, contain sections that are not needed for their biological activity.

In order to function optimally, a hormone molecule needs to possess some other properties consonant with its physiological role. For adequate control, hormonal responses often need to be rapidly terminated. The excitant can either be readily excretable in the urine or bile, or, by virtue of the presence of chemical groups that can be changed by metabolic processes, be converted to an inactive form. The synthesis of hormones is not always rapid enough to meet the immediate demands for their release, so that their accumulation and storage in glandular tissues may be necessary. In this instance the molecule should possess a considerable measure of innate stability and be able to interact with cellular (or even extracellular; as for thyroxine) binding proteins that facilitate this storage. Related to such a process is an ability to undergo rapid mobilization from such storage sites so that the hormone can be released into the blood.

Clearly, hormones are highly specialized molecular structures incorporating (or programmed for) several important interrelated properties. They are not just 'keys' that fit various 'locks'. In order to function as a hormone a molecule must also exhibit a variety of other physical and chemical properties consistent with the hormone's synthesis, storage, release, transport, and removal from the body.

Structural differences between hormones are tentatively assumed on the basis of differences in their biological actions. They are confirmed by the demonstration of variations in their chemical behavior and ultimately by the determination of their molecular structure. It is usually a comparatively simple procedure to show that two hormones differ from each other. Tests for biological activity; for instance changes in blood glucose levels, an ability to alter blood pressure, decrease urine flow and so on, are reasonably straightforward laboratory procedures. Broad chemical differences in even very impure preparations can often be seen when, for instance, one compares their stabilities at different temperatures and pH's, solubilities in different solvents, relative rates of destruction when incubated with various enzymes and so on. Such biological and chemical characterization can be used to identify and measure the relative quantities of the hormonal material present in an extract. Determination of chemical composition and structure is a more complex procedure and, before this can be done, highly purified preparations must be made.

Differences in the composition of related homologous hormones in distinct (or even on occasion the same) species are more difficult to detect. The molecules may only exhibit quite small quantitative, in contrast to qualitative, differences in a common biological activity.

Several procedures are available to help us make such distinctions.

(1) Biological activities can be compared. A simple cross-test between two or more species can be informative. For instance, a comparison between the action of extracts of the neurohypophyses of frogs with those from mammals indicated the presence of different but analogous hormones. Heller (in 1941) found that extracts from the neurohypophyses of European frogs (*Rana temporaria*) were about 20 times as active in eliciting water retention in frogs than a comparable amount of a hormone extract from the mammalian pituitary. The two extracts (frog and mammal) were each standarized by their ability to contract the guinea-pig uterus and to increase the blood pressure of cats. Equal amounts of activity, as measured in these ways, showed that the homologous frog hormone was much more active in frogs than the mammalian one. As we shall see, this change reflects a single amino acid substitution in the octapeptide molecule of mammalian antidiuretic hormone. Similar comparisons have been made between many species and usually the more distantly related they are the greater their effects differ. Even two closely related species may show distinct differences; thus pituitary growth hormone from animals, including other primates, is completely ineffective in man even though that from man is active in many other species. Hormones often exhibit a variety of biological actions and, by comparing them in several assay systems, a pharmacological profile or 'fingerprint' can be made. This can be used to characterize some hormones with a considerable degree of accuracy.

(2) The chemical behavior of hormones can be shown to differ. In closely related molecules this may be difficult to detect. It may depend on differences in electrophoretic or chromatographic mobilities in different solvent systems. A correlation (isopolarity), however, does not necessarily confirm the chemical identity of two molecules. Steroid molecules are often

chemically altered, for instance by methylation, and their chromatographic behavior is again compared. If it still corresponds, the evidence of identity is much stronger. The absorption spectra of extracts can also be compared. This measurement may involve the use of ultraviolet and infrared light when the hormones are dissolved in different solvents. Mass spectrometry and nuclear magnetic resonance spectrometry are effective but expensive methods for identifying many hormonal materials. Using such procedures a chemical 'fingerprint' can be obtained for comparison with standard preparations of known structure. Such chemical methods are particularly useful in aiding the identification of steroid hormones (see Brooks *et al.*, 1970; Sandor and Idler, 1972).

(3) Immunological responses are used to predict the differences in the structure of protein hormones. Antibodies (or antisera) to purified hormones can be made following their injection into another species, often a rabbit. The degree of interaction of protein hormones with such antisera can be followed in a precipitin test or by comparing their abilities to displace known radio-iodinated hormones from binding with the antibodies. Apart from indicating similarities and differences between hormones, such immunological procedures are now widely used to measure the quantities of hormones by radioimmunoassay. This procedure is not confined to protein hormones, as antibodies to other hormones including steroids and thyroid hormones, can be made by conjugating them to a protein molecule and using this to elicit formation of antibodies.

(4) The chemical structure of a hormone may be determined. This is ultimately desirable and allows one to relate the structure of the molecule to its biological actions and even predict the genetic basis of its formation and possible evolution. As some hormones are very complex structures containing as many as 200 amino acids, knowledge of their precise chemical structure has been somewhat slow in coming. A special relationship with immunological procedures also exists, as these can be used to confirm the chemical proposals. Initial chemical analyses of human growth hormone and prolactin indicated that they had identical chemical structures but this was not confirmed immunologically. Presumably small, as yet undetected, differences exist in man, as has been shown between these hormones in other species.

While the chemical structure of many hormones is known, this knowledge is mainly confined to the mammals, especially with respect to the larger protein hormones. In addition, although the disposition of chemical groups and the sequence of amino acids may be known, little information is available as to their three-dimensional (tertiary) arrangement. Such data will ultimately be required if we are to understand properly how the hormones work.

Steroid hormones

Steroids are chemical compounds derived from cholesterol. They consist of a series of carbon rings, the basic unit being the cycloperhydrophen-

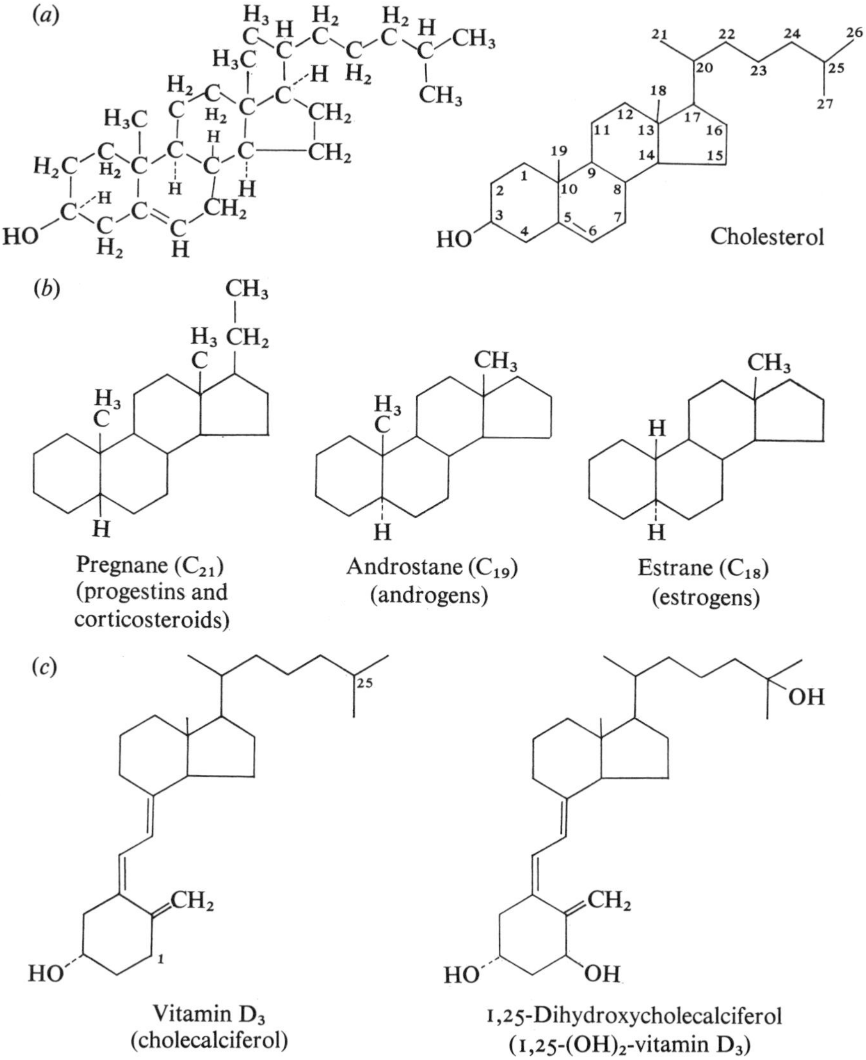

(a)

Cholesterol

(b)

Pregnane (C_{21})
(progestins and
corticosteroids)

Androstane (C_{19})
(androgens)

Estrane (C_{18})
(estrogens)

(c)

Vitamin D_3
(cholecalciferol)

1,25-Dihydroxycholecalciferol
(1,25-$(OH)_2$-vitamin D_3)

Fig. 3.1. (*a*) The chemical structure of cholesterol and the conventional manner of numbering the carbon atoms.

(*b*) The parent steroid compounds for the progestins and corticosteroids (C_{21}), androgens (C_{19}), and estrogens (C_{18}).

(*c*) Vitamin D_3 and its active metabolite 1,25-dihydroxycholcalciferol.

anthrene nucleus. Such compounds occur widely in nature and are not confined to the animal kingdom. Plants contain many steroids and some of these may even exhibit activities reminiscent of those of the mammalian hormones: for example, catkins of the pussywillow plant contain steroids that have the effects of female sex hormones (estrogenic activity). Steroids obtained from plants are indeed often used as starting materials for the preparation of vertebrate-like hormones in the laboratory.

Several different types of steroids function as hormones in vertebrates. These and the parent cholesterol molecule are shown in Fig. 3.1*a*, *b*. They are often classified in the following manner: (1) Those based on pregnane and containing 21 carbon atoms (C_{21}). These include the adrenocortical steroids and progesterone which, apart from being a metabolic intermediate in the formation of most steroid hormones, also acts as a sex hormone, especially during pregnancy. (2) Androstane compounds with 19 carbons (C_{19}) and which include the androgens that have the actions of male sex hormones. (3) Estrane (C_{18}) compounds that have actions of female sex hormones (estrogenic). (4) Vitamin D (Fig. 3.1*c*), a group of sterols the precursors of which are commonly obtained in the diet and which can be converted (see Chapter 6) into hormones that influence calcium metabolism.

The hormones from the gonads and adrenal cortex are all derived from cholesterol compounds. In the instance of C_{18}, C_{19}, and C_{21} steroids, various metabolic pathways, usually involving several hydrolase enzymes, lead to the formation of the ultimate hormone product (Fig. 3.2). These include the female sex hormones, called *estrogens* (C_{18}) – estradiol-17β, estrone and estriol; the *androgens* (C_{19}) mainly testosterone but also its metabolic precursor androstenedione; *progestins* (C_{21}), progesterone and the *adrenocorticosteroids* (C_{21}) cortisol, corticosterone, aldosterone and 1α-hydroxycorticosterone. Some other adrenocorticosteroids, such as cortisone, are also sometimes found in the blood.

Many other steroids are found in the steroidogenic tissues where they constitute intermediates of the hormones, while others may represent products of steroid catabolism. The chemical structures of these hormones are shown in Fig. 3.2.

The C_{18}, C_{19} and C_{21} steroids have been identified in tissues and also often in the blood of all the main groups of vertebrates. The compounds present are, however, not identical in all of these while the evidence of their precise identity in some (for instance cyclostomes) has been noted as 'tentative' or 'only suggestive' (see Idler, 1972). Steroids of a hormonal nature, nevertheless, undoubtedly have a wide phyletic distribution among vertebrates.

The sex hormones show a remarkable uniformity; testosterone, progesterone and estradiol-17β are common throughout the vertebrates. This possibly reflects the 'conservative' nature of the sexual process and the early evolution of a mechanism of such efficiency that little subsequent endocrine modification of the hormonal excitants could be advantageous. It can be seen in Fig. 3.2 that, when the structures of the steroid hormones are compared, the chemical differences appear surprisingly minor. Nevertheless each molecule exerts distinct effects. A high degree of specificity based on such simple structural differences probably allows little room for subsequent successful evolutionary 'experiments'.

Among the adrenocorticosteroids different molecules have emerged and these often have a distinct systematic distribution (Fig. 3.3). There is some doubt as to whether such corticosteroid hormones exist in the cyclostome fishes (see Weisbart and Idler, 1970). The steroid 1α-hydroxycorticosterone is widespread in the adrenal tissues and blood of the Chondrichthyes. This hormone is, however, only present in the Selachii (sharks and rays) and not the Holocephali (chimaeroids) which instead have cortisol (see Idler and Truscott, 1972). Among the Actinopterygii (including the Holostei, Chondrostei and Teleostei) cortisol is the predominant corticosteroid in the blood. Cortisone, aldosterone and corticosterone have also been identified in teleosts but the quantities appear to be small. The criteria for such identifications are, however, sometimes in doubt. The Dipnoi (lung-fishes) possess cortisol, like other bony fishes, and recently a fascinating discovery has been the additional identification of aldosterone in the South American lungfish *Lepidosiren paradoxa* (Idler, Sangalang and Truscott, 1972). Aldosterone has been identified in the blood of representatives of all the tetrapod groups so that its presence in the Dipnoi, but not apparently in most other fish (though it has been found in the Atlantic herring) is consistent with their suggested phylogenetic relationships to tetrapods. A second major corticosteroid, corticosterone, is also present in amphibians, reptiles and birds. This hormone is also the major corticosteroid in some mammals while in others cortisol is predominant.

The ratio of cortisol to corticosterone varies among the mammals. Rats, rabbits and mice secrete little or no cortisol from their adrenal cortices; corticosterone (aldosterone is also present) predominates. Other mammals secrete a mixture of cortisol and corticosterone, usually with the former predominant. It was at one time suggested that the ratio cortisol: corticosterone may be a characteristic of a species and therefore be determined genetically. It has, however, been subsequently found that this ratio can vary considerably, even in a single animal, depending on the physiological conditions. Nevertheless, the inability of the rat to form cortisol reflects the absence of an enzyme, 17α-hydroxylase, and it seems likely that this, at least, is genetic. Most mammals, including placentals and marsupials, secrete more cortisol than corticosterone. An interesting exception is the

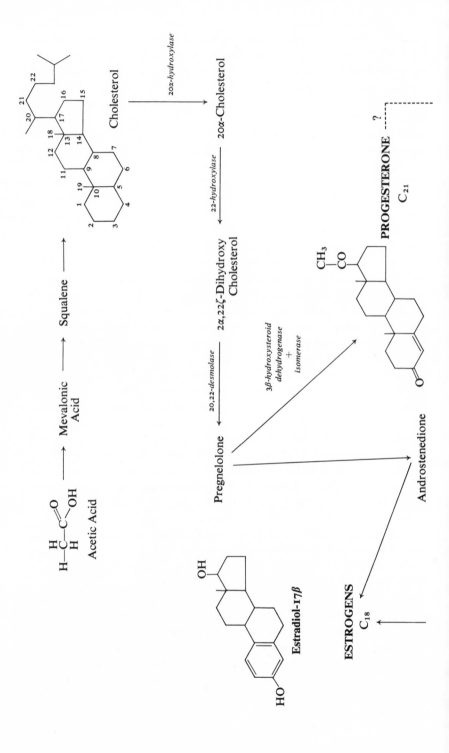

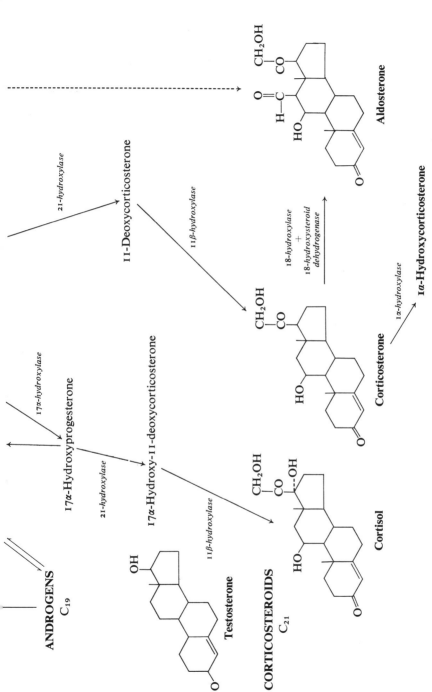

Fig. 3.2. Interrelationships and formation of the steroid hormones.

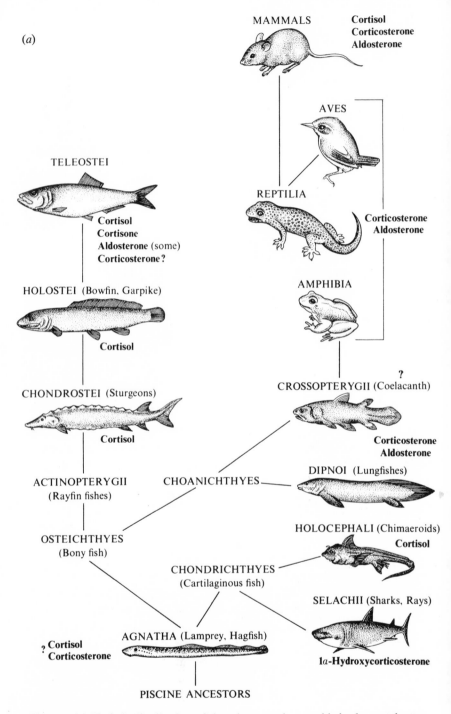

(a)

MAMMALS **Cortisol**
Corticosterone
Aldosterone

AVES

REPTILIA **Corticosterone**
Aldosterone

TELEOSTEI

Cortisol
Cortisone
Aldosterone (some)
Corticosterone?

AMPHIBIA

HOLOSTEI (Bowfin, Garpike)

Cortisol

?

CROSSOPTERYGII (Coelacanth)

Corticosterone
Aldosterone

CHONDROSTEI (Sturgeons)

Cortisol

DIPNOI (Lungfishes)

ACTINOPTERYGII
(Rayfin fishes)

CHOANICHTHYES

HOLOCEPHALI (Chimaeroids)

Cortisol

OSTEICHTHYES
(Bony fish)

CHONDRICHTHYES
(Cartilaginous fish)

SELACHII (Sharks, Rays)

AGNATHA (Lamprey, Hagfish)

? Cortisol
Corticosterone

1*a*-Hydroxycorticosterone

PISCINE ANCESTORS

Fig. 3.3. (*a*) Phyletic distribution of the adrenocorticosteroids in the vertebrates.

(b)

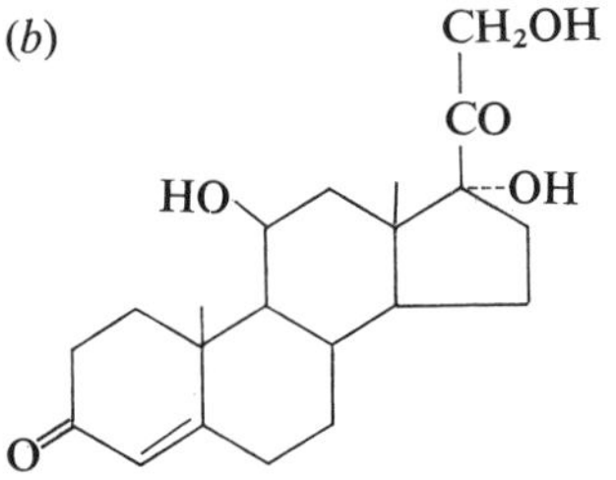

Cortisol

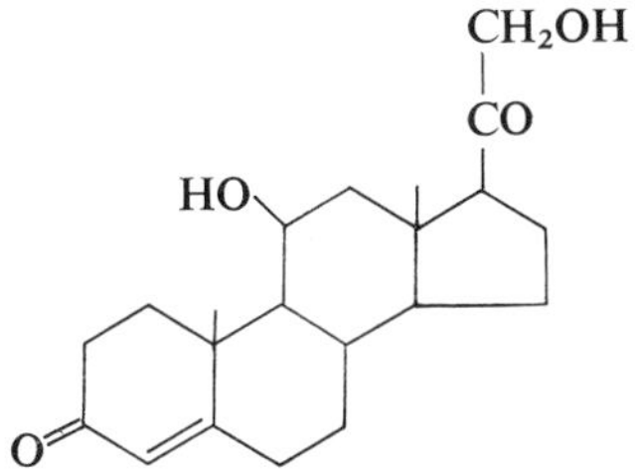

Corticosterone

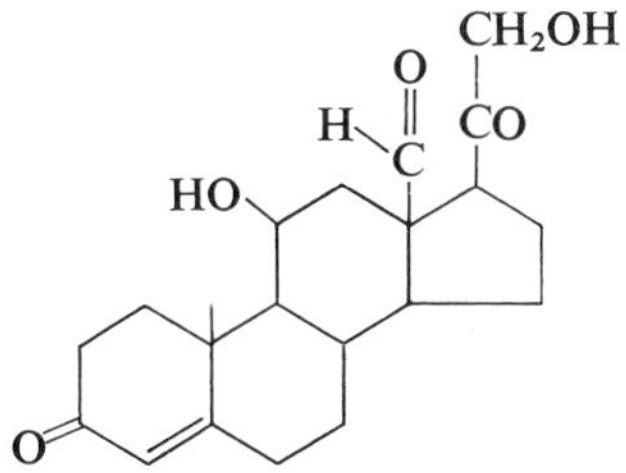

Aldosterone

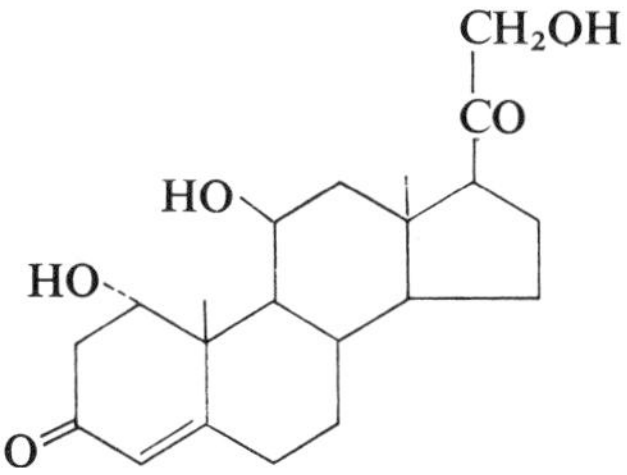

1α-Hydroxycorticosterone

(b) The principal structures of the corticosteroid hormones.

echidna *Tachyglossus aculeatus*, a monotreme in which corticosterone predominates (Weiss and McDonald, 1965). This pattern is more like that in reptiles and birds than that in most other mammals.

The corticosteroids, in contrast to the sex steroids, display different chemical structures that probably reflect evolutionary changes. Sex is a relatively uniform process but the roles of corticosteroids show some variation. This may be reflected in the different structures of these steroids. The role of 1α-hydroxycorticosterone in the Selachii is unknown. In other vertebrates, cortisol and corticosterone influence intermediary metabolism, which is a basic function in all vertebrates. Aldosterone, and to a lesser extent corticosterone, exert a prominent effect on sodium and potassium metabolism in tetrapods. These animals have special osmotic problems not faced by their piscine ancestors, so that it is conceivable that the solutions to them were not only accompanied by the evolution of special effector mechanisms but also hormones to fit them.

Hormones made from tyrosine

Catecholamines

The adrenal medulla and other chromaffin tissues secrete two hormones, adrenaline and noradrenaline (Fig. 3.4). These are amine derivatives of

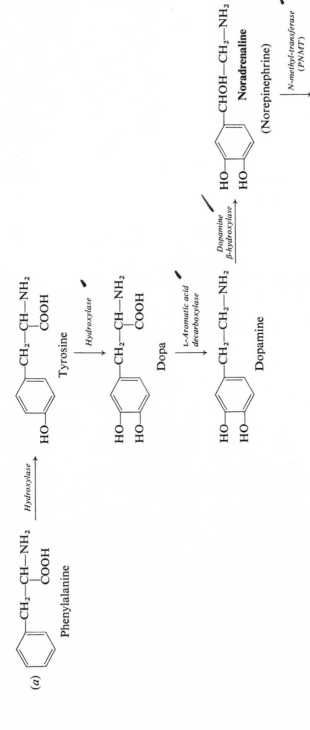

(a)

(b)

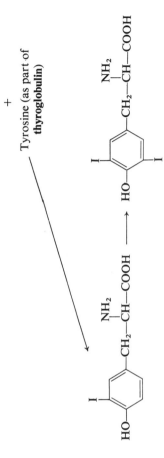
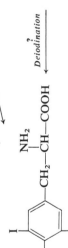

Fig. 3.4. Chemical structure and biological synthesis of (a) the catecholamine hormones; (b) the thyroid hormones.

catechol; hence their name catecholamines. Such compounds are found in all vertebrates where they also act as neurotransmitters in the sympathetic nervous system. They are also present in many invertebrates and even in the ciliated protozoan *Tetrahymena* where they influence metabolism in a manner reminiscent of that in more sophisticated metazoan animals (Blum, 1967).

Using phenylalanine, then tyrosine, as substrates, noradrenaline (or norepinephrine) is formed in chromaffin tissue. This may, under the influence of a methyl transferase enzyme PNMT (phenylethanolamine-*N*-methyl-transferase) have a methyl group added to become adrenaline (or epinephrine). These two hormones have differing actions though a cross-over of their effects occurs. Their actions are separately classified as α-adrenergic and β-adrenergic that can each separately be prevented by specific drugs. β-Adrenergic actions include stimulation of the heart, relaxation of the bronchi, dilatation of certain blood vessels and increase in blood glucose levels. These are principally seen in response to adrenaline rather than noradrenaline which does, however, affect the heart. The constriction of peripheral blood vessels and the sphincter muscles in the gut and bladder are α-adrenergic effects. Both noradrenaline and adrenaline are effective at such sites but adrenaline is more versatile and also has β-adrenergic effects.

The ratio of adrenaline to noradrenaline in the adrenal chromaffin tissue varies considerably among vertebrates. This is illustrated in Table 3.1

TABLE 3.1. *Noradrenaline as per cent of total catecholamines in the adrenals of various species of vertebrates.* (Based on West, 1955)

	Noradrenaline as % total catecholamines
Whale	83
Domestic fowl	80
Dogfish	68
Turtle	60
Pigeon	55
Frog	55
Toad	55
Pig	49
Sheep	33
Ox	26
Man	17
Rat	9
Rabbit	2
Guinea-pig	2

where it can be seen that noradrenaline makes up about 80% of the total catecholamines in whales and the domestic fowl (an ill-assorted phyletic pair) and as little as 2% in rabbits and guinea-pigs. No phyletic pattern in the distribution can be seen. Young and fetal animals possess a predominance of noradrenaline because of the relative lack of the methylating enzyme (see West, 1955).

Thyroid hormones

These are unique hormones as they contain, as part of their structure, the halogen iodine. The formation of thyroxine (tetra-iodothyronine or T_4) and tri-iodothyronine (T_3) is shown in Fig. 3.4. The thyroid gland also contains, probably mainly as metabolic intermediates, mono- and di-iodotyrosine. Thyroxine and tri-iodothyronine differ quantitatively in their effects; those of the former are much slower in onset but longer in duration than those of the latter. It has been suggested that in order to act, T_4 must be converted to T_3 but this is uncertain. T_4 can be bound more strongly to plasma proteins in a complex which may contribute to the difference in the time course of its effects from T_3. In mammals, T_4 is probably secreted at about five times the rate of T_3.

Biologically-active iodothyronine compounds occur throughout the vertebrates and have also been identified in a number of protochordates. Claims as to their presence in non-chordates have not been substantiated though iodotyrosines undoubtedly exist. Iodine readily combines with proteins containing tyrosine (*in vitro*) so that the natural occurrence of such compounds, especially in iodine-rich solutions like sea-water, is perhaps not surprising. Their transformation to iodothyronine compounds, however, seems to depend on specialized metabolic pathways and conditions such as those that occur in the thyroid gland, which has a unique ability to trap and oxidize iodide. Nevertheless, this process can be imitated *in vitro* providing the appropriate amounts of iodine and tyrosine-containing proteins are incubated together.

The spontaneous occurrence of thyroxine compounds in nature, even before the origin of the thyroid gland, is not inconceivable. Whether such compounds did arise and acquire a usefulness as hormonal excitants is sheer conjecture. If this did occur, subsequent specializations may have led to the hormones' more efficient formation in the thyroid gland.

Finally, we may consider the question of why iodine is present in the thyroid hormone molecules. It has been suggested that it plays a vital part in the initiation of the hormonal effect, the hormone acting as a 'carrier' moving iodine into the cell. Thyroxine-like molecules that contain no iodine have been made, however (Taylor, Tu and Barker, 1967). These contain bromine and *iso*propyl groups instead of iodine and yet they still exhibit considerable biological activity. Iodine thus does not appear to have a unique role in the action of thyroid hormones. Presumably, however, it was available in the environment and this, along with its ready chemical reactivity with ty.osine, resulted in its utilization.

The thyroid and catecholamine hormones are clearly very 'conservative' with respect to evolutionary changes of their chemical structure. This would seem to be, at least partly, because of their small size which limits the possibility for change in their molecules, and the fact that they are made from simple precursors that are abundant in nature. The catecholamines and thyroid hormones (and to a slightly lesser extent the steroid hormones) provide us with an illustration of the dictum that 'it is not the hormones that have evolved but the uses to which they have been put'. As will become particularly apparent in the succeeding sections, this is not always true.

The peptide hormones of the neurohypophysis

Two chemically related hormones are usually secreted by the neurohypophysis. These are peptides containing eight amino acids. They are arranged in a 5-membered ring, joined by a disulfide bridge (contributed by two half-cystine residues) and a side chain with three amino acids (Fig. 3.5

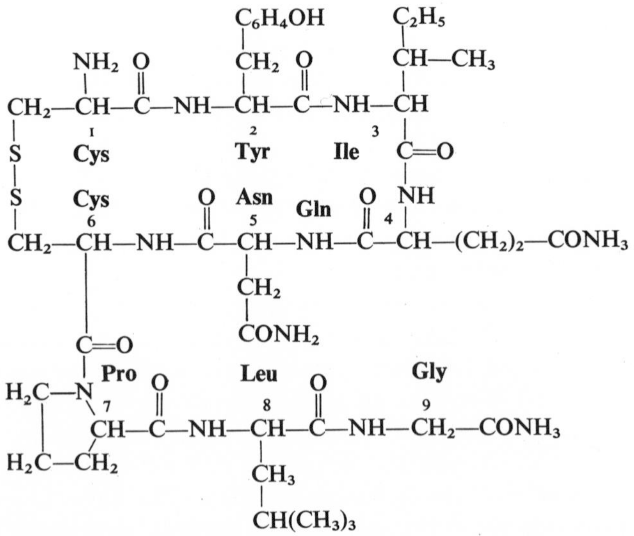

Fig. 3.5. The structure of oxytocin showing the conventional numbering of the amino acids.

TABLE 3.2. *Amino acid sequences of known neurohypophysial hormones.* (Heller, 1974)

Common structure (Variations in positions 3, 4, and 8 indicated by (X)	1 Cys	2 Tyr	3 (X)	4 (X)	5 Asn	6 Cys	7 Pro	8 (X)	9 Gly(NH$_2$)

Basic peptides	Amino acids in position		
	3	4	8
Arginine vasopressin (AVP)	Phe	Gln	Arg
Lysine vasopressin (LVP)	Phe	Gln	Lys
Arginine vasotocin (AVT)	Ile	Gln	Arg
Neutral (=Oxytocin-like) peptides			
Oxytocin	Ile	Gln	Leu
Mesotocin	Ile	Gln	Ile
Isotocin (= ichthyotocin)	Ile	Ser	Ile
Glumitocin	Ile	Ser	Gln
Valitocin	Ile	Gln	Val
Aspartocin	Ile	Asn	Leu

and Table 3.2). In most mammals the two hormones are arginine-vaso-pressin (AVP, also called antidiuretic hormone or ADH) and oxytocin. These differ by two amino acid substitutions; vasopressin has phenyl-alanine and arginine at positions 3 and 8 in the molecule, where oxytocin has *iso*leucine and leucine. This change confers considerable differences in biological activity; vasopressin enhances water reabsorption across the renal tubule, and so reduces urine flow, while oxytocin can contract the uterus and initiate 'milk let-down' from the mammary glands. There is little cross-over in their actions.

Homologous hormones have been identified in the neurohypophyses of representatives of all the systematic groups of vertebrates. Considerable differences in chemical structure however exist so that, so far, nine such peptides have been identified in nature. Amino acid substitutions occur at the 3, 4 and 8 positions in the molecule (Table 3.2). The occurrence of these natural analogues has a well-defined systematic distribution (Fig. 3.6). For example, arginine-vasopressin is confined to mammals while arginine-vasotocin (a combination of the ring of oxytocin and the side chain of vasopressin) is present in all other vertebrates. The second, oxytocin-like (or neutral) peptide in non-mammals exists in five variant forms; mesotocin (*iso*leucine instead of leucine at position 8) is present in birds, reptiles, amphibians and lungfishes; isotocin (*iso*leucine at 8, serine instead of glutamine at 4) is found in all the myriad of bony fishes except lungfishes.

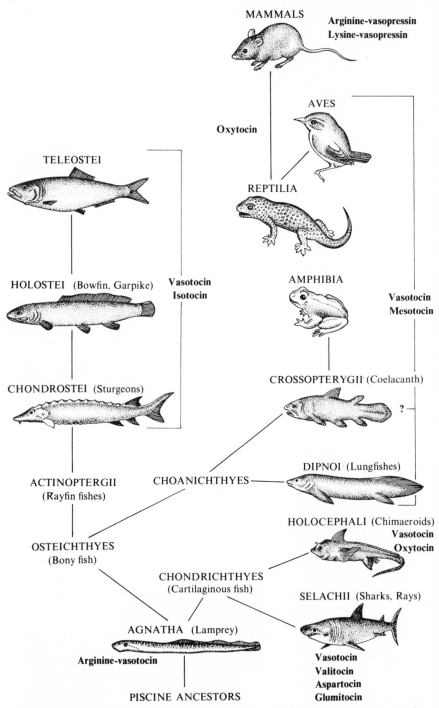

Fig. 3.6. The phyletic distribution of the neurohypophysial hormones among the vertebrates.

The chondrichthyeans exhibit more variability; vasotocin and oxytocin being present in the Holocephali, and vasotocin as well as glumitocin, valitocin and aspartocin are distributed among the Selachii (see Fig. 3.7). While the physiological roles of vasopressin and oxytocin in mammals, and vasotocin in tetrapods, are reasonably well understood, the functions of the other peptides remain unknown, particularly in fish. They are nevertheless present and, from our knowledge of extant species, apparently have persisted for about 500 million years since the first cyclostomes evolved (see Fig. 3.7).

Such polymorphism of the hormones is genetically determined. It can be examined more closely among mammals where a variant of arginine-

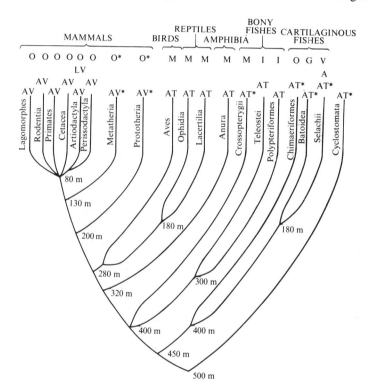

Fig. 3.7. Evolution of the neurohypophysial hormones.
Letters represent hormones that have been identified in extant species from each group. O, oxytocin; AV, arginine–vasopressin; LV, lysine–vasopressin; A, aspartocin; V, valitocin; G, glumitocin; AT, arginine–vasotocin; M, mesotocin. 500 m indicates 500 million years since divergence, and so on.
*, Identification is pharmacological, not chemical. (From Acher, Chauvet and Chauvet, 1972.)

vasopressin occurs. Many pig-like mammals (Suiformes; including the true pigs, Suidae, the peccaries, Tayassuidae and the hippopotamus, Hippopotamidae) possess a vasopressin with lysine instead of arginine present in the 8 position. This probably arose as a result of a single-step mutation from arginine-vasopressin. Its present distribution suggests that this transformation occurred in an ancestor of the Suiformes before the Hippopotami broke away from the pig–peccary stock (Ferguson and Heller, 1965). The change occurred in the Eocene, about 60 million years ago. The neurohypophyses of domestic pigs only contain lysine-vasopressin (and oxytocin) but among other Suiformes, such as peccaries, warthogs and hippopotami, both arginine- and lysine-vasopressin may be present in the same individual. The homozygotes contain one such peptide, the heterozygotes both. The evolutionary persistence of lysine-vasopressin seems to reflect the fact that its biological potency is only a little less than that of its arginine-containing relative so that it is not appreciably disadvantageous. In addition, an adaptive increase in sensitivity of the kidney to lysine-vasopressin may occur (Stewart, 1973). It is, of course, possible that the presence of this hormone also confers adaptive advantages which we do not know about.

Mutations that result in such changes in the amino acid composition of the neurohypophysial peptides may occur periodically. A strain of mice descended from some wild Peruvian specimens has been shown to contain lysine-vasopressin which is not usual in this species (Stewart, 1968). Also, a strain of laboratory rats (Brattleboro) has been found to lack vasopressin altogether and so cannot regulate its urine flow normally (Valtin, Sawyer and Sokol, 1965). When these rats are bred with normal rats the pituitary glands of the heterozygotes contain reduced amounts of vasopressin. Such genetic changes may reflect the complete absence of such a peptide or a mutational 'mutilation' of the molecule so that it has no biological action. The physiological result to the animal would be the same.

A number of biologists have played genetic games with the neurohypophysial peptides. Forearmed with the structures of the natural analogues and the genetic code it is possible to construct feasible lines for their evolution. These are usually calculated on the basis of the minimum number of mutations needed to produce a change from one amino acid at a certain position in a molecule, to another. Four such schemes (that do not include the diverse chondrichthyean hormones) are shown in Fig. 3.8. Most consider vasotocin to be the 'parent' or original, ancestral molecule.

The chemists have made more than 600 analogues of the neurohypophysial hormones that have not been identified in nature (see Berde and Boissonnas, 1968). By looking at the biological activity of these we can speculate about why the nine identified natural analogues have arisen.

Mutations have been perpetuated at only three positions, 3, 4 and 8, in these molecules. With regard to the 8 position, basic amino acids like arginine and lysine endow it with the most activity on the mammalian kidney and non-mammalian effectors such as the frog and toad skin and urinary bladder (osmotic water transfer across these preparations is increased by such peptides). Less-basic amino acids such as ornithine and histidine are much less effective. Similarly leucine at the 8 position in oxytocin results in a hormone that is most potent in its ability to contract the mammalian uterus and stimulate milk let-down from the mammary gland. Substituents at positions other than 3, 4 or 8 usually result in drastic reductions in biological activity. It seems that of all the thousands of possibilities available, nature has, in the course of time, provided the hormones with a structure optimal to that of the receptors. There is, however, an interesting exception. If threonine is substituted for glutamine in the 4 position of oxytocin a peptide is formed that is about four times as effective on the mammalian uterus as oxytocin itself (Manning and Sawyer, 1970). Why then is this not found in nature? The transition to this molecule from oxytocin would require two successive mutations, the first being the substitution of lysine or proline in the 4 position. Such analogues have a very low activity and so may not survive in nature. The succeeding mutation to threonine may thus not have been possible.

The quantities of the hormones stored in the neurohypophysis as well as the ratios of their concentrations are also determined genetically. Storage of these peptides in the neurohypophysis of fishes is more than ten times less than that normally seen in tetrapods. Five inbred strains of mice have been shown to exhibit a two-fold range of variation in the stores of vasopressin and oxytocin in their neurohypophyses (Stewart, 1972). The molecular ratio of vasopressin to oxytocin (V/O ratio) however remained a steady 1.5. As seen in Table 3.3 systematic variations in this V/O ratio occurs among the mammals. The marsupials have a much higher V/O ratio than most other mammals. It is also interesting that two geographically-separated members of the Tylopoda, the camel (from Asia) and the llama (from South America) both have a V/O ratio of about 3, which is higher than that observed in other placentals. The nature of such hereditarily determined differences could be in the ratio of neurosecretory fibers, if one type made oxytocin and another vasopressin. Alternatively if the two hormones are made together, in the same neuron, this synthesis may be regulated by a control system of genes.

The neurohypophysial peptides are stored in granules in the secretory neurons and are associated with proteins (with molecular weights about 10 000) called neurophysins. Two (or possibly even more) such neurophysins (I and II) occur in mammals and it is possible that one of these is

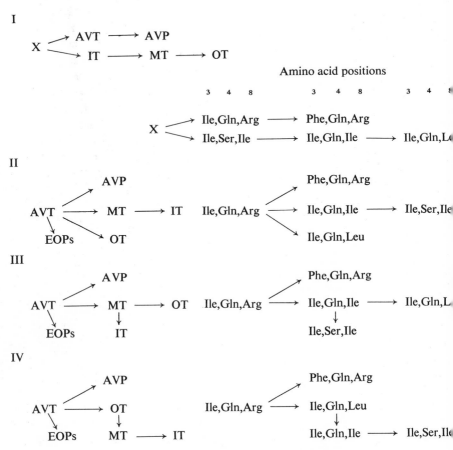

Fig. 3.8. Schemes that have been suggested to represent the successive steps in the molecular evolution of the neurohypophysial peptide hormones (from Geschwind, 1969).

Each transition has been represented as a single amino acid replacement and in some instances it can be seen that two successive changes must be proposed to account for a hormones evolution from the 'parent' molecule; usually considered to be 8-arginine-vasotocin.

Such changes in amino acid composition can be described according to codon base changes in the genetic code. These transformations are often consistent with a single base change (one-step mutation) but in other instances, two base replacements would be needed; such as the transition from isotocin to mesotocin. In this instance the intermediate could be 4-proline, 8-isoleucine-oxytocin but this peptide has not been identified in nature. At the time that the proposed schemes were advanced the structures of the chondrichthyean neurohypophysial peptides (EOPs; glumitocin, aspartocin and valitocin, see Fig. 3.5) were unknown but it was suggested that they also arose from vasotocin. This is still thought to be likely but the transitions may involve more than a single unknown intermediate

TABLE 3.3. *Examples of the distribution of mole ratios of neurohypophysial hormones in mammalian taxonomic groups. The ratio of arginine-vasopressin to oxytocin is shown.* (From Heller, 1966)

Order or suborder	Species	V/O
Order: Marsupialia	American opossum	2.9
	Australian opossum	6.2
	Wallaby	3.8
	Red kangaroo	4.8
Order: Perissodactyla	Horse	0.93
	Zebra	0.44
	Tapir	0.72
	Black rhinoceros	0.77
Order: Artiodactyla		
Suborder: Ruminantia	African buffalo	0.9
	Kongoni	1.5
	Topi	1.4
	Blue wildebeeste	1.3
	Kob	1.4
	Bushbuck	1.4
Suborder: Tylopoda	Llama	3.1
	Camel	3.6

associated with the storage and synthesis of oxytocin and the other with vasopressin. Domestic pigs and warthogs that have lysine-vasopressin have been shown to possess a neurophysin (II) that shows chemical and immunological differences from that in other mammals (Uttenthal and Hope, 1972). The mutational changes in the hormone's structure thus appear to be accompanied by changes in its binding neurophysin. This association may reflect a common synthetic pathway for both the hormone and neurophysin.

Arginine-vasotocin is present in the neurohypophyses of non-mammals but not in those of adult mammals; it was fascinating to find that vasotocin occurs in the fetuses of sheep and seals (Vizsolyi and Perks, 1969).

The neurohypophysial peptide hormones are apparently not confined to the neurohypophysis, as vasotocin has also been identified in the pineal

'hormone'. 8-Lysine-vasopressin (present in the Suina) probably evolved from arginine-vasopressin by a single base replacement.

AVT = arginine-vasotocin, AVP = arginine-vasopressin, IT = isotocin, MT = mesotocin, OT = oxytocin, EOPs = the chondrichthyean neurohypophysial hormones and X = a 'parent' molecule of unknown composition.

gland (see Pavel *et al.*, 1973) and probably the teleostean urophysis (see Chapter 8). Vasopressin has also been identified in some cells from human tumors from non-endocrine tissues.

The neurohormones of the median eminence

The median eminence, at the base of the brain, is the site of release of several hormones that pass into the portal blood vessels of the adenohypophysis (see Schally, Arimura and Kastin, 1973). There is reasonable evidence to indicate that at least nine such hormones (Table 3.4) exist which stimulate or inhibit the release of the adenohypophysial hormones.

TABLE 3.4. *Hypothalamic hormones believed to control the release of pituitary hormones.* (Based on Schally, *et al.*, 1973. Copyright © 1973 by the American Association for the Advancement of Science)

Hypothalamic hormone (or factor)	Abbreviation
Corticotrophin (ACTH)-releasing hormone	CRH or CRF
Thyrotrophin (TSH)-releasing* hormone	TSH-RH or TRH or TRF
Luteinizing hormone (LH)-releasing* hormone	LH-RH or LH-RF
Follicle-stimulating hormone (FSH)-releasing* hormone	FSH-RH or FSH-RF
Growth hormone (GH)-releasing* hormone	GH-RH or GH-RF
Growth hormone (GH) release-inhibiting hormone–somatostatin	GH-R-IH or GIF
Prolactin release-inhibiting hormone	P-R-IH or PIF**
Prolactin-releasing hormone	PRH or PRF***
Melanocyte-stimulating hormone (MSH) release-inhibiting hormone	MSH-R-IH or MRIH or MIF
Melanocyte-stimulating hormone (MSH)-releasing hormone	MRH or MRF

* Or regulating hormone.   ** ? Dopamine.   *** ? TRH.
The evidence for the presence of some of these hormones is still equivocal.

These are, like the neurohypophysial hormones, thought to be secretory products of neurons present in the hypothalamus. Similarly, they are, chemically, usually thought to be peptides (it is possible that prolactin-R-IH is dopamine, MacLeod and Lehmeyer, 1974) and are called releasing- or inhibiting-hormones (-RH or -IH). For instance, the one that releases corticotrophin is called corticotrophin-releasing hormone or CRH, and that which inhibits the release of melanocyte-stimulating hormone (MSH),

Thyrotrophic-stimulating hormone-releasing hormone, TRH or TSH-RH

(Pyro) (Glu)(His)(Pro)—(NH₂)

Luteinizing hormone/follicle-stimulating hormone-releasing hormone,
LH/FSH-RH, Gn-RH

(Pyro)(Glu)(His)(Trp)(Ser)(Tyr)(Gly)(Leu)(Arg)(Pro)(Gly)—NH₂

Growth hormone-releasing hormone, GRH or GHRH

H—(Val)(His)(Leu)(Ser)(Ala)(Glu)(Glu)(Lys)(Glu)(Ala)—OH

Growth hormone-release-inhibiting hormone (somatostatin)

H—(Ala)(Gly)(Cys)(Lys)(Asn)(Phe)(Phe)(Trp)(Lys)(Thr)(Phe)(Thr)(Ser)(Cys)—OH

Melanocyte-stimulating hormone-releasing -inhibiting hormone,
MSH-R-IH or MRIH

(Pro)(Leu)(Gly)—NH₂

Melanocyte-stimulating hormone-releasing hormone, MRH

H—(Cys)(Tyr)(Ile)(Gln)(Asn)—OH

Fig. 3.9. The amino acid sequences of some of the hormones from the median eminence that influence the release of adenohypophysial hormones. The structures of GH-RH and MSH-RH are tentative, they exhibit such activity but have not been positively identified in the median eminence. Some doubt has even been expressed about MSH-R-IH.

melanocyte-stimulating hormone-release-inhibiting hormone or, much more conveniently, MSH-R-IH.

The amino acid sequence in some of these hormones has been elucidated. Thyrotrophin-releasing hormone, TRH, is a tripeptide containing glutamine, histidine and proline; luteinizing hormone-RH and follicle stimulating hormone-RH (LH/FSH-RH) is a decapeptide (see Fig. 3.9). This RH stimulates release of both the gonadotrophins and so is also called gonadotrophin-releasing hormone or Gn-RH. It is uncertain whether the structures proposed for growth hormone-RH (GH-RH) and MSH-RH are the same as those that exist in the median eminence and the identity of MSH-R-IH has also been questioned.

Little is known about the presence or chemical identities of such hormones in the median eminence of non-mammals but sporadic evidence suggests that they are present. Mammalian LH/FSH-RH promotes ovulation in the domestic fowl and a hormone that also has this effect has been found in the fowl's median eminence (van Tienhoven and Schally, 1972; Smith and Follett, 1972). Using an immunoassay, TRH has been identified in the hypothalamus of a variety of non-mammals, including the domestic fowl, a reptile, an amphibian, a teleost fish and even from the brain of a larval cyclostome and the head region of a protochordate *Amphioxus* (Jackson and Reichlin, 1974). It is also interesting that TRH has been identified in other parts of the brain, apart from the hypothalamus, in both mammals and non-mammals, suggesting that it may have a more widespread role as a neurotransmitter.

Polymorphism among these hormones has not been documented. Porcine and ovine TRH are identical. As this hormone is a tripeptide there is little opportunity for change and substituted synthetic analogues have little activity. It is, however, noteworthy that while mammalian TSH can stimulate the thyroid gland in the African lungfish, mammalian TRH is ineffective in these fish (Gorbman and Hyder, 1973). This observation suggests that, as in frogs, a different mechanism or molecular variant of TRH may be present in non-mammals. LH/FSH-RH is similar in pigs and sheep. The biological activity of this molecule is usually decreased when amino acid substitutions are made in it but when the terminal glycinamide (see Fig. 3.9) is replaced by structures lacking an electrical charge, a much more potent substance results (Rippel *et al.*, 1973). It would thus appear that the chemist in his laboratory can improve somewhat on nature's hormone. MSH-R-IH is an interesting tripeptide as its composition seems to be identical to that of the side chain of oxytocin (Celis, Taleisnik and Walter, 1971; Celis, Hase and Walter, 1972). It can be formed from this neurohypophysial hormone as a result of the action of an enzyme present in the median eminence. Tripeptides with the side chain of lysine- and arginine-vasopressin also have MSH-R-IH activity, though somewhat less than that of the side chain of oxytocin. Another remnant of the oxytocin molecule, a 5-membered pentapeptide, has been found to increase the release of MSH (MSH-RH?) but this molecule has not been positively identified in the median eminence. The activity of those peptides formed from oxytocin, which thus acts as a pro-hormone, suggests that comparable median eminence hormones in other vertebrates may have structures which reflect that of the particular neurohypophysial peptide they possess.

It is even possible that other fragments of these molecules may influence the release of other adenohypophysial hormones.

The neurohypophysial hormones themselves have also been identified in the median eminence and the injection of vasopressin has been shown to stimulate the release of corticotrophin and growth hormone. It is thus possible that these peptide hormones may also influence the normal release of the adenohypophysial hormones and indeed vasopressin has recently been identified, in high concentration, in the blood of the hypophysial portal vessels of monkeys (Zimmerman *et al.*, 1973).

The renin–angiotensin system

It has been known since the turn of the century that saline extracts of the mammalian kidney, when injected into mammals, produce a large increase in the blood pressure. This effect is due to the interaction of an enzyme present in the kidney called renin which, as described in the last chapter, is formed by the juxtaglomerular cells. Renin interacts with an α-2 globulin in the blood plasma to form angiotensin I which is converted by a 'converting' enzyme, to angiotensin II which is the hormone that actively constricts the peripheral blood vessels. Angiotensinogen is the substrate.

Renin has a wide phyletic distribution in bony fishes and tetrapods. It is, however, absent in cyclostomes and chondrichthyean fishes (Nishimura *et al.*, 1970). Renin has also been tentatively identified in the corpuscles of Stannius of some bony fishes. It thus appears to have made a slightly later entry into the vertebrates than most other hormones.

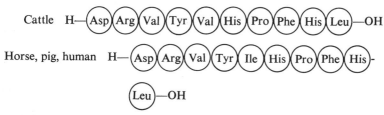

Fig. 3.10. The amino acid sequence of angiotensin I derived from the ox and horse, and pig and man. The active octapeptide, angiotensin II, is formed by removal of the histidine and leucine residues by a 'converting enzyme'.

Angiotensin I is a decapeptide (Fig. 3.10) which is activated by conversion to an octapeptide on the removal of histidine and leucine from one end of the molecule. Species differences in structure exist as can be seen in Fig. 3.10. There is also evidence indicating that marsupial angiotensinogen (from kangaroos) differs from a placental (sheep) substrate (Simpson and

Renins

	Fish	Amphi-bian	Reptile	Aves	Mammal
Mammal	—	—	—	—	+
Aves	—	—	—	+	—
Reptile	—	$-$ $+$	+	+	—
Amphi bian	—	+	+	+	—
Fish	+	+	+	+	—

Angiotensinogens

Fig. 3.11. The interactions of renin and angiotensinogen, from different verte-brates, to form angiotensin. On the ordinate are the renins from the different vertebrates and on the abscissa, the angiotensinogens. +, Angiotensin (or angiotensin-like) formed; –, no interaction. (From Nolly and Fasciola, 1973. Reprinted with permission of Pergamon Press.)

Blair-West, 1972). The structures of angiotensin have not been determined in non-mammals but differences in their chemical and pharmacological behavior indicate that structural variation is widespread (Sokabe and Nakajima, 1972).

Considerable differences have also been observed in the interactions between renins and angiotensinogens from the various major phyletic groups (Fig. 3.11). Fish (teleost) renin interacts with angiotensinogens (to form angiotensin) from all the tetrapods except mammals. Mammalian renin will not interact with the plasma substrate from non-mammals. Bird renin is also very specific while that of other vertebrates is less so. Amphibian renin interacts with the angiotensinogens from birds and reptiles and reptilian renin reacts with those of birds and amphibians.

Parathormone and calcitonin

Parathormone and calcitonin are, respectively, the peptide hormones originating in the parathyroids and ultimobranchial bodies (or in mammals

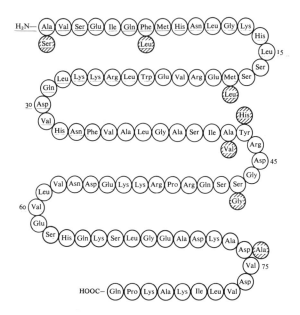

Fig. 3.12. Amino acid sequences of parathyroid hormone from the ox (main structure) and the substitutions that occur in that from the pig (shaded residues). (From Aurbach *et al.*, 1972.)

the thyroid 'C' cells). Parathormone increases calcium levels in the plasma and calcitonin decreases them. Parathormone contains 84 amino acids and is present in all tetrapods but not fishes. Variations in its structure undoubtedly exist but have not been chemically elucidated in non-mammals. The porcine and bovine hormones differ from each other by seven amino acid substitutions (Fig. 3.12) which can each be accounted for genetically by a single base change. When tested in the rat (*in vivo*, blood calcium levels) these two parathyroid hormones do not exhibit different biological activities, though human parathormone is only about $\frac{1}{3}$ as active. When bovine and porcine parathormone are compared *in vitro* (by their ability to activate renal adenyl cyclase) the porcine hormone is less effective. This observation reflects differences in their rates of inactivation by the kidney tissue *in vitro* (Aurbach *et al.*, 1972). It is likely that the active hormone at the effector site only represents a portion of the whole polypeptide molecule and that the hormone, once released, is converted at some peripheral site into an active fragment. The complete molecule is not essential for the exertion of a biological effect as it has been shown that a portion, the amino acids 1 to 34 at the amino-terminal of the bovine hormone, has a similar activity to the complete molecule with its 84 amino acids (Tregear *et al.*,

TABLE 3.5. *Calcitonin concentration in glands from various vertebrates.* (From Copp, 1969)

Class and species	Units (MRC)/g fresh gland wt			
	Thyroid	Ultimo-branchial	Internal parathyroid	Unit/kg body wt
Mammalia				
Man, *Homo sapiens*				
Normal thyroid	0.4	–	0.1–0.5	0.16
Medullary cell carcinoma of thyroid	17	–	–	–
Rat, *Rattus rattus*	5–15	–	–	0.2–0.6
Hog, *Sus scrofa*	2–5	–	–	0.4–0.8
Dog, *Canis familiaris*	1–4	–	1.5–3.3	0.25–0.50
Rabbit, *Oryctolagus cuniculus*				
Lower pole	1.5–2	–	2.1–2.5	–
Upper pole	a	–	–	–
Aves				
Domestic fowl, *Gallus domesticus*	a	30–120	–	0.5–0.8
Turkey, *Meleagris gallopavo*	a	60–100	–	0.5–0.9
Reptilia				
Turtle, *Pseudemys concinna suwaniensis*	a	3–9	–	0.002–0.006
Amphibia				
Bullfrog, *Rana catesbeiana*	–	0.5–0.8	–	0.001–0.002
Teleosti				
Chum salmon, *Oncorhynchus keta*	–	25–40	–	0.4–0.6
Gray cod, *Gadus macrocephalus*	–	10–20	–	0.2–0.4
Elasmobranchii				
Dogfish shark, *Squalus suckleyi*	a	25–35	–	0.25–0.40

a, No detectable hypocalcemic activity.

1973). This observation suggests that considerable polymorphism of the parathormone molecule is possible.

Calcitonin activity has been measured in all vertebrates except the cyclostomes (Table 3.5). The hormones contain 32 amino acids. Chemical analysis indicates that considerable differences in their sequence occur that result in quantitative differences of biological activity. The amino acid sequence of the calcitonins in four mammals and a teleost fish (salmon) are

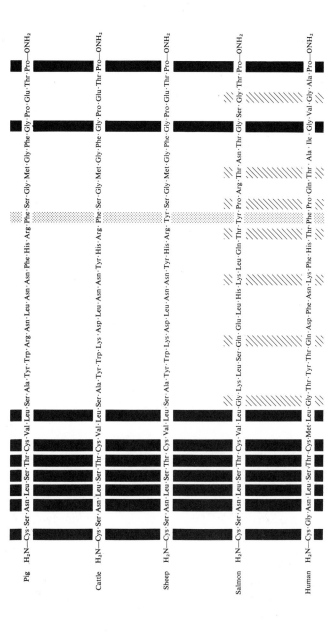

Fig. 3.13. A comparison of the amino acid sequence of calcitonin in four mammals and the salmon. The solid bars indicate the amino acids are homologous in all species. It can be seen that extensive differences exist, especially in the central parts of the molecules. Cross-hatched bars indicate homologies between salmon and human; stippled bar indicates comparable hydrophobic residues. (From Potts *et al.*, 1972.)

shown in Fig. 3.13. Only nine amino acid positions are commonly shared by all five species. The differences, however, can nearly all be accounted for by single-base changes in the genetic code (Potts *et al.*, 1972),

The salmon calcitonins are of special interest. Three variants have been identified among four different species of salmon with amino acid substitutions at four or five positions (Fig. 3.14). All species have a common hormone, calcitonin I, but others, calcitonin II or III, may also be present.

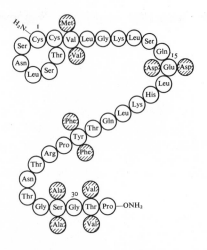

Fig. 3.14. The three variant forms of salmon calcitonin. The main amino acid sequence is that of calcitonin I. Calcitonin II differs by only four residues; valine is still present at position 8. Calcitonin III has 5 substitutions. Among the four species of salmon examined all form calcitonin I, the Chum, Pink and Sockeye salmon also form calcitonin II while the Coho salmon has calcitonin III. (From Potts *et al.*, 1972.)

Salmon calcitonin is much more active when tested in mammals (20–100-fold) than is the natural (homologous) hormone. This is probably a unique situation and is due to two factors: a slow rate of destruction of the piscine hormone as well as a greater affinity for the receptor in the kidney and in bone (Marx, Woodward and Aurbach, 1972). This is a very interesting situation theoretically and shows that a species need not necessarily have evolved a hormone with a structure that has the maximal possible biological activity. It is nevertheless conceivable that other factors may be involved in the 'choice' of such a molecule and sheer persistency in the circulation may even be a disadvantage. Calcitonin could even have other roles in mammals that are not reflected in its ability to increase calcium concentrations in the bioassay system.

The hormones of the Islets of Langerhans and the gastrointestinal tract

Insulin

This hormone has an important role in controlling several processes in intermediary metabolism that involve levels of glucose, fatty acids and proteins. It has been identified in all vertebrate groups and, in addition, in extracts of the gut, and its associated tissues, of many invertebrates. Some coelenterates, crustaceans, molluscs and protochordates have been shown to possess insulin-like substances. This activity has been demonstrated in biological assays in mammals and also by the interaction with antibodies to ox insulin (see Falkmer and Patent, 1972). The chemical nature of the pre-vertebrate 'insulin' is uncertain and it may be a large molecule, a pro-insulin, that can react with vertebrate antibodies and from which biologically active fragments may be split off. It has been suggested (Steiner *et al.*, 1972) that insulin may have originated from a large proteolytic digestive enzyme, a proto-pro-insulin, in such a manner that when it was absorbed into the blood it became associated with certain metabolic processes in the body.

Several of the vertebrate insulins have been described chemically. The molecule consists of two main parts, an A-chain with 21 amino acids and a B-chain with 31. These are joined by two disulfide bridges contributed by four cysteine residues (Table 3.6). Among the 20 species so far examined amino acid substitutions have been recorded at 29 of the 51 positions in the insulin molecules. The A-chain is identical in the insulin present in man, pigs, rabbits, dogs and sperm whales while the B-chain is the same in pig, horse, ox, dog, sheep, goat, sperm whale and sei whale. The intact insulin in the pig, dog, sperm whale and fin whale are identical. Most of the differences that occur in mammals are localized at three positions (8, 9, 10) in the A-chain and in one position (30) in the B-chain. In the guinea-pig, however, changes have occurred, compared to the pig, at 17 positions in the molecule.

Fish (teleost) insulins show a number of distinct differences from those of mammals (Fig. 3.15). Amino acid substitutions occur at more than 15 loci. The B-chain also has an additional amino acid at its *N*-terminal end while the mammalian *C*-terminal amino acid is missing.

The laboratory rat and mouse and some fishes each have two insulins (see Smith, 1966 and Fig. 3.15). The two rodents have insulins which differ by two amino acids (in the B-chain) and both hormones are present in the same individuals. This is thought to be a homozygous condition which nevertheless could be the result of a gene duplication so that two genes are present, each controlling the form of one insulin. It is unknown whether the intraspecific polymorphism among the fish insulins is similar to that in the rodents or whether each hormone is present in separate, individual, fish.

Despite the chemical differences in the structure of the vertebrate insulins there is surprisingly little demonstrable variation in their specific biological activities when they are tested on mammalian preparations.

TABLE 3.6. *Amino acid sequence in vertebrate insulins.* (From Humbel, Bosshard and Zahn, 1972)

The italicized amino acids indicate the principle differences
(a) Amino acid sequences of insulin A chains

Type of insulin	1	2	3	4	5	6	7	8	9	10
Human*	Gly	Ile	Val	Glu	Gln	Cys	Cys	Thr	Ser	Ile
Sei whale	Gly	Ile	Val	Glu	Gln	Cys	Cys	*Ala*	Ser	*Thr*
Horse	Gly	Ile	Val	Glu	Gln	Cys	Cys	Thr	*Gly*	Ile
Beef	Gly	Ile	Val	Glu	Gln	Cys	Cys	*Ala*	Ser	*Val*
Sheep, goat	Gly	Ile	Val	Glu	Gln	Cys	Cys	*Ala*	*Gly*	*Val*
Elephant	Gly	Ile	Val	Glu	Gln	Cys	Cys	Thr	*Gly*	*Val*
Rat, mouse (I and II)	Gly	Ile	Val	*Asp*	Gln	Cys	Cys	Thr	Ser	Ile
Guinea-pig	Gly	Ile	Val	*Asp*	Gln	Cys	Cys	Thr	*Gly*	*Thr*
Chicken, turkey	Gly	Ile	Val	Glu	Gln	Cys	Cys	*His*	*Asn*	*Thr*
Cod	Gly	Ile	Val	*Asp*	Gln	Cys	Cys	*His*	*Arg*	*Pro*
Tuna (II)	Gly	Ile	Val	Glu	Gln	Cys	Cys	*His*	*Lys*	*Pro*
Angler fish	Gly	Ile	Val	Glu	Gln	Cys	Cys	*His*	*Arg*	*Pro*
Toadfish (I)	Gly	Ile	Val	Glu	Gln	Cys	Cys	*His*	*Arg*	*Pro*
Toadfish (II)	Gly	Ile	Val	Glu	Gln	Cys	Cys	*His*	*Arg*	*Pro*

11	12	13	14	15	16	17	18	19	20	21
Cys	Ser	Leu	Tyr	Gln	Leu	Glu	Asn	Tyr	Cys	Asn
Cys	Ser	Leu	Tyr	Gln	Leu	Glu	Asn	Tyr	Cys	Asn
Cys	Ser	Leu	Tyr	Gln	Leu	Glu	Asn	Tyr	Cys	Asn
Cys	Ser	Leu	Tyr	Gln	Leu	Glu	Asn	Tyr	Cys	Asn
Cys	Ser	Leu	Tyr	Gln	Leu	Glu	Asn	Tyr	Cys	Asn
Cys	Ser	Leu	Tyr	Gln	Leu	Glu	Asn	Tyr	Cys	Asn
Cys	Ser	Leu	Tyr	Gln	Leu	Glu	Asn	Tyr	Cys	Asn
Cys	*Thr*	*Arg*	*His*	Gln	Leu	Glu	*Ser*	Tyr	Cys	Asn
Cys	Ser	Leu	Tyr	Gln	Leu	Glu	Asn	Tyr	Cys	Asn
Cys	*Asp*	*Ile*	*Phe*	*Asp*	Leu	*Gln*	Asn	Tyr	Cys	Asn
Cys	*Asn*	*Ile*	*Phe*	*Asp*	Leu	*Gln*	Asn	Tyr	Cys	Asn
Cys	*Asn*	*Ile*	*Phe*	*Asp*	Leu	*Gln*	Asn	Tyr	Cys	Asn
Cys	*Asp*	*Ile*	*Phe*	*Asp*	Leu	*Gln*	*Ser*	Tyr	Cys	Asn
Cys	*Asp*	*Lys*	*Phe*	*Asp*	Leu	*Gln*	*Ser*	Tyr	Cys	Asn

(*b*) *Amino acid sequences of insulin B chains*

Type of insulin	−1	1	2	3	4	5	6	7	8	9
Pig**		Phe	Val	Asn	Gln	His	Leu	Cys	Gly	Ser
Man, elephant		Phe	Val	Asn	Gln	His	Leu	Cys	Gly	Ser
Rabbit		Phe	Val	Asn	Gln	His	Leu	Cys	Gly	Ser
Rat, mouse (I)		Phe	Val	*Lys*	Gln	His	Leu	Cys	Gly	*Pro*
Rat, mouse (II)		Phe	Val	*Lys*	Gln	His	Leu	Cys	Gly	Ser
Guinea-pig		Phe	Val	*Ser*	*Arg*	His	Leu	Cys	Gly	Ser
Chicken		*Ala*	*Ala*	Asn	Gln	His	Leu	Cys	Gly	Ser
Cod	Met	*Ala*	*Pro*	*Pro*	Gln	His	Leu	Cys	Gly	Ser
Tuna (II)	*Val*	*Ala*	*Pro*	*Pro*	Gln	His	Leu	Cys	Gly	Ser
Angler fish	*Val*	*Ala*	*Pro*	*Ala*	Gln	His	Leu	Cys	Gly	Ser
Toadfish (I)	Met	*Ala*	*Pro*	*Pro*	Gln	His	Leu	Cys	Gly	Ser
Toadfish (II)	Met	*Ala*	*Pro*	*Pro*	Gln	His	Leu	Cys	Gly	Ser

10	11	12	13	14	15	16	17	18	19	20	21	22
His	Leu	Val	Glu	Ala	Leu	Tyr	Leu	Val	Cys	Gly	Glu	Arg
His	Leu	Val	Glu	Ala	Leu	Tyr	Leu	Val	Cys	Gly	Glu	Arg
His	Leu	Val	Glu	Ala	Leu	Tyr	Leu	Val	Cys	Gly	Glu	Arg
His	Leu	Val	Glu	Ala	Leu	Tyr	Leu	Val	Cys	Gly	Glu	Arg
His	Leu	Val	Glu	Ala	Leu	Tyr	Leu	Val	Cys	Gly	Glu	Arg
Asn	Leu	Val	Glu	*Thr*	Leu	Tyr	*Ser*	Val	Cys	(*Gln*	*Asp*	*Asp*)
His	Leu	Val	Glu	Ala	Leu	Tyr	Leu	Val	Cys	Gly	Glu	Arg
His	Leu	Val	*Asp*	Ala	Leu	Tyr	Leu	Val	Cys	Gly	*Asp*	Arg
His	Leu	Val	*Asp*	Ala	Leu	Tyr	Leu	Val	Cys	Gly	*Asp*	Arg
His	Leu	Val	*Asp*	Ala	Leu	Tyr	Leu	Val	Cys	Gly	*Asp*	Arg
His	Leu	Val	*Asp*	Ala	Leu	Tyr	Leu	Val	Cys	Gly	*Asp*	Arg
His	Leu	Val	*Asp*	Ala	Leu	Tyr	Leu	Val	Cys	Gly	*Asp*	Arg

23	24	25	26	27	28	29	30
Gly	Phe	Phe	Tyr	Thr	Pro	Lys	Ala
Gly	Phe	Phe	Tyr	Thr	Pro	Lys	*Thr*
Gly	Phe	Phe	Tyr	Thr	Pro	Lys	*Ser*
Gly	Phe	Phe	Tyr	Thr	Pro	Lys	*Ser*
Gly	Phe	Phe	Tyr	Thr	Pro	*Met*	*Ser*
Gly	Phe	Phe	Tyr	*Ile*	Pro	Lys	*Asp*
Gly	Phe	Phe	Tyr	*Ser*	Pro	Lys	Ala
Gly	Phe	Phe	Tyr	*Asn*	Pro	Lys	
Gly	Phe	Phe	Tyr	*Asn*	Pro	Lys	
Gly	Phe	Phe	Tyr	*Asn*	Pro	Lys	
Gly	Phe	Phe	Tyr	*Asn*	Pro	Lys	
Gly	Phe	Phe	Tyr	*Asn*	Ser		

* Sequence is identical in man, rabbit, dog, pig and sperm whale.
** Sequence is identical in pig, horse, ox, dog, sheep, sperm whale and sei whale.

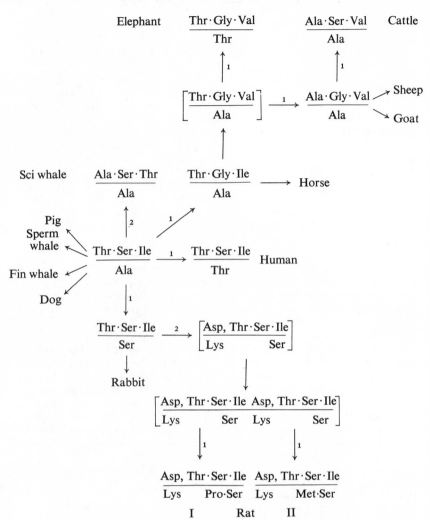

Fig. 3.15. A scheme for the evolution of the mammalian insulins. The sequences of the A-chain are given above the line and the B-chain below it. Numbers alongside the arrows are the minimum number of base changes (mutations) required for the amino acid substitution. The sequences given in brackets are postulated intermediates that have not been identified in nature. The amino acids in the A-chain are those at positions 8, 9 and 10 or in the rat 4, 8, 9, 10.

In the B-chain they refer to the 30 position or in the rat 3, 29 and 30. The rat produces two insulins (I and II), a process that may be due to a gene duplication and a mutation that occurred subsequently in one of the gene-pairs. (From L. F. Smith, 1966 and personal communication.)

Attempts to relate the similarities and differences in the amino acid composition of the insulins to the closeness of the relationship and systematic position of the species have not been very successful. Guinea-pig insulin differs from that in man by 17 amino acid substitutions (16 in a close relative, the elephant). That of the chicken and man (and the elephant) have only six such disparities. Nevertheless an attempt has been made to trace the evolution of the mammalian insulin. The 'parent', or prototype, may be the most common form; which is found in the pig and others. The successive mutations that would be required to produce the hormones present in other mammals can then be traced (Fig. 3.15). As with the neurohypophysial hormones this must be considered a 'game' which is fun and may even be partly correct.

Glucagon

This hormone is a smaller molecule than insulin, consisting of a single chain of 29 amino acids. It has been identified in teleost fishes and all the main tetrapod groups. Glucagon has not been identified in the Chondrichthyes or Cyclostomata though immunological evidence suggests its possible presence in invertebrates including some molluscs and protochordates (see Falkmer and Patent, 1972). The teleost's 'glucagon', while biologically effective (it has a hyperglycemic action) in fish, has no effect in the rabbit. This fact suggests structural differences from the homologous mammalian hormone(s). Failure to detect glucagon in other fishes could reflect even greater structural disparities.

Glucagon-like activity has also been identified in various segments of the mammalian gut (stomach, duodenum, jejunem and colon). Immunological behavior indicates that these excitants from the gut differ structurally from pancreatic glucagon (Samols *et al.*, 1966; Heding, 1971) and are called entero-glucagon.

Secretin

Glucagon and another gut hormone, secretin (which stimulates the exocrine pancreatic secretions), have a remarkable number of similarities in their structure (Weinstein, 1968, Fig. 3.16). Porcine secretin has two less amino acids than porcine glucagon but shares 15 amino acids at identical positions in the molecule. This strongly suggests a common origin. The number of genetic base-changes necessary for the difference involve one or two mutations in each of the disparate amino acid positions. From information about spontaneous mutation rates it has been calculated that gene duplication (of the parent molecule) would have occurred 200 million years ago,

Glucagon

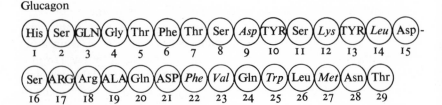

Secretin

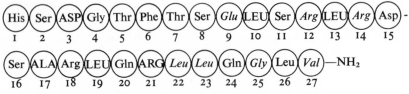

Base changes: glucagon versus secretin

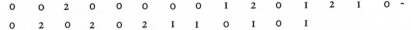

Fig. 3.16. The amino acid sequence of pig glucagon and secretin. They are aligned for direct correspondence between the amino acid positions. When no substitutions have occurred ordinary letters are used, with 'conservative' substitutions capitals are used, and 'radical' changes are italicized. (From Weinstein, 1968.)

early in the Mesozoic period. If glucagon really exists in teleost fish, this genetic event may have occurred somewhat earlier. The 'parent' molecule could, however, possess some glucagon-like activity. Weinstein considers it most unlikely that glucagon could have arisen from an insulin-like molecule as has been suggested by some.

Gastrin

This hormone (which stimulates gastric acid secretion) has been found in the D-cells of the pancreatic islets and in the pyloric region of the stomach. Its distribution in non-mammals has not yet been systematically explored. Early reports have indeed failed to confirm its presence. Several gastrins have been identified in mammals. The structures are shown in Fig. 3.17. Amino acid substitutions (accounted for by single mutations) occur at three positions in the molecule and result in considerable variation in their biological activities. When tested by their ability to alter gastric juice secretion (in cats) the ratio of activities in ovine:porcine:human:canine gastrins were 1.8:1.0:0.5:0.3 (see Bromer, 1972).

(a) GASTRIN I

| | 1 | 2 | 3 | 4 | 5 | 6 | 7 | 8 | 9 | 10 | 11 | 12 | 13 | 14 | 15 | 16 | 17 |

Human   Pyr-Gly-Pro-Trp-Leu-Glu-Glu-Glu-Glu-Glu-Ala-Tyr-Gly-Trp-Met-Asp-Phe—NH_2

Pig   Pyr-Gly-Pro-Trp-*Met*-Glu-Glu-Glu-Glu-Glu-Ala-Tyr-Gly-Trp-Met-Asp-Phe—NH_2

Cat   Pyr-Gly-Pro-Trp-Leu-Glu-Glu-Glu-Glu-Glu-*Ala*-Ala-Tyr-Gly-Trp-Met-Asp-Phe—NH_2

Dog   Pyr-Gly-Pro-Trp-*Met*-Glu-Glu-Glu-*Ala*-Glu-Glu-Ala-Tyr-Gly-Trp-Met-Asp-Phe—NH_2

Cattle, sheep Pyr-Gly-Pro-Trp-*Val*-Glu-Glu-Glu-Glu-Glu-*Ala*-Ala-Tyr-Gly-Trp-Met-Asp-Phe—NH_2

(b)   Cholecystokinin–Pancreozymin

Lys · (Ala$_1$, Gly$_1$, Pro$_1$, Ser$_1$) · Arg · Val · (Ileu$_1$, Met$_1$, Ser$_1$) · Lys ·

Asn · (Asx$_1$, Glx$_1$, His$_1$, Leu$_2$, Pro$_1$, Ser$_2$) · Arg · Ileu · (Asp$_1$, Ser$_1$) ·

                          SO_3H
                          |

Arg · Asp · Tyr · Met · Gly · Trp · Met · Asp · Phe · NH_2

Terminal Octapeptide

               SO_3H
               |

Asp · Tyr · Met · Gly · Trp · Met · Asp · Phe · NH_2

Fig. 3.17. (a) Amino acid sequences of some mammalian gastrins. Residues in italics indicate differences from the hormone present in man. (From Bromer, 1972.)

(b) Tentative amino acid sequence of porcine cholecystokinin–pancreozymin. The active C-terminal octapeptide is also shown. It can be seen that the last five amino acids in the gastrin and cholecystokinin–pancreozymin molecules are identical. (From Rubin et al., 1969.)

Gastrin has not been identified in birds but a polypeptide, which stimulates acid secretion from the avian proventriculus, and so has a 'gastrin-like' action, has been found in the pancreas (Hazelwood *et al.*, 1973). This has been identified in 10 species of birds and in the circulation of the chicken so that it could have an hormonal role. It may thus be a third pancreatic hormone in birds (in addition to insulin and glucagon) and has been dubbed 'APP', which stands for avian pancreatic polypeptide. The chemical structure of APP is quite unlike that of mammalian gastrin as it consists of 36 amino acids (molecular weight, 4200) and contains glycine at its *N*-terminus and tyrosine-amide at its *C*-terminus. It is interesting, however, that the pancreas of several mammals has been shown to contain a similar peptide which shares a common sequence of 16 amino acids with APP. Such a polypeptide has not been identified in amphibians or snakes and it does not act like gastrin in mammals. The structural similarities between the avian and mammalian pancreatic polypeptides, however, suggest that they may have a common ancestry and that, in the birds, the molecule has been used as a hormone (like gastrin) that stimulates the secretion of acid in the gut.

Enterogastrone

On physiological grounds a hormone called enterogastrone, that inhibits the secretion of acid by the stomach, is thought to be released from the upper part of the intestine. This 'hormone' has, however, defied attempts to isolate it chemically so that its existence as a distinct hormone remains in doubt.

Cholecystokinin–pancreozymin

Cholecystokinin is a gastrointestinal hormone that is released from the mucosal cells in the upper regions of the intestine. Pancreozymin was originally thought to be a separate hormone but the two are now known to be identical (Jorpes and Mutt, 1966) and the hormone is called cholecystokinin–pancreozymin. Materials that behave in a similar biological and chemical manner have also been isolated from the intestines of cyclostomes (lampreys, *Lampetra fluviatilis* and *Petromyzon marinus*), a teleost (the eel, *Anguilla anguilla*), and a chondrichthyean (the holocephalian fish *Chimaera monstrosa* (Barrington and Dockray, 1970, 1972; Nilsson, 1970). This hormone thus appears to have a long phylogenetic history. It is a single chain polypeptide containing 33 amino acids, the tentative sequence of which is given in Fig. 3.17. The *C*-terminal octapeptide part of the molecule possesses all of the biological actions of the intact molecule (Rubin *et al.*, 1969). The *C*-terminal pentapeptide (= pentagastrin) is identical to the *C*-terminal sequence in gastrin and this fragment of the molecule exerts all the important actions of gastrin (Tracy and Gregory, 1964). These similarities in the structures of the active sections of these gastrointestinal hormones contribute to the cross-over that is often observed in their actions and suggests that they may have had common ancestral origins.

Adrenocorticotrophin (ACTH), melanocyte-stimulating hormone (MSH) and β-lipotrophin

These pituitary hormones are polypeptides that share a number of common features. They all possess a common core of seven amino acids while the ones adjoining it are also often similar (Table 3.7). They possess, with a few variations, common sequences of up to 20 amino acids. This is reflected in a cross-over in their biological activities; all, for instance, exhibit an ability to stimulate amphibian melanophores.

MSH (Table 3.7) exists in many polymorphic forms though the chemical structures of only a few are known. α-MSH has the same structure in many species (13 amino acids in a chain) but the associated β-MSH (usually 18 amino acids but 22 in man) shows considerable variation. They all, however share the same 'core'. Dogfish MSH has more recently been described chemically and is similar to α-MSH but contains two less amino acids (Lowry and Chadwick, 1970). Comparison of the chromatographic, immunological and biological behavior indicates that many variants of MSH exist among the vertebrates. The differences are seen in such diverse species as the codfish, *Gadus morhua*, the frogs, *Rana catesbeiana* and *R. pipiens*, the lizard, *Anolis carolinensis*, and the domestic fowl, *Gallus domesticus*, to name but a few (see Burgers, 1963; Shapiro *et al.*, 1972). The polymorphism exhibited by MSH is not only an interspecific one. In the sheep, for instance, three chemically distinct MSHs have been found within a single pituitary gland. Apart from indicating considerable genetic variation the reasons for their perpetuation are not apparent, especially as the role of such peptides in mammals is unknown.

ACTH, which stimulates the adrenal cortex to secrete certain steroid hormones consists of a chain of 39 amino acids in mammals. While variations in configuration undoubtedly occur, the chemical structure of ACTH of a non-mammal has not yet been described. The ability to stimulate adrenocortical tissue resides in the first 19 to 23 amino acids at the amino-terminus. Synthetic peptides with such structures have been made and exhibit a similar biological potency to the intact hormone. Differences between the amino acid sequences of ACTH have been described in man,

TABLE 3.7. *Amino acid sequences of meloncyte-stimulating hormones (MSH), adrenocorticotrophin (ACTH) and lipotrophin. The relative MSH activity (as tested on amphibian color change) of some of the materials is shown in the last column*

Hormone	Structure	Relative MSH potency *in vitro*
α-MSH: Pig, cattle, horse, man, monkey	Acetyl-Ser-Tyr-Ser-Met-Glu-His-Phe-Arg-Trp-Gly-Lys-Pro-Val-NH$_2$ 1 4 10 13	1.00
MSH: Dogfish	Ser-Met-Glu-His-Phe-Arg-Tyr-Gly-Lys-Pro-Met-NH$_2$ 2 8	–

β-MSH: Man	Ala-Glu-Lys-Lys-Asp-Glu-Gly-Pro-Tyr-Arg-
	  1
	Met-Glu-His-Phe-Arg-Trp-Gly-Ser-Pro-Pro-Lys-Asp
	  11 22
		0.23

Monkey	Asp-Glu-Gly-Pro-Tyr-Arg-Met-Glu-His-Phe-Arg-Try-Gly-Ser-Pro-Pro-
	  1 13
	Lys-Asp
	  18
		–

Pig, sheep	Asp-Glu-Gly-Pro-Tyr-Lys-Met-Glu-His-Phe-Arg-Trp-Gly-
	  1 13
	Ser-Pro-Lys-Asp
	  18
		0.26

Cattle, sheep	Asp-Ser-Gly-Pro-Tyr-Lys-Met-Glu-His-Phe-Arg-Trp-Gly– 1 · · · · · · · 7 · · · · · · · · 13 Ser-Pro-Pro-Lys-Asp 18	0.66
Horse	Asp-Glu-Gly-Pro-Tyr-Lys-Met-Glu-His-Phe-Arg-Trp-Gly– 1 · · · · · · · 7 · · · · · · · · 13 Ser-Pro-Arg-Lys-Asp	–
ACTH:	Ser-Tyr-Ser-Met-Glu-His-Phe-Arg-Trp-Gly-Lys-Pro-Val-Gly– 1 · · · 4 · · · · · · · · · · · · 10 *Lys*······Phe 15 · · · · 39	0.01
Lipotrophin	Ala-Glu-Lys-Lys-Asp-*Ser*-Gly-Pro-Tyr-*Lys*-Met-Glu-His-Phe-Arg-Trp– 37 · · · · · · · · · · · · · · · · 47 Gly-Ser-Pro-Pro-Lys-Asp······Gln 53 · · · 58 · · · · 90	–

Italicized amino acids indicate different from human MSH.

ox, sheep and pig. The substitutions occur in the less important part of the hormone between positions 25 and 33. The last six amino acids are identical in these mammals.

A large polypeptide, which has an ability to mobilize lipids in the body, called β-lipotrophin, is present in the adenohypophysis of sheep, cattle, pigs and possibly man (Li *et al.*, 1965; Desranleau, Gilardeau and Chrétien, 1972). In the sheep, it contains 90 amino acids. Immunological comparisons indicate that this polypeptide is structurally similar in sheep, cattle and pigs but that in man it is different. The structure of this molecule is of special interest in relation to that of MSH and ACTH. As can be seen in Table 3.7, it shares a common sequence with them, at positions 47 to 53. In addition, with two exceptions, positions 37 to 58 are identical to that of β-MSH in man. These molecules, MSH, ACTH and β-lipotrophin thus share many common chemical features. It has been suggested that they could arise in the pituitary from a single parent protein though this is difficult to reconcile with the dispersion of the morphological sites where they are found within the gland. It remains likely, nevertheless, that such chemical similarities reflect their evolution from a common ancestral molecule.

The pituitary glycoprotein hormones: luteinizing hormone (LH), follicle-stimulating hormone (FSH) and thyrotrophic hormone (TSH)

The gonadotrophins, LH and FSH, and TSH originate in the adeno-hypophysis. They exert trophic actions on the gonads and thyroid gland, stimulating their development and growth as well as the formation and secretion of their hormones.

These three hormones are large molecules (with a molecular weight of about 30 000) composed of amino acids and certain carbohydrate moieties (14 to 18 % of their weight). The precise chemical structure of bovine and ovine LH and TSH is now known, while considerable information is also available about FSH. They are each composed of two subunits termed α- and β- (or sometimes CI and CII) (see Papkoff, 1972). The subunits TSH-α and LH-α (or ICSH-α) each contain 96 amino acids and are virtually identical (Fig. 3.18). The differences in the molecules, and hence their biological specificity, reside in the structure of their β-subunits. LH-β contains 120 amino acids while TSH-β has 113. These alone have little or no action in mammals. If THS-α is joined to LH-β the LH activity is restored and in the same way if LH-α and TSH-β are combined the molecule has the usual TSH activity.

The precise chemical constitution of these adenohypophysial molecules is only known in mammals. Extracts of the pituitaries of other vertebrates have similar biological actions on the gonads and thyroid which are seen when homologous extracts are injected into a species or even when such preparations are tested on other vertebrates. Some gonadotrophins and TSH in non-mammals have been chemically identified as glycoproteins. An amino acid analysis has been made on a highly purified preparation of a teleost gonadotrophin (the carp) and this shows considerable differences from that of mammalian LH and FSH (Burzawa-Gerard and Fontaine, 1972). Chemical behavior (principally the ability to separate the biological activities chromatographically or electrophoretically) of extracts from non-mammals suggests that the gonadotrophic activities may reside in a single molecule which incorporates both FSH- and LH-like activities. A distinct LH and a FSH have been isolated in mammals, in birds and a chelonian but in other reptiles, amphibians and fishes a single gonadotrophin seems to be present which can react with more than one type of effector and so fulfill the dual functions of FSH and LH.

The principal evidence that vertebrate gonadotrophin(s) and TSH exist in various polymorphic forms comes from comparative measurements of their activities using bioassay preparations derived from different species. The ratio of the biological activity of two glandular extracts can be initially compared in one type of assay system and then repeated in other, different, types of preparations. If the ratio differs in the two or more systems used for the measurements it suggests that the activities result from hormones that differ somewhat in their chemical structures. For the measurement of gonadotrophic activity such assay preparations include stimulation of the uptake of radioactive phosphorus (^{32}P) by the testes of day-old chicks or eels, spermiation in amphibians and teleosts, maintenance and development of the gonads and secondary sex characters in lizards and ovulation in amphibians and mammals. Comparison of TSH activity depends on measurements of ^{131}I uptake by the thyroid gland, the release of thyroxine and the histological appearance and size of the cells in the thyroid. Phyletically diverse species have been used for these comparative assays ranging from mammals to teleosts and chondrichthyean fishes.

'There are no clear cut, well documented cases of species specificity of gonadotrophic hormones' (Nalbandov, 1969). Different species invariably show *some* response to heterologous gonadotrophins. There are, however, considerable variations in the biological potency of such hormones indicating that polymorphic variations exist. Mammalian gonadotrophins exhibit some activity in all vertebrates. In teleost fish, mammalian LH (and human chorionic gonadotrophin, HCG) are sometimes effective but FSH is inactive (Burzawa-Gerard and Fontaine, 1972). Teleost gonadotrophin, on the other hand, while being very active in teleosts, has little effect in mammals. Amphibians, reptiles and birds show considerable responsiveness

(a) H—(Phe)(Pro)(Asp)(Gly)(Glu)(Phe)(Thr)(Met)(Gln)(Gly)(Cys)(Pro)(Glu)(Cys)-
                                                        10

(Lys)(Leu)(Lys)(Glu)(Asn)(Lys)(Tyr)(Phe)(Ser)(Lys)(Pro)(Asp)(Ala)(Pro)-
                              20

(Ile)(Tyr)(Gln)(Cys)(Met)(Gly)(Cys)(Cys)(Phe)(Ser)(Arg)(Ala)(Tyr)(Pro)-
        30   40

                                                                CHO
                                                                 |
(Thr)(Pro)(Ala)(Arg)(Ser)(Lys)(Lys)(Thr)(Met)(Leu)(Val)(Pro)(Lys)(Asn)-
                              50

(Ile)(Thr)(Ser)(Glu)(Ala)(Thr)(Cys)(Cys)(Val)(Ala)(Lys)(Ala)(Phe)(Thr)-
              60   70

                                                  CHO
                                                   |
(Lys)(Ala)(Thr)(Val)(Met)(Gly)(Asn)(Val)(Arg)(Val)(Glx)(Asn)(His)(Thr)-
                                        80

TSH-α: (Cys)(Ser)

(Glu)(Cys)(His)(Ser)(Cys)(Thr)(Cys)(Tyr)(Tyr)(His)(Lys)(Ser)—OH
                        90

                                                          CHO
                                                           |
(b) H—(Ser)(Arg)(Gly)(Pro)(Leu)(Arg)(Pro)(Leu)(Cys)(Glu)(Pro)(Ile)(Asn)(Ala)-
                                                    10

(Thr)(Leu)(Ala)(Ala)(Glu)(Lys)(Glu)(Ala)(Cys)(Pro)(Val)(Cys)(Ile)(Thr)-
                        20

(Phe)(Thr)(Thr)(Ser)(Ile)(Gly)(Ala)(Tyr)(Cys)(Cys)(Pro)(Ser)(Met)(Lys)-
        30   40

(Arg)(Val)(Leu)(Pro)(Val)(Pro)(Pro)(Leu)(Ile)(Pro)(Met)(Pro)(Gln)(Arg)-
                              50

(Val)(Cys)(Thr)(Tyr)(His)(Gln)(Leu)(Arg)(Phe)(Ala)(Ser)(Val)(Arg)(Leu)-
              60   70

(Pro)(Gly)(Pro)(Cys)(Pro)(Val)(Asp)(Pro)(Gly)(Met)(Val)(Ser)(Phe)(Pro)-
                              80

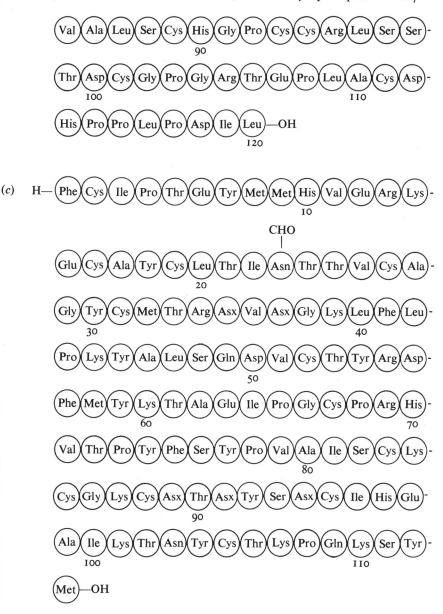

Fig. 3.18. Amino acid sequences the α- and β- chains of mammalian thyro-trophic hormone (TSH) and luteinizing hormone (LH, *or* ICSH).
(*a*) Ovine ICSH-α and bovine TSH-α; (*b*) ovine ICSH-β; (*c*) bovine TSH-β. (From Papkoff, 1972.)

to gonadotrophins from teleost, chondrichthyean and dipnoan fishes as well as those from mammals. The avian hormones are more effective in lizards than in mammals (Burzawa-Gerard and Fontaine, 1972; Donaldson *et al.*, 1972; Licht and Stockell Hartree, 1971; Scanes, Dobson, Follett and Dodd, 1972). While a reptilian gonadotrophin preparation, from the snapping turtle, is ineffective *in vivo* on mammalian test-preparations it stimulates the ovarian granulosa cells of a monkey *in vitro* and is also active in birds, amphibians and other reptiles (Channing *et al.*, 1974; Licht and Papkoff, 1974*a*). No gonadotrophin preparations thus would appear to be completely species specific but they exhibit considerable differences in their activity when tested on preparations from other phyletic groups. Such differences are assumed to reflect variations in their molecular structures.

Immunological cross-reactions are also indicative of chemical relationships between hormones. Antibodies to highly purified preparations of LH from the domestic fowl have been tested on a variety of vertebrates (Scanes, Follett and Goos, 1972). Such antiserum reacted with pituitary gland extracts and plasma from 10 other species of birds as well as three species of reptiles and a dogfish. Non-parallel dose–response reactions were observed with preparations from three amphibians, a lungfish and one teleost (the goldfish) while no reaction could be seen in another teleost (the carp). One cannot trace the phylogenetic history of hormones in this way but it emphasizes the variations between them. It should be emphasized that similarities and differences in biological and immunological responses of hormones need not parallel each other, as the associated changes in the molecules may have evolved independently for each type of activity.

In 1940 Gorbman found that the goldfish thyroid tissue was stimulated by pituitary extracts from a teleost fish, two amphibians, a bird and a mammal. This suggested that TSH had a wide phyletic distribution. Subsequent measurements using a greater variety of species to compare the activity of such glandular extracts (in addition to the goldfish, a salamander, lizard and guinea-pig were used) indicated the hormones present in the various species were not identical, though they exerted the same general biological effects. While mammalian TSH preparations are active in teleosts, teleost TSH has little activity in mammals (Fontaine, 1969*a*, *b*). As the phyletic scale is ascended it is found that TSH preparations from a lungfish (*Protopterus*), amphibians, reptiles and birds can exert well-defined effects on the thyroid of both mammal (mouse) and teleost fish (trout). The thyroid of chondrichthyean fish (the stingray, *Dasyatis sabina*) while responding to its own, homologous TSH shows no response to the mammalian or even teleost hormones (Jackson and Sage, 1973). Chondrichthyean TSH, on the other hand, stimulates the mammalian thyroid

(Dodd and Dodd, 1969). Mammalian TSH increases thyroidal activity in the Pacific hagfish, *Eptatretus stouti* (Cyclostomata), though TSH-activity (when tested in mammals) has not been demonstrated in the pituitary of the Atlantic hagfish, *Myxine glutinosa* (Kerkof, Boschwitz and Gorbman, 1973; Dodd and Dodd, 1969). One obviously cannot construct an ordered story from these observations but they serve to show that TSH, like other hormones, has suffered, during evolution, changes in its structure.

The mammalian gonadotrophins have a thyrotrophic effect when they are injected into teleost fishes, an action that was initially attributed to the presence of a distinct heterothyrotrophic factor, or HTF, in the mammalian pituitary (Fontaine, 1969*a*). This cross-over in the actions of gonado-trophins and TSH presumably reflects similarities in the chemical structure of teleost TSH, and mammalian FSH and LH. It has been proposed (Fontaine, 1969*a*) that the gonadotrophins and TSH in extant species evolved from a common ancestral molecule, probably by a process of gene duplication and subsequent genetic change. It is interesting that there is no clear evidence for the presence of a distinct TSH in cyclostome fish where hypophysectomy (in the lamprey) has no effect on thyroidal activity (see Sage, 1973). It has also been observed that reproductive rhythms in fish are associated with parallel changes in the activity of the thyroid tissue. It has thus been suggested that the ancestral, or parent, molecule had a gonadotrophic role which was extended to a thyrotrophic one when the thyroid gland assumed a role in reproduction in fishes.

Growth hormone, prolactin and human chorionic somatomammotrophin

These three hormones are proteins containing about 190 amino acids and they show many structural and functional homologies to one another. Growth hormone and prolactin are formed in the adenohypophysis, the former being concerned with the regulation of growth and the latter with diverse processes ranging from lactation in mammals to osmoregulation in some fish. Human chorionic somatomammotrophin (or human placental lactogen) has been isolated from the placenta of man, and some other primates, and exerts some of the effects of both its adenohypophysial analogues.

The amino acid sequence of the hormones has been described. That of human growth hormone is shown in Fig. 3.19. Human chorionic somato-mammotrophin and ovine prolactin have similar structures to this hormone. Human prolactin has not been isolated but sensitive bioassay and immuno-logical techniques indicate that it exists as a distinct entity, apart from growth hormone, although this was once in doubt. The amino acid sequence of the three hormones has been compared (Fig. 3.20) and considerable

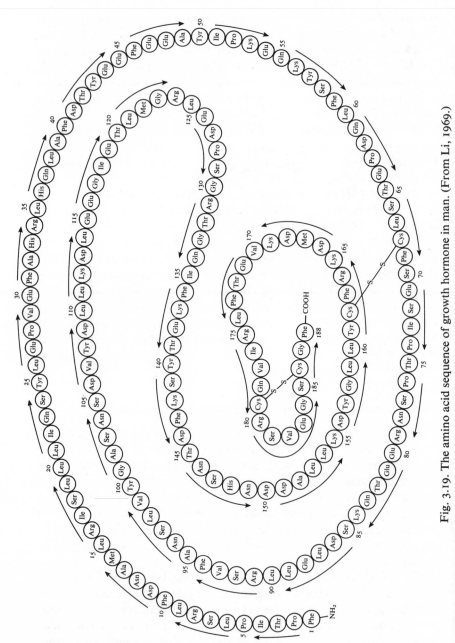

Fig. 3.19. The amino acid sequence of growth hormone in man. (From Li, 1969.)

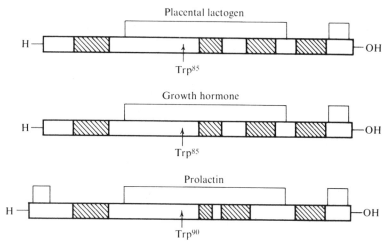

Fig. 3.20. Diagrammatic representation of the structures of placental lactogen (chorionic somatomammotrophin) and growth hormone from man, and prolactin from sheep. The cross-hatched areas represent regions of internal homology in the sequence of the amino acids. Other similarities can be seen in the presence of disulfide bridges (narrow lines) and the tryptophan residues at position 85 in placental lactogen and growth hormone and 90 in prolactin. (From Niall *et al.*, 1971.)

homologies exist. There are thus 160 (out of 190) identical residues in human growth hormone and chorionic somatomammotrophin while, of the remainder, only seven positions are occupied by what are considered 'non-homologous' amino acids (Li, 1972). Such similarities have led to the suggestion (Bewley and Li, 1970; Niall *et al.*, 1971) that the three hormones may have arisen from a common ancestral molecule. Various segments of each hormone molecule also bear considerable similarities to each other (internal homologies). The ancestral molecule may have been a smaller peptide of 25 to 50 amino acids that, by a process of genetic reduplication in a 'tandem' manner, led to an increase in the chain length of the hormones.

The chemical similarities in the molecules are reflected in their biological activities. Apart from the dual effects of chorionic somatomammotrophin on growth and lactation, prolactin exhibits considerable growth hormone-like activity while growth hormone has (though more limited) prolactin-like actions.

Growth hormone and prolactin are present throughout the vertebrates (with the possible exception of the cyclostomes).

Growth hormone can be measured by its ability to increase growth of the tibia of young, hypophysectomized rats. Pituitary extracts from all the groups of tetrapods exhibit this effect but it is not manifested by those of teleost or chondrichthyean fish (see Geschwind, 1967). Extracts from teleosts, however, have a comparable effect on growth when injected into another teleost, the killifish, *Fundulus kansae*. Pituitary extracts from the lungfish (Dipnoi), the bowfin and garpike (Holostei) and sturgeon and paddlefish (Chondrostei), all stimulate growth of the rat's tibia (Hayashida and Lagios, 1969; Hayashida, 1971). It would thus seem that the teleost growth hormone has a greater degree of structural dissimilarity from the homologous rat hormone than do growth hormones from other bony fishes. Other bioassay test systems are responsive to a phyletic range of growth hormone preparations. Such tests include the incorporation of sulfate into the cartilage of embryonic chicks as well as growth in toads (Meier and Solursh, 1972; Zipser, Licht and Bern, 1959), lizards (Licht and Hoyer, 1968) and turtles (Nichols, 1973). These tests also show that not only the hormone, but a similar biological response to it, occurs in many vertebrates.

Growth hormones from all species do not always exhibit an effect when injected into an heterologous species. This is seen very clearly in man who is unresponsive to all animal growth hormones, including those from other primates. As indicated earlier, this is an observation of some practical significance as the supply of growth hormone for administration to man has to be obtained from human cadavers and is thus limited.

Immunological evidence also emphasizes the differences, and similarities, among growth hormones from different vertebrates. Growth hormones show varying activities as antigens, depending on the species of the donor and the recipient. Rat growth hormone is not antigenic in rabbits but is very effective when injected into monkeys. Primate growth hormone is, however, antigenic in rabbits. Rabbit antiserum to human growth hormone has been shown to react (as measured by complement-fixation) with hormone preparations from other primates and in nine such species (Fig. 3.21) the degree of these interactions was closely correlated with their phyletic relationships.

A wider survey has shown that pituitary growth hormone extracts from most vertebrates can react with monkey antiserum that is formed in response to injected rat growth hormone (Hayashida and Lagios, 1969; Hayashida, 1970, 1971, 1973). Measurements of the relative ability of such pituitary extracts to antagonize the interaction of rat growth hormone with its antisera, in radioimmunoassays, has shown that the wider the phyletic distance between the species the less effective this antagonism (or ability to react with the antisera) becomes (Fig. 3.22a). No significant interaction was seen in glandular extracts from teleost fishes (which also lacks activity in the rat-tibia bioassay test). Hormones from other vertebrates, including the lungfish, have an interaction in this system which

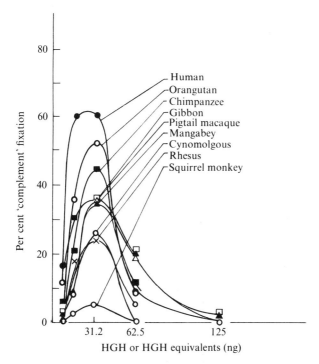

Fig. 3.21. The interactions between rabbit antiserum to human growth hormone and growth hormone from 9 different species of primates. The values given are for complement-fixation curves. The equivalence points of all the curves have been aligned with the antigen-concentration of human growth hormone which gave the maximum value. The degree of the immunological relationship is directly proportional to the amount of complement fixed at equivalence by that of the human and animal growth hormones. (From Tashjian, Levine and Wilhelmi, 1965.)

increases as the phyletic scale is ascended. It is also notable (Fig. 3.22*b*, *c*) that pituitary extracts of chondrichthyean, holostean and chondrosten fishes show abilities to antagonize the reaction of rat growth hormone with its antiserum. These reactions parallel their abilities to stimulate growth in the rat-tibia test.

Prolactin increases milk secretion in mammals and this response can be used to measure the hormone's activity even in the low concentrations that appear in the plasma (Frantz, Kleinberg and Noel, 1972). Pituitary extracts from birds, reptiles and amphibians all promote this response but not those from fishes. Pigeons secrete a milk-like paste (pigeons' milk) from their crop-sac, with which they feed their young. This response is stimulated by prolactin from tetrapods *and* lungfishes but that from other fishes is

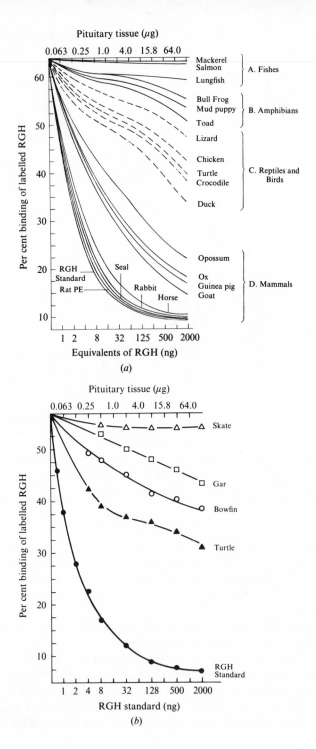

Pituitary tissue (µg)

0.063 0.25 1.0 4.0 15.8 64.0

Per cent binding of labelled RGH

60

50

40

30

20

10

Mackerel
Salmon } A. Fishes
Lungfish

Bull Frog
Mud puppy } B. Amphibians
Toad

Lizard

Chicken

Turtle
Crocodile

Duck

} C. Reptiles and Birds

Opossum

Ox
Guinea pig
Goat

} D. Mammals

RGH
Standard
Rat PE

Seal

Rabbit

Horse

Equivalents of RGH (ng)

1 2 8 32 125 500 2000

(a)

Pituitary tissue (µg)

0.063 0.25 1.0 4.0 15.8 64.0

Per cent binding of labelled RGH

50

40

30

20

10

Skate

Gar

Bowfin

Turtle

RGH
Standard

RGH standard (ng)

1 2 4 8 32 128 500 2000

(b)

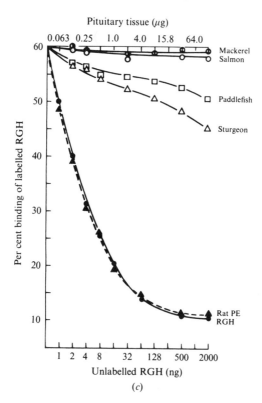

Fig. 3.22. Diagrams showing the immunochemical relationships of different vertebrate growth hormones.

The curves represent the relative abilities of these preparations to reduce the binding of ^{131}I-labelled rat growth hormone to monkey antiserum (to rat growth hormone). Thus rat growth hormone (RGH) nearly completely displaced the labelled rat growth hormone, while that from the mackerel, salmon and skate was completely ineffective. Growth hormone from other sources fell between these extremes. (*a* and *b* from Hayashida, 1970, 1971; *c*, Hayashida and Lagios, 1969.)

ineffective. Prolactin, when injected into certain newts (*Notophthalmus* (*Diemictylus*) *viridescens*) at a particular stage in their life cycle, causes them to seek water preparatory to breeding. This is called the 'eft (or newt) water-drive response' and can be initiated by prolactin from all the principal groups of vertebrates [except the cyclostomes which seem to lack a prolactin hormone (see Bern and Nicoll, 1968)]. This response cannot be mimicked by any other pituitary hormone and has been used to demonstrate the presence of an analogous prolactin-like secretion throughout the vertebrates. Further evidence of the occurrence of this hormone in fishes has followed the recent discovery that certain teleost fishes, when in fresh water, usually die following removal of the pituitary gland, and this is due to excessive losses of sodium. When injected with mammalian prolactin they retain sodium and survive. Teleosts's pituitaries contain a 'hormone' that also has the latter effect and which, as a reflection of its difference from mammalian prolactin, has been called 'paralactin'. The phyletic distribution of all these effects follows a precise pattern which is shown in Fig. 3.23.

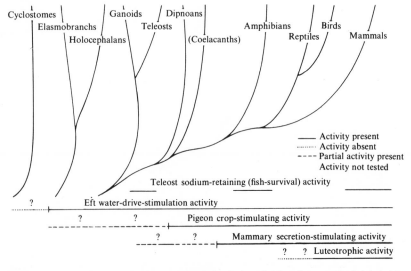

Fig. 3.23. Distribution of some of the biological activities that can be initiated by mammalian prolactin and prolactin-like hormones from other vertebrates. (From H. A. Bern, personal communication in Bentley, 1971.)

The observations described above suggest two things: first, that the prolactin hormone is not identical in all vertebrates and has been subject to evolutionary change and second, it seems likely that it has assumed diverse biological roles.

The precise chemical structure of prolactins from non-mammals has not yet been described. The similarities and differences in their biological effects (see Table 3.8) nevertheless indicate that, while they are basically analogous, differences exist in their structure. The 'ancestral' molecule may have been relatively less specific in its action than, for instance, that present in contemporary mammals. This may be reflected in the one-way specificity of mammalian prolactin which acts in fish while fish prolactin fails to have an effect in mammals. With the origin of the tetrapods, changes occurred in the molecule which are illustrated by its ability to stimulate the pigeon

TABLE 3.8. *Distribution of several prolactin activities in vertebrate pituitaries. (Group in which pituitary has been examined for prolactin activity; from Bern and Nicoll, 1969)*

Prolactin activity	Cyclostomes	Chondrichthyes	Teleosts	Lungfish	Amphibians	Reptiles	Birds	Mammals
Osmoregulatory (in teleosts)			+/−					+
Water-drive-inducing (in efts)	−?	+	+		+	+	+	+
Growth-stimulating (?) (in tadpoles)			−		+	+	+	+
Crop-sac-stimulating (in pigeons)	−	−*	−*	+	+	+	+	+
Mammotrophic (in mice)		**	−**	−	+	+	+	+
Luteotrophic (in mice)						−?	+?	+

* Partial activity has been reported from these groups; this is considered to be minimal and not fully crop-sac-stimulating in the manner seen with lungfish and amphibians.

** Partial activity has been reported from these groups, which is distinguished from the 'fully effective' response seen with tetrapods.

crop-sac and mammary gland. This evidence is, of course, derived from extant species and, if it indeed did occur in that long-past time, the particular effects were not then of contemporary biological significance. The pigeon crop-sac and mammary gland were not to appear for many millions of years. It seems quite likely that prolactin had other roles at that time.

As described above, human chorionic somatomammotrophin has both prolactin- and growth hormone-like action in mammals. When tested in non-mammals, however, this hormone behaves differently to the mammalian pituitary hormones as it exhibits no 'eft water-drive' activity, does not increase growth in tadpoles and fails to promote sodium retention in a teleost fish, *Tilapia mossambica* (Gona and Gona, 1973; Clarke *et al.*, 1973). Thus, while human chorionic somatomammatrophin shares some actions with mammalian prolactin and growth hormone, these do not necessarily extend to the effects of the last two in non-mammals.

Tetrapod prolactins have been shown to exhibit differing chemical behavior. The electrophoretic mobilities (R_f), reflecting size and electrical charge, show considerable differences (Fig. 3.24) from each other and cross an almost five-fold range. No phyletic order is apparent; the ox, duck and frog all have a similar R_f, while the rat, quail, turtle and toad are all much higher. Interspecific differences, nevertheless, can be seen to exist. Growth hormones can also be separated in this way and, while they also exhibit differences in electrophoretic mobility, they are quite distinct from the prolactins (Fig. 3.24). The two hormones may still exhibit immunochemical similarities as seen in amphibians where the electrophoretically separated

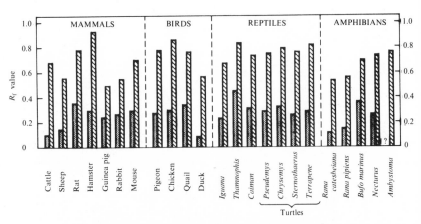

Fig. 3.24. A comparison of the electrophoretic mobilities (R_f) of prolactins and growth hormones from different species of tetrapod vertebrates, stippled bars, growth hormone; cross-hatched bars, prolactin. (From Nicoll and Licht, 1971.)

hormones both compete with rat growth hormone for binding to rat growth hormone antiserum (Hayashida, Licht and Nicoll, 1973).

The prolactins exhibit a multitude of different biological effects. These (Bern and Nicoll, 1968) have been classified into several categories including actions related to reproduction and parental care, osmoregulation, growth, metabolism and the integument. In all, over 60 different effects have been described. It is unlikely that these all reflect physiological roles but they do illustrate the considerable biological reactivity of the prolactin molecule. This may provide it with an adaptability, and propensity, to be utilized in a great diversity of physiological roles, only some of which have already been identified, or even applied. Prolactin would seem to be a hormone whose structure, as well as physiological role, have both evolved. Its diverse biological actions suggest that it is the most versatile vertebrate hormone.

Evolution of hormones: general comment

While it is difficult to make any generalizations about the directions and effects of the evolution of the structure of hormones, a certain amount of speculation on this subject has been made. It has been observed that, while the homologous hormones of 'higher' vertebrates may interact with receptors, and elicit a response, in 'lower' vertebrates the reverse is seen less often. There are, however, many exceptions and one must be cautious about accepting such occurrences as a 'rule'. The one-way phenomenon is, nevertheless, often apparent though it does not always follow phyletic lines strictly. For example: the growth hormone of the teleostean fishes does not elicit effects in mammals, but that from the phyletically less-advanced holostean and chondrosteans is effective. Also, chondrichthyean thyro-trophic hormone is effective in mammals but that from mammals, or teleosts, has no action in chondrichthyeans. With such exceptions to the one-way rule in mind, we may nevertheless consider what the structural basis for the differing interspecific effects of hormones may be and what effects such changes may have on the actions of the hormones themselves.

As indicated earlier, in order to elicit a response, hormones interact with a specific chemical entity which has been called the receptor. The precise nature of such receptors is still largely unknown but they are thought of as being a group of chemical properties arranged in such a manner as to be able to combine, or interact, with molecules of a certain specific form. Several types of interactions may occur and not all of these will necessarily initiate a response or even equal responses. A 'successful' combination (that results in a biological effect) in some manner, not well understood, triggers a reaction(s) that initiates an effect. This process could be the activation of an enzyme, the supplying of an essential substrate or cofactor, or a structural change in the cell membrane such as may alter its permeability.

Properties that may determine the ability of a hormone to interact with such a receptor site include its size, shape, electrical charge, disposition of hydrophobic and hydrophilic groups, the presence of acidic or basic moieties, its ability to form hydrogen bonds, ionic bonds and so on. The hormone and its receptors thus must exhibit a complex affinity for each other that is complementary.

Changes in the structure of a hormone, if it is to remain effective, need to be accompanied by changes in the receptor. The evolution of both must therefore occur in some sort of harmony. By utilizing such parallel changes, the specificity of the hormones can be maintained, or altered in such a direction as may be advantageous (or possibly disadvantageous) to the animal.

Changes in the structure of a hormone, especially a large polypeptide, can influence, in either of two ways, its ability to combine with a receptor (see for instance Fontaine, Y-A., 1964):

(1) Each group of phyletically related hormones may possess a 'common-core' in which the attachment sites for the receptor reside. These structural features are primarily essential for the hormone's interaction with its receptor. Modifications that survive in nature may involve other, less essential, parts of the molecules that 'mask' the 'common-core' in such a way that it cannot interact with its receptor unless aligned in a special way. Alternatively, it may be necessary to 'activate' the molecule by removing part of it as the result of the action of an enzyme. The receptor necessarily must possess a complementary arrangement to insure that such effects are possible. This hypothesis tends to gain support from observations of common amino acid sequences in phyletically diverse molecules and the retention of biological activity by fragments of these.

(2) The activity of a hormone molecule may be intrinsically related to its over-all structure. The entire configuration of the molecule could contribute to its interaction with the receptor.

A parent, or ancestral, hormone is sometimes thought of as being more 'simple' in structure than its evolutionary descendants. As a result of mutational changes, it may be altered in several ways. Single-base changes at individual gene loci may occur that result in amino acid substitutions. Depending on the nature of the substituent amino acid this may influence its reaction at the receptor; it could facilitate it, make little difference, decrease or even make such an interaction impossible. If the change were not too disadvantageous it is possible that compensatory changes in the receptor may also arise subsequently, or that the molecule may even be able to act at hitherto inaccessible sites. Complex changes of the hormones' structure may result from genetic reduplication, like those of the 'tandem' type proposed for prolactin and growth hormone and which result in an increase in the size of the molecule. Evidence for this type of change among extant species is equivocal as there is little indication of dramatic differences in the size of analogous hormones in different species.

Changes in a hormone's structure may, providing receptors can adapt to it, make possible a greater specificity of its action in the body which could be advantageous. There is much evidence available from the examination of cross-reactivity in extant species to suggest that such changes in hormones and their receptors has occurred during the course of evolution. This may also contribute to an explanation of 'one-way specificity' in the effects of hormones. Thus the hormones and their receptors that are present in some phyletic groups may be less fastidious as compared to those in others. A mammalian hormone, such as a gonadotrophin or thyrotrophin, may be able to interact (though possibly not as readily) with a less discriminating receptor in, for instance, a teleost fish. On the other hand, a more specific receptor (such as may have evolved in conjunction with the homologous hormones in a mammal) can no longer react with the teleost hormones. This theory is still, however, highly speculative and, as we have seen, one cannot make generalizations from the evidence obtained from extant species.

Conclusions

An examination of the chemical, biological and immunological behavior, as well as the chemical structure, of homologous hormones from different species suggests that many of these may have been subject to an orderly evolutionary change. This possibility is also indicated by the similarities that persist between each hormone from closely, in contrast to distantly, related species and within the principal systematic groups of the vertebrates. In some instances, a genetic background for such changes has been described. It is nevertheless noteworthy that some hormones display little or no difference in their structure even when they are present in such distantly related species as a lamprey and a man. On some occasions it even appears that a completely 'new' hormone that lacked a genetic homologue in its ancestors has evolved.

While the evolution of 'new' hormones has potential importance in novel processes of coordination in the body, the functional significance of alterations in the structure of 'old' hormones is less clear. Such changes may take place at chemical sites on the hormone molecule that are not essential for its action and so have little or no effect on its functioning.

However, if more important sites are involved the hormone's activity may be altered and this could have important results in the animal. These changes may include a virtual absence of its effect, differences in the quantities of the hormone required to mediate the response or an alteration in its relative ability to influence different processes within the same animal (specificity). As we shall see in the succeeding chapters, the roles of hormones in the body have often changed completely during the course of evolution. Such a modification of a hormone function is not necessarily accompanied by an alteration in its chemical structure, but it often is; though we are uncertain as to how important this change may be in the transition.

4. The life history of hormones

The use of hormones for the purpose of coordination involves a complex series of physiological events. Such a life history begins with the formation of the excitant by the endocrine glands and concludes with the response of a target, or effector tissue, and the hormone's ultimate destruction or its excretion from the body. The events that determine the action of a hormone are shown in Fig. 4.1. This basic pattern persists throughout the vertebrates, though, as will be described, certain differences exist.

The formation of hormones

While the formation of all hormones is determined at the genetic level it can be either a relatively direct translational procedure or, alternatively, occur as a result of the prior formation of enzymes that mediate synthesis.

Translational formation of hormones

It seems likely that the sequences of amino acids in the polypeptide and protein hormones directly reflect genetic translation via messenger RNA. Some of the small peptides, however, like thyrotrophin-releasing hormone (TRH), which is a tripeptide, may be assembled separately, in this instance as a result of the action of a TRH synthetase enzyme present in the hypothalamus (Mitnick and Reichlin, 1972). Even when a direct translation of genetic material into the amino acid sequence of a hormone is made, the active hormone product does not always result. Many peptide hormones, including glucagon, corticotrophin, MSH, the neurohypophysial peptides and calcitonin, contain relatively few amino acids and may be formed as a result of a process of disassembly from larger protein molecules. In addition, some polypeptide hormones are not simple strings of amino acids but consist of subunits as seen in insulin, the gonadotrophins and TSH. The assemblage of these subunits into a hormone must occur subsequent to the formation of the individual parts. The formation of polypeptide hormones thus often involves the initial formation of a parent molecule called a *pro-hormone*. As a result of post-translational changes, such as cleavage by enzymes, the pro-hormone is broken up to form the hormone itself. There

[123]

PROCESS   MOLECULAR MECHANISM

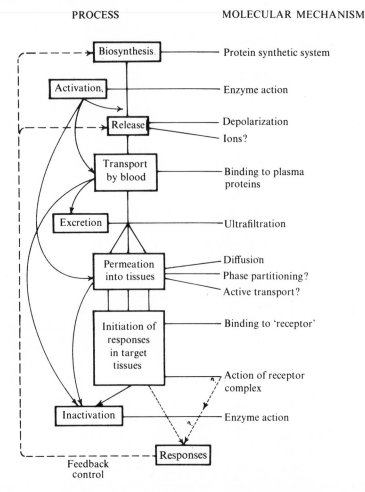

Fig. 4.1. Diagrammatic summary of the life history of a hormone commencing with its biosynthesis and concluding with the response and its inactivation. (From Rudinger, 1968. Reprinted by permission of the Royal Society.)

is good evidence to indicate that this occurs in the formation of parathormone, the neurohypophysial peptides, and insulin, and probably other examples occur.

Such a process of hormone formation is well illustrated in the production of insulin (Steiner *et al.*, 1972). Tumors with B-cells of the Islets of Langerhans contain a molecule that exhibits insulin-like activity, both biologically and immunologically, but which is about 1.5 times larger than insulin itself. A small quantity of this material has been found in normal

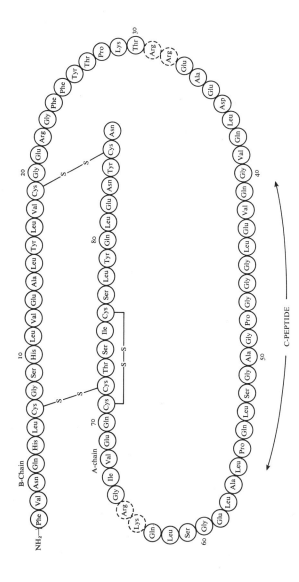

Fig. 4.2. The proposed amino acid sequence in proinsulin from man. The basic residues indicated by the broken circles have been assigned as they are known to occur in bovine and porcine proinsulin. (From Oyer *et al.*, 1971.)

B-cell extracts as well as in the 'principal islets' tissue of a teleost fish (the cod). When this pro-hormone molecule is treated with a proteolytic enzyme, trypsin, a product is formed that behaves similarly to true insulin. It thus appears that the latter is formed post-translationally from a large parent molecule. The chemical structure of human pro-insulin is shown in Fig. 4.2. It incorporates the B-chain of insulin at the amino-terminus which is followed by the A-chain. Interposed between these sections is a segment of amino acids called the 'C' peptide. The transposition to insulin takes place in the B-cells, as a result of the action of two enzymes; one like trypsin and the other like carboxypeptidase B. The final products of the actions of these enzymes are shown in Fig. 4.3. The folding arrangement of the pro-

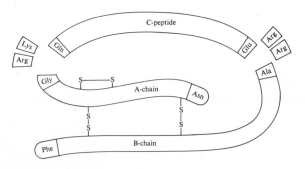

Fig. 4.3. Products formed as a result of the conversion of proinsulin to insulin in the pancreatic B-cells. (From Steiner *et al.*, 1972.)

hormone is an important feature that ensures the correct alignment of the cysteine residues to form the disulfide bridges that are present in the hormone proper. The amino acid sequence of the 'C' peptide can be seen to differ in various species (Fig. 4.4).

Many polypeptide hormones are stored in granules present in the endocrine cells. These organelles are bounded by membranes and are 0.1 to 0.4 μm in diameter. They appear to originate in the Golgi apparatus of the cell. The precursor, or pro-hormone, becomes associated with the granules and it seems likely that conversion to the hormone takes place in these. The granules can travel to the peripheral regions of the cell, and, in response to releasing-stimuli, combine with the plasma membrane and discharge their contents into the region of blood vessels. A summary of this process as it is thought to occur for insulin is shown in Fig. 4.5.

Such granules apparently furnish sites for the formation and storage of many hormones. If released into the cytoplasm of the cell the hormones may be destroyed as has been observed for the catecholamines when they are exposed to the mitochondrial enzyme monoamine oxidase (MAO). In

	1	2	3	4	5	6	7	8	9	10	11	12	13	14	15	16
Human	Glu	Ala	Glu	Asp	Leu	Gln	Val	Gly	Gln	Val	Glu	Leu	Gly	Gly	Gly	Pro
Monkey	Glu	Ala	Glu	Asp	Pro	Gln	Val	Gly	Glx	Val	Glu	Leu	Gly	Gly	Gly	Pro
Pig	Glu	Ala	Glu	Asn	Pro	Gln	Ala	Gly	Ala	Val	Glu	Leu	Gly	Gly	Gly	Leu
Cattle	Glu	Val	Glu	Gly	Pro	Gln	Val	Gly	Ala	Leu	Glu	Leu	Ala	Gly	Gly	Pro

	17	18	19	20	21	22	23	24	25	26	27	28	29	30	31
Human	Gly	Ala	Gly	Ser	Leu	Gln	Pro	Leu	Ala	Leu	Glu	Gly	Ser	Leu	Gln
Monkey	Gly	Ala	Gly	Ser	Leu	Gln	Pro	Leu	Ala	Leu	Glu	Gly	Ser	Leu	Gln
Pig	Gly	—	Gly	—	Leu	Gln	Ala	Leu	Ala	Leu	Glu	Gly	Pro	Pro	Gln
Cattle	Gly	Ala	Gly	—	—	—	—	—	Gly	Leu	Glu	Gly	Pro	Pro	Gln

Fig. 4.4. A comparison of the amino acid sequences of the human, monkey, porcine and bovine C-peptides. The solid bars indicate residues that are identical in all species. (From Steiner et al., 1972.)

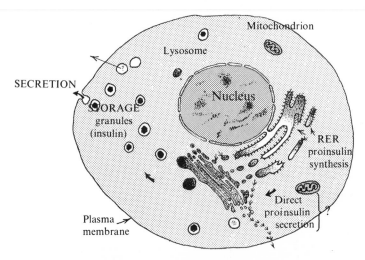

Fig. 4.5. Subcellular organization of the processes involved in the biosynthesis of insulin in the cell. The proinsulin is formed in the ribosomes of the rough endoplasmic reticulum (RER). This polypeptide rapidly folds and assumes its normal conformation during its transfer to the Golgi region. The storage granules are formed in this part of the cell from the Golgi apparatus and these travel to the peripheral regions of the cells. The contents of these granules may be secreted following fusion of the granule to the plasma membrane. (From Steiner *et al.*, 1972.)

addition, storage granules may afford convenient vehicles in which hormones can be transported for considerable distances along nerve cells.

Some neurons form hormones by a process called *neurosecretion*. These are like ordinary nerve cells and consist of a cell body with an extended axon and they can also be depolarized and so convey electrical information. The axon instead of terminating at another neuron or an effector tissue, like a gland or muscle, lies near to a capillary into which it can discharge certain of its products (Fig. 4.6). These products may be hormones, the formation of which is initiated some distance away in the cell body. The hormones, parceled up in their granules, travel along the nerves to the peripheral sites in the axon where they can be released into the blood.

Hormones that are formed as a result of neurosecretion are those of the neurohypophysis and median eminence including vasopressin, oxytocin and the various releasing-hormones that control the adenohypophysis. In mammals, vasopressin and oxytocin are formed in the supraoptic and paraventricular nuclei which are situated at the base of the brain. These hormonal products pass down the axons in granules to the neural lobe. Inside the granules they are attached to protein molecules of neurophysin,

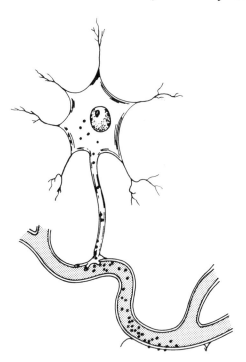

Fig. 4.6. A neurosecretory cell. The hormonal products are transported from the cell body down to the axon from which they can be released into capillaries. In contrast, ordinary nerve cells have axons that abut onto other neurones (instead of capillaries). (From R. Guillemin and R. Burgus, *The Hormones of the Hypothalamus*. Copyright © 1972 by Scientific American, Inc. All rights reserved.)

the synthesis of which seems to be closely associated to that of the hormone. What relationship neurophysin has to a pro-hormone is, however, as yet unknown. Amphibians and fishes have a single preoptic nucleus where the neurohypophysial hormones originate. The putative hormones of the urophysis in fishes (see Chapter 2) are also apparently formed by a process of neurosecretion.

The formation of hormones by enzymically controlled synthesis

Thyroid hormones

The endocrine secretions of the thyroid, the adrenal medulla, the gonads and the adrenal cortex, are the result of biosynthetic processes controlled by enzymes. While the enzymes themselves are the result of genetic translational processes, the hormones are assembled as a result of chemical reactions controlled by the enzymes. This synthesis may involve many reactions, some of which have been summarized in Fig. 3.2 (for steroid hormones) and Fig. 3.4 (for thyroid and catecholamine hormones).

Thyroid hormones contain iodine and the thyroid gland has a special ability to concentrate inorganic iodide from the blood. This ability to transport iodide actively against an electrochemical gradient is shared by some other tissues, including the intestine and salivary glands. This ability may be controlled by a single gene: in man a congenital inability to accumulate iodide in the thyroid is accompanied by a parallel deficiency at the other iodide transport sites. The accumulated iodide is oxidized to iodine, which combines with tyrosine to form the precursor of thyroxine and tri-iodothyronine. The latter reaction occurs with the tyrosine residue that is part of a large protein, thyroglobulin (molecular weight 670 000), that is stored extracellularly in the thyroid follicles. Each molecule of thyroglobulin binds two of thyroxine and on the average, less than one tri-iodothyronine. Thyroglobulin itself does not appear to be a particularly remarkable protein; it contains about 30 molecules of tyrosine (or about 3% by weight), and about 0.5% iodine. It nevertheless provides a site for the synthesis and storage of the thyroid hormones. The biosynthetic process for the thyroid hormones appears to be common to all vertebrates and was apparently attained early in their evolution. Nevertheless most of our information has been derived from studies of mammals.

Thyroglobulins have been identified in thyroid tissues of species from most groups of vertebrates, even including larval cyclostomes (lampreys) where it is present in the subpharyngeal gland or endostyle (Suzuki and Kondo, 1973). These proteins exhibit many similarities with respect to their molecular size (though a few differences have been observed), as determined by centrifugation in sucrose gradients, but their amino acid constitutions may differ. Thyroglobulins also exhibit different immunological behavior. Antibodies to specific thyroglobulins have been prepared and these react, *in vitro*, with the homologous protein, which can be radioactively labelled. Thyroglobulins from different species may compete with this labelled protein for binding to its antibodies. The relative ability to do this suggests the degree of immunological similarity to the homologous thyroglobulin. Considerable interspecific differences have been observed in such radioimmunoassays (Torresani *et al.*, 1973). Sheep thyroglobulin readily displaces its labelled form from anti-sheep thyroglobulin antibodies but thyroglobulins from other mammals, such as pigs and rabbits, are much less effective. Thyroglobulin from a python and a crocodile also compete with the homologous labelled protein for such binding but this is also much less so than that for the sheep protein. Bird thyroglobulin, from ducks, had no ability to bind with the sheep antibodies. While it is tempting to construct phylogenetic trees with such information the paucity of species examined makes such predictions of doubtful significance but the measurements nevertheless illustrate the diversity that can occur among thyroglobulins from different species.

Catecholamines

Adrenaline and noradrenaline are formed in chromaffin tissues. These hormones are present not only in the adrenal gland but also are associated with nervous tissue in other parts of the body. Noradrenaline is also formed in certain nerve endings in the sympathetic nervous tissue and the brain. The original precursor of these catecholamines is tyrosine which by a series of enzymically controlled reactions is converted to 3,4-dihydroxy-phenylalanine, or dopa, and thence to dopamine. These reactions occur in the cell's cytoplasm. The dopamine is accumulated by storage granules in which it is converted, under the influence of dopamine β-hydroxylase to noradrenaline. Noradrenaline can be N-methylated to adrenaline under the influence of the enzyme phenylethanolamine-N-methyl-transferase (PNMT) which, in mammals (see Chapter 3), can be induced in the presence of high concentrations of corticosteroids. There is evidence to suggest that noradrenaline and adrenaline are stored in different granules and even different cells in the adrenal medulla. This could be determined by regional differences in the access of corticosteroids to the medullary tissue thus influencing the local levels of PNMT (Pohorecky and Wurtman, 1971).

Steroid hormones

The formation of the steroid hormones also appears to be basically the same in all vertebrates (Sandor, 1969). All of these are formed from cholesterol which is present in high concentrations in the steroidogenic endocrine glands. This parent molecule may be formed *in situ* from acetate (Fig. 3.2) or be accumulated from the plasma. It has been suggested that the synthesis of some steroid hormones may not involve cholesterol as a precursor but this is difficult to prove. The enzymic conversion of cholesterol to pregnelolone and progesterone is common to all of the steroidogenic endocrine glands. Sandor (1969) has suggested that the use of steroids as hormones may have been determined by a primeval mutation which invented the enzyme systems that determine this transformation of cholesterol. The conversion of progesterone to the androgen, estrogen and adrenocorticosteroid hormones involves the successive actions of diverse enzymes that hydroxylate, oxidize and reduce the steroid at some of the

21 carbon positions present. The ability of a species to synthesize steroid hormones with differing structures depends on the presence or absence of the enzymes that mediate these changes. The chondrichthyean fishes that secrete 1α-hydroxycorticosterone possess an enzyme, 1α-hydroxylase (that converts corticosterone to the hormone), that has not been found in other vertebrates. The formation of aldosterone in tetrapods is determined by the presence of 18-hydroxylase (it converts corticosterone to aldosterone). Rats and mice cannot form cortisol and lack 17α-hydroxylase (that converts progesterone to 17α-hydroxyprogesterone) in their adrenal cortex. Mutations may also arise that influence the ability of a species to form certain hormones. In man, a congenital condition known as the adrenogenital syndrome is due to the complete or partial block of the 21-hydroxylating

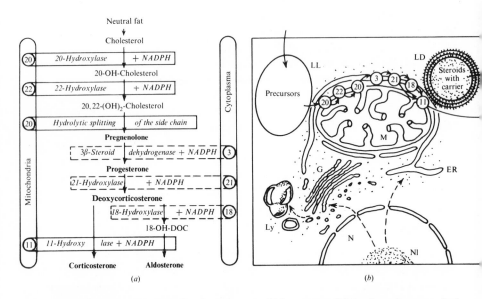

Fig. 4.7. The biosynthesis of corticosteroid hormones (*a*) in the adrenocortical tissue of a frog. This is illustrated with reference to the cell organelles (*b*). Corticosterone can also act as a precursor for aldosterone. In some species, 18-hydroxylase may be a mitochondrial rather than a microsomal enzyme. In vertebrates that form cortisol another microsomal enzyme, 17α-hydroxylase, active on progesterone and leading to 17α-OH-progesterone, deoxycortisol and cortisol, is present.

ER, endoplasmic reticulum; G, Golgi apparatus; N, nucleus; LD, electron dense lipid droplet; LL, electron lucid lipid droplet; Ly, lysosomes; M, mitochondrion; Nl, nucleolus.

Solid arrows are pathways of steroid synthesis from precursors to steroid bound to a carrier; broken arrows indicate cellular responses activated by ACTH. (From Lofts and Bern, 1972.)

system (the conversion of progesterone to deoxycorticosterone and 17α-hydroxyprogesterone to 17α-hydroxy-11-deoxycorticosterone). Such enzymic differences determine the presence or absence of the various steroid hormones and furnish the raw materials of evolutionary change.

The steroidogenic enzymes are associated in the cytoplasm of the cell and the mitochondria. The pathways and sites of the enzymes determining the formation of adrenocorticosteroids in the frog's interrenal are shown in Fig. 4.7.

The release of hormones from endocrine glands

Nature of the stimuli

The role of the endocrine glands in the regulation of bodily functions is dependent on the release of their secretions on appropriate occasions. Secretion is initiated upon the receipt, by the gland, of a suitable stimulus which may increase or decrease, the discharge of its hormone. The message may arrive either by way of a nerve or be carried in the blood that perfuses the tissue. The primary event that initiates this stimulus may arise either from the external environment, exteroceptive stimulus, or inside the body, interoceptive stimulus.

Exteroceptive stimuli that may affect the endocrine glands include the receipt of light, a change in temperature or of the osmotic concentration (of an aqueous environment) and the acquisition of food, water and salts. Social situations such as the proximity of prey, a predator, a mate or the young may evoke psychogenically-mediated responses in the endocrine glands. Climatic events such as rain, temperature and even, possibly, humidity and atmospheric pressure can also influence a hormone's release. The receipt of and endocrine response to such external stimuli helps the animal to maintain an equitable relationship with the events that happen around it. Exteroceptive stimuli are especially useful in providing cues that are involved in reproduction.

Interoceptive stimuli are those that result from changes in the physico-chemical conditions within the body. Ultimately they may reflect the external conditions: for instance a lack of drinking water and a hot, dehydrating environment will lead to an increase in the osmotic concentration of the body fluids. Internal stimuli include: changes in the concentration of salts, such as sodium, potassium and calcium, in the body fluids, alteration of the hydrostatic pressure of the blood vascular system, oscillations in the levels of nutrients, like glucose, amino acids and fatty acids, as well as changes in the body temperature. The physiological factors influencing release of hormones are summarized in Table 4.1.

TABLE 4.1. *Principal stimuli influencing the release of hormones*

Hormone	Releasing stimuli
Adrenaline	Neural stimuli (mediated by acetylcholine)
Aldosterone	Low plasma Na concentration, angiotensin
Angiotensin	Renin
Calcitonin	Hypercalcemia
Cortisol and corticosterone	Corticotrophin
Enterogastrone	Fats and oils in intestine
Entero-glucagon	Feeding
Estrogens	FSH
FSH	External stimuli, such as light, low estrogen levels
Gastrin	Feeding (vagal reflex; local reflex from food in stomach)
Glucagon	Hypoglycemia, gastrin, pancreozymin, high amino acids and low fatty acids in plasma, exercise
Growth hormone	Sleep, exercise, apprehension, hypoglycemia
Insulin	Hyperglycemia and amino acids in plasma, glucagon, growth hormone, in ruminants high levels of propionic and butyric acid, vagal stimulation, pancreozymin and secretin. Inhibition by adrenaline
LH or ICSH	External stimuli, sexual excitement (male), estrogen 'surge' (female), low progesterone or testosterone levels
MSH	Light on retina, low plasma corticosteroids levels, inhibition by neural stimuli and MSH-R-IH
Melatonin	Darkness (adrenergic neural stimulation)
Oxytocin	Suckling, parturition
Pancreozymin–cholecystokinin	Digestive products in the upper intestine
Parathormone	Hypocalcemia
Pituitrophins: CRH, TRH, etc.	Hypothalamic neuronal stimuli (dopamine and mono-amine transmitters), inhibited by negative-feedback mechanisms carried by hormones and metabolites
Progesterone	LH, chorionic gonadotrophin, prolactin
Prolactin	Diurnal rhythm (sleep), suckling, parturition, plasma osmotic concentrations (low in fish, high in mammals?), estrogens
Renin	Low Na in plasma, hemorrhage, reduced renal blood flow, nerve stimulation (β-adrenergic), increased osmotic concentration in renal blood supply
Secretin	Acid in upper intestine
Testosterone	LH or ICSH
Thyroid hormones	TSH
TSH	Low thyroxine, temperature reduction
Vasopressin	Increased osmotic concentration of plasma
$1,25$-$(OH)_2$-vitamin D_3	Low Ca and phosphate levels in plasma; parathormone

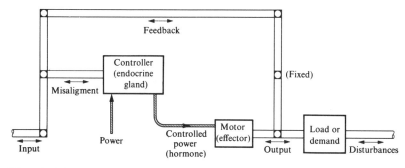

Fig. 4.8. Schematic diagram of a servo-system showing the analogies of the classic engineering control mechanism of physiological coordination by the endocrines. The double arrows indicate that the links can move to and fro. Stimuli (both input and feedback) are fed into the controller (endocrine gland in the physiological analogy) through a misalignment detector which is sensitive to changes from a certain set-point. It is conceivable that the latter may be reset in certain physiological conditions (such as hibernation). In response the controller varies its power output (or hormone secretion) to the motor or effector. The latter adjusts the physiological needs and tends to restore equilibrium (of metabolites, salts, water, other hormones, temperature, etc.) the degree of which is transmitted back to the controller via the feedback arc. (From Bentley, 1971.)

Hormones can exert trophic effects on other endocrine glands and the terminal secretions, once released, may travel back to the region where the trophic hormone originated and inhibit its further release. This last effect completes the cycle of events that closes the loop of a *negative-feedback system* that plays a vital role in regulating the endocrine system. Such a control-system is well known to engineers and its action is illustrated in Fig. 4.8. The release of hormones from the median eminence, that in turn controls the formation and discharge of the trophic hormones of the adenohypophysis, is regulated in this manner. The hormones that exert the inhibitory effects in the median eminence may alter the thresholds for stimulation of the neurosecretory cells. In addition, such a negative-feedback can also act directly on the adenohypophysis as seen with the action of thyroid hormones which inhibit the release of thyrotrophic hormone in this way. The interrelations of the hormones of the hypothalamus, adenohypophysis and more peripheral endocrine glands are shown in Fig. 4.9.

The feedback mechanisms involving the action of the peripheral endocrine secretions on the hypothalamus is called a *long-loop feedback*. There is also evidence suggesting that the adenohypophysial secretions may exert a similar action on the hypothalamus by what is termed a *short-loop feedback*. It should be noted that peripheral hormones do not necessarily initiate a

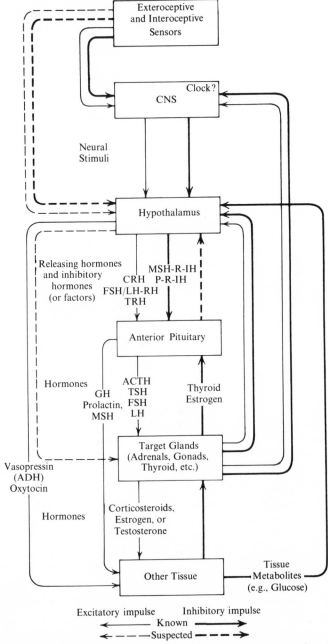

Fig. 4.9. Factors controlling the release of hormones from the anterior pituitary (Adenohypophysis). This illustrates the use of negative-feedback inhibition of secreted hormones and tissue metabolites in influencing further hormonal release.

negative-feedback inhibition on the hypothalamus. High estrogen levels can stimulate the release of LH/FSH-RH which initiates the events that result in ovulation. Such a positive-feedback has also been shown in the hypothalamus of the goldfish where thyroxine stimulates the release of a thyrotrophin-inhibiting hormone (Peter, 1971). The sporadic information available from experiments on non-mammals suggests that feedback control working through the hypothalamus, and, in the instance of thyrotrophin, the adenohypophysis, is widespread. More information is needed and there is some doubt as to the importance of such effects in cyclostomes (Larsen and Rosenkilde, 1971; Fernholm, 1972).

A negative-feedback inhibition of hormone secretion also results from changes in the concentrations of the products of the hormone's actions. The retention of water or sodium, as a result, respectively, of the actions of antidiuretic hormone and aldosterone, reduces the further release of these hormones. Comparable mechanisms exist involving glucose levels and the regulation of insulin, glucagon and growth hormone, as well as calcium and parathormone, and calcitonin.

The endocrines, apart from influencing each other's release through trophic and feedback mechanisms, may also interact with each other and so modify their secretory activity. Adrenaline can inhibit the release of antidiuretic hormone from the neurohypophysis and that of insulin from the Islets of Langerhans. This inhibition is an α-adrenergic effect which contrasts with the β-adrenergic effect which increases the release of insulin. Glucagon promotes the release of insulin and growth hormone directly; such effects do not depend on changes in blood glucose levels. Excesses of growth hormone are diabetogenic and inhibit the formation of insulin, though smaller amounts apparently stimulate its release.

Many less-precise and less-specific stimuli than those described above can initiate the discharge of hormones from endocrine glands. Such stimuli may contribute to the homeostatic process though they may also confuse it. No endocrine gland can exist and be uninfluenced by events outside what we may like to think of as its homeostatic area of influence. Non-specific stimuli, especially if strong enough, can elicit a discharge of many hormones. This is sometimes referred to as 'stress' and is particularly likely to occur in experimental situations that contribute to the confusion of the perpetrating scientists.

The trophic effects of hormones on the secretion of other endocrine glands can contribute to the processes of biological amplification in the

---

As can be seen, the hypothalamus (and its associated median eminence) plays a central role in this process. Stimulatory effects are shown by thin lines and inhibitory processes by thick lines. (From Krieger, 1971.)

body. An initial stimulus may only produce a change that involves a very small amount of energy while the energetic demands of the response may be relatively immense. The quantities of the excitants necessary to initiate an effect and the amount of the products formed (on a weight for weight basis) are illustrated in Fig. 4.10. We have no information as to the quantity

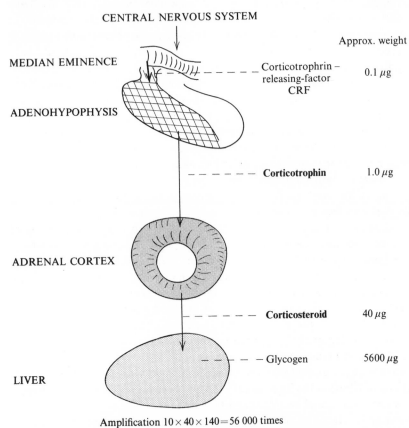

Fig. 4.10. An illustration of biological amplification as shown in the endocrine system. Corticotrophin-releasing hormone initiates the release of corticotrophin (ACTH) which leads to the deposition of glycogen in the liver.

of neural transmitters necessary to trigger a release of corticotrophin-releasing hormone from the median eminence but quantitative changes have been estimated for the subsequent physiological events. The final release of cortisol promotes the formation of 5600 μg of glycogen. The initial amount of CRH required to do this is about 0.1 μg so that the final response represents an amplification of 56 000 times.

Cyclical release of hormones

Many hormones are released periodically: at certain well-defined hours of the day (diurnal or circadian rhythms), during the reproductive cycle, or at certain seasons and times of the year (circannual rhythms). Such timing may be especially important in coordinating the events of the reproductive cycle and insuring that this occurs during the times of the year most appropriate to the survival of the young. A predictably functioning release mechanism also insures that adequate hormone levels, necessary for the animal's optimal daily activities, are available. Such release of hormones is usually controlled by centers in the brain (sometimes called 'biological clocks') which are programmed by stimuli that include the length of the daily period of light, the external temperature, changes in the seasons, as well as certain interoceptive stimuli.

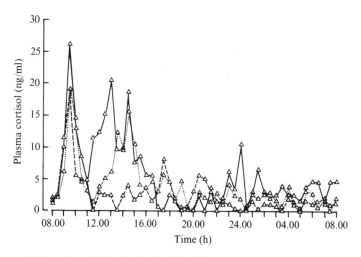

Fig. 4.11. Diurnal variation in the levels of plasma cortisol in three sheep. It can be seen that the highest concentrations were recorded during the early hours of daylight. The concentrations of cortisol in the plasma were not uniform but showed continual oscillations suggesting that it is released in sudden 'spurts'. (From McNatty, Cashmore and Young, 1972.)

The levels of corticosteroids in the blood vary in a distinct pattern during the course of the day. In mammals, this is well known in primates including man, and dogs, rats and mice. Release is related to the incidence of light and the corticosteroids are lowest in concentration during the night and reach a distinct peak after various periods of daylight. In man and sheep (Fig. 4.11), this is seen in the morning hours, soon after dawn, though in laboratory rats, which are nocturnal, it is delayed until the early evening hours (Fig. 4.12). Comparable changes in the plasma corticosteroid concentrations have been observed in some teleost fishes [the channel catfish, *Ictalurus punctatus* (Boehlke *et al.*, 1966), and the gulf killifish, *Fundulus grandis* (Srivastava and Meier, 1972)] where peak concentrations occur about eight hours after the onset of light. In another teleost (the eel *Anguilla rostrata*), diurnal variation of plasma cortisol levels does not seem to occur (Forrest *et al.*, 1973*a*). Prolactin and growth hormone in man are released in greatest amounts during the period of sleep (Fig. 4.13) and does not appear to be directly dependent on light.

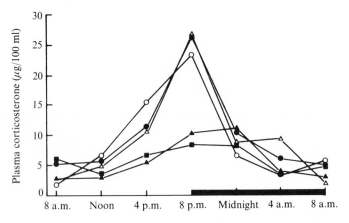

Fig. 4.12. The circadian pattern in the concentration of corticosteroids in the plasma of rats. The effects of the administration of exogenous corticosteroids to young, developing rats on the subsequent circadian periodicity in the endogenous corticosteroid levels. It can be seen that the administration of dexamethasone or hydrocortisone (cortisol) on days 2 to 4 after birth suppressed the rhythmical release. However, when dexamethasone was given on days 12 to 14 after birth no effect was seen: ●, control; ○, saline, 0.1 ml, day 2–4; ■, hydrocortisone acetate, 500µg, day 3; ▲, dexamethasone PO₄, 1 µg, day 2–4; △, dexamethasone PO₄, 1 µg, day 12–14. (From Krieger, 1972. Copyright © 1972 by the American Association for the Advancement of Science.)

The release of gonadotrophins from the adenohypophysis may also occur at precise times of the day. In the Japanese quail (*Coturnix coturnix japonica*), the pituitary is depleted of gonadotrophin over a period of four hours commencing about 16 hours after dawn (Follett and Sharp, 1969). This release is seen in birds kept on a long-day cycle of light, 20 hours light/4 hours of dark. When they are exposed to a short-day of 6 hours light/18 hours dark such oscillation is not observed. This long-day photoperiodic pattern is similar to that which quail experience in nature where

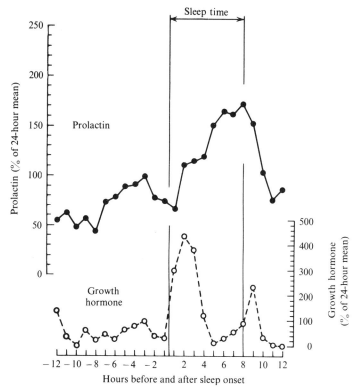

Fig. 4.13. The diurnal pattern in the release of prolactin and growth hormone in man. It can be seen that the release of both these hormones was accentuated during sleep. (From Sassin *et al.*, 1972. Copyright © 1972 by the American Association for the Advancement of Science.)

it is associated with the onset of breeding. Comparable precise patterns in the daily release of pituitary gonadotrophins have been observed in brook trout (*Salvelinus fontinalis*) and rainbow trout (*Salmo gairdneri*) as well as in leopard frogs (*Rana pipiens*) (O'Connor, 1972).

The pineal gland of the rat shows an interesting cyclical pattern in activity during the day which may be related to reproduction. As described in Chapter 2, the weight of this gland is lowest at the end of the daylight hours and rises during the hours of darkness. The enzyme, hydroxy-indole-O-methyl transferase (HIOMT) is essential for the synthesis of melatonin (which is secreted by the pineal) and its activity also increases at night (Fig. 4.14). Noradrenaline is present in the pineal and it attains higher levels in rats during darkness. Blinding the rats, cutting the sympathetic nerve supply to the pineal or placing them in continuous light prevents this

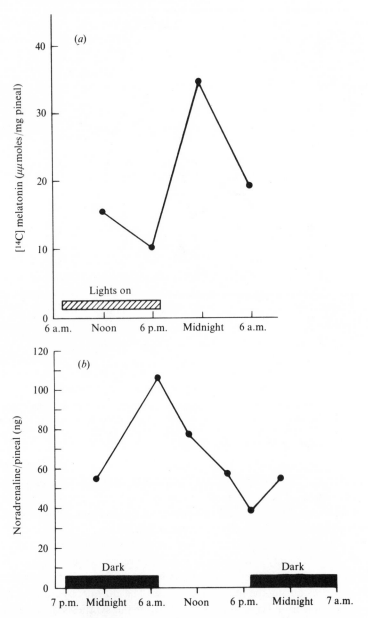

Fig. 4.14. Diurnal changes in the activity of the rat pineal gland.

(*a*) Increases in the levels of the enzyme HIOMT during the hours of darkness. (From Axelrod *et al.*, 1965.)

(*b*) The rise in the glandular content of noradrenaline during the night and the decline in the daytime. (From Wurtman and Axelrod, 1966.)

circadian rhythm of activity. The precise events involved in this cycle in the pineal appear to be as follows (Axelrod, 1974). During the hours of darkness there is an increase in the release of the neurotransmitter noradrenaline and a rise in the sensitivity of its pineal receptor. These changes result in the synthesis of the enzyme serotonin-N-acetyltransferase. This response is a β-adrenergic one involving cyclic AMP. The N-acetyl-serotonin that is formed from serotonin is then converted to melatonin under the influence of HIOMT. This rhythm in pineal activity is thought to be controlled by a biological clock that is situated in the hypothalamus, near the suprachiasmatic nucleus, which can be inhibited as a result of the receipt of light by the retina. In some birds and baby rats, blinding does not affect the rhythm as light can apparently impinge through their thin skulls onto photoreceptors in the brain.

In many vertebrates, melatonin inhibits growth of the gonads and this can be correlated with the inhibitory effects of darkness on this process. It is unknown how widespread this nocturnal rhythm in the activity of the pineal is. Periods of prolonged darkness or light, however, have been shown to change the activity of the pineal in other vertebrates. In mammals this activity, and the melatonin levels, are enhanced by darkness but depressed by light, but at this time the observations appear to be confined to rats. In the domestic fowl, however, melatonin levels still rise at night, just like in rats and this increase (a 10-fold one) is related to elevated levels of N-acetyltransferase though HIOMT declines (Binkley *et al.*, 1973). Changes in HIOMT concentration do not appear to be essential for formation of melatonin. In another bird, the Japanese quail, and an amphibian, the frog *Rana pipiens*, no change in the synthesis of melatonin could be produced by altering the lighting conditions (see Wurtman, Axelrod and Kelly, 1968). It has been suggested that the pineal, by responding to a biological clock, functions like the median eminence as a neuroendocrine-transducer that modifies the actions of other endocrine glands. It thus may provide an important pathway whereby light provides cues to the endocrine system.

The activity of the endocrine glands and the release of their hormones often shows profound changes that are associated with the season of the year. The diurnal patterns of release just described may be initiated at such times or be superimposed on these gross changes in activity. Such changes have most often been observed in relation to the breeding season which, especially in animals from non-equatorial regions, usually only occurs at certain times of the year. Reproduction may be dictated by predictably favorable seasons or in less-favored areas, like deserts, by the sudden appearance of rain. Apart from pituitary and gonadal sex hormones, cyclical changes in the activity of the thyroid gland and the adrenal cortex and medulla have been observed. The thyroid of the Japanese quail follows a pattern of activity that parallels that of the gonads (Follett and Riley, 1967). Thyroid activity (Leloup and Fontaine, 1960) is increased in fishes undergoing seasonal migrations. In the African lungfish (*Protopterus annectens*), thyroid activity is lowest during estivation at the time of seasonal drought. In toads and frogs, adrenaline attains its highest concentration in the plasma during autumn and winter (Donoso and Segura, 1965; Harris, 1972). Stores of growth hormone are highest in the pituitary of the perch (*Perca fluviatilis*) in June, a few weeks prior to the summer rapid-growth period (Swift and Pickford, 1965).

The cyclical and episodic release of hormones is usually controlled through the brain. This is especially apparent when light provides the cues for these changes. External temperature may also be involved in controlling reproductive and possibly other cycles, especially in poikilotherms (see Licht, 1972). It is likely that such heat stimuli act through the central nervous system but they could also exert more direct effects on the pituitary and gonads and even influence the rates of a hormone's metabolism. The principal photoreceptors are the eyes but there is evidence that in some vertebrates others may exist, such as those of the pineal gland. Temperature impinges on receptors present in the skin but may also directly influence the brain.

The diurnal rhythm of activity of glands such as the pineal and the adrenal cortex is not present at birth. In rats, the rhythmical release of corticosteroids from the latter usually arises about 21 days after birth while in man it takes two to eight years to appear. The adequate development of a biological clock is thought to require the presence of optimal levels of hormones. If newborn rats are injected with cortisol two to four days after they are born, the subsequent normal pattern of diurnal release of the corticosteroids is suppressed (Fig. 4.12). Similarly, the normal pattern in the cyclical activity of the gonads is prevented when newborn female rats are injected with androgens. If newborn male rats are castrated soon after birth the hypothalamus develops along the female pattern. When ovaries and a vagina are grafted into these animals they undergo cyclical changes typical of the female. Such endocrine manipulation can also contribute to behavioral abnormalities.

Systematic differences in releasing stimuli

There are many instances among the vertebrates in which the physiological roles of analogous hormones exhibit systematic differences. Such a change in the use of a hormone necessarily results in an altered responsiveness to excitatory stimuli that prompt the endocrine gland to discharge its secretion. When certain teleost fish move from fresh water to sea-water they lose water and accumulate salt so that the concentration of sodium in their body fluids rises. This initiates a release of corticosteroids. Tetrapods, on the other hand, usually release such analogous hormones in response to declines in the sodium concentration of the body fluids. Prolactin and oxytocin are released during suckling in mammals. Other vertebrates lack mammary glands so that this represents a unique and phyletically novel stimulus. On the other hand, in certain teleost fish a prolactin-type hormone is released when the fish migrate from the sea into fresh water. In homeotherms, thyroid hormones are discharged following a decline in the temperature of the blood flowing through certain areas in the hypothalamus but there is no indication that this happens in poikilotherms. Indeed thyroid secretion usually increases in cold-blooded vertebrates exposed to elevated temperatures. Changes in glucose concentrations usually determine insulin release, but in ruminants (sheep and cattle) fatty acids are more important. In fishes, amino acids may be the major stimulant for secretion of insulin. Such phyletic differences in the propensity of endocrine glands to respond to certain stimuli are necessary for the evolution of a hormone's physiological role.

Conduction of stimuli to the endocrines

The conveyance of a stimulus to an endocrine gland may involve a complex series of events that take place along rather circuitous pathways. These are consistent with, and indeed are dictated by, the particular physiological requirements of the animal. The initial stimulus is usually translated into another form that may be a chemical compound or an electrical event or both. It travels to the endocrine gland, in such a modified form, by routes of varying complexity and may suffer further translation on the way. During this voyage, the stimulus may be modulated and interpolated with other information that is already available, and other stimuli that also impinge on that particular communication pathway. This may take place in the brain and endocrine glands that are temporally proximal to the gland that is destined to receive, eventually, the final message. Such intermediary substations may involve neural areas in the brain, interconnecting endocrine glands, like the median eminence and the pineal, as well as the pituitary gland.

These events can be illustrated by summarily following the effects of light on reproduction. The stimulus is received by the eye where it is translated by the retinal receptors to electrical impulses that travel along nerves within the brain. These messages, after further translation to other transmitter substances (such as acetylcholine, noradrenaline and dopamine) eventually reach the median eminence and, probably also, in some species, pass through the pineal gland which may release melatonin. The median eminence modulates its release of LH/FSH-RH that crosses, in the short,

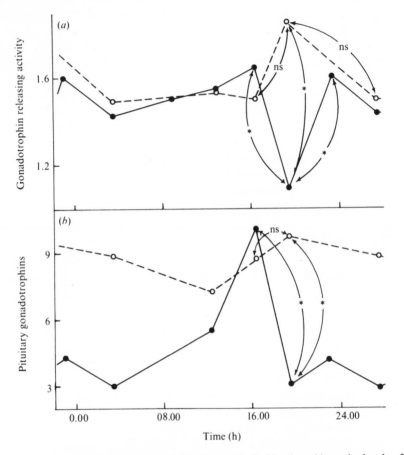

Fig. 4.15. The effects of different daily photoperiods (day-length) on the levels of hypothalamic gonadotrophin-releasing activity (*a*) and pituitary gonadotrophins (*b*) in the quail (*Coturnix coturnix*). The length of the day was either 20 hours light/4 hours dark (●——●) for ten days, or 6 hours light/20 hours dark also for 10 days (○----○).

The values are expressed as gonadotrophic-releasing activity, μg equivalents NIH standard/mg/pituitary tissue/hour; gonadotrophin, μg equivalents NIH standard/gland; ns, not significantly different; *, difference statistically significant.

It will be noted that no significant changes occurred in the birds on the 'short day' but in those on long day there was a periodic (and corresponding) release of both hormones in the late afternoon. (From Follett and Sharp, 1969.)

portal blood vessels to the adenohypophysis. The receipt of this information will be further interpreted (in a process which possibly involves the formation of cyclic AMP) in terms of an appropriate release of FSH and/or LH which are carried in the blood to the ovaries or testes. Such an effect is illustrated by the levels of gonadotrophin-releasing hormones and the gonadotrophins in the hypothalamus and pituitary of the Japanese quail. As can be seen in Fig. 4.15, there is a drop in the concentration of gonadotrophin-releasing hormone which corresponds to the release of adenohypophysial gonadotrophins. The gonadotrophins may influence such events as ovulation and the formation and release of estrogens and progesterone as well as testosterone. These latter steroids, in turn, are carried back to the brain (in the vicinity of the median eminence) where they provide information about the current hormone levels that will be used to modify stimuli that subsequently pass through this tissue. The process is, in detail, undoubtedly even more complex than that which has been described.

In other instances, the mechanism of the hormone's release may be simpler. The discharge of vasopressin (or antidiuretic hormone) from the neurohypophysis can occur in response to small increases in the osmotic concentration of plasma. This is thought to induce changes in osmoreceptors in the region of the supraoptic nucleus, possibly by releasing small amounts of acetylcholine, which initiates a wave of depolarization along the axons of the supraopticohypophysial tract. This results in the release of ADH from the storage granules at the terminus of the nerve. Neurohypophysial hormones may also be released in other circumstances. This release may involve non-specific stimuli, often termed 'stress', that pass through higher centers in the brain to the nerve cells of the gland. Oxytocin is discharged in response to suckling in mammals, the initial receptor is in the nipple of the mammary gland from which it is transmitted along nerves to the paraventricular nucleus which contains the cell bodies of the oxytocinergic neurons of the neurohypophysis.

Even simpler processes, not involving nerve pathways, may exist and determine the release of hormones in the body. The release of insulin can be demonstrated in isolated, perfused, pieces of pancreatic tissues containing the Islets of Langerhans. Elevated glucose and certain amino acid concentrations in the perfusate initiate a release of insulin. In a similar way, glucagon is discharged when the blood glucose concentration is depressed. Although it is difficult to be completely certain (as these preparations of Islets' tissue are not pure B-cells) it is considered that glucose reacts with a 'glucoreceptor' associated with the B-cells. This event, which results in the formation of a metabolite, then initiates the release of the hormone. Even at this level the process may be quite complex and involve several intermediate metabolites, ionic changes (especially that of calcium), and possibly cyclic AMP. Glucagon potentiates the release of insulin and the α-adrenergic effects of catecholamines can inhibit it. The events that occur near or in the endocrine cell itself are summarized in Fig. 4.16a and in the next section.

The mechanism of release of the hormones

Upon receiving a stimulus an endocrine gland may release its hormones from their storage sites. Many hormones, such as catecholamines, neuro-hypophysial peptides and insulin are spewed from their storage granules, steroid hormones are released from lipid droplets and thyroid hormones are detached from thyroglobulins. In some instances, this is preceded by the formation of cyclic AMP as in the thyroid gland, the adrenal cortex, the ovary and possibly the adenohypophysis and pancreatic B-cells.

The release of hormones from intracellular storage granules has been studied in detail in the instances of the secretion of catecholamines from the adrenal medulla and vasopressin and oxytocin from the neurohypophysis (Douglas, 1968; Kirshner and Viveros, 1972). Hormones that are stored in granules exist there as a non-diffusible complex with proteins and adenine nucleotides. Their release from the cell involves the process of emiocytosis (exocytosis or reverse pinocytosis) across the cell membranes. The nature of the events that result in such a release of hormones can be studied *in vitro* and is as follows:

(i) As a result of nerve stimulation the cell membrane is depolarized and ions enter the cells. *In vitro*, this can be performed by exposing the tissue to high concentrations of potassium or by stimulating it electrically.

(ii) An increase in the concentration of intracellular Ca^{2+} occurs, as a result of its uptake across the cell membrane and probably also its dissociation from binding within the cell. No release of hormone occurs *in vitro* following depolarization if the external media contains no Ca^{2+} which thus appears to be vital for excitation–release coupling.

(iii) Excitation–release coupling involves a migration of the hormone storage granules towards the cell membrane; a process that may involve the cell microtubular system.

(iv) When contact between the cell membrane and the granules is made they fuse and Ca^{2+} may be important in the structural links. The entire contents of the granule, hormones, proteins and adenine nucleotides are then extruded and pass into the capillaries. The empty granule may then be reconstituted and return to the cell cytoplasm.

Modifications of this process may occur in different endocrine glands. Insulin is released as a result of nerve stimulation but glucose and amino acids have direct effects and these apparently also result in the admission

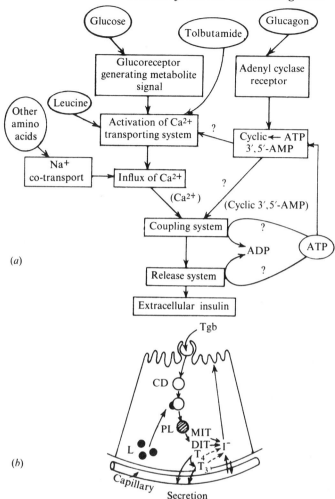

Fig. 4.16. (*a*) A diagrammatic representation of the events that result in the release of insulin from the B-cells of the Islets of Langerhans.

Changes in calcium concentration in the cell have a central role and 'couple' the primary stimuli (glucose, amino acids, etc.) to the secretion of insulin from its storage granules. β-Adrenergic stimulation, which results in increased concentrations of cyclic AMP, can also mediate a release, probably via such a calcium mechanism, while α-adrenergic excitation opposes this effect and reduces hormone secretion. (From Randle and Hales, 1972.)

(*b*) The secretion of thyroid hormones. Thyroglobulin (Tgb) is transported by endocytosis from the follicles into the thyroid cells where it appears as the colloid droplets (CD) that fuse with lysosomes (L) to form phagolysosomes (PL) in which proteolytic digestion of the thyroglobulin occurs. The iodothyronines (MIT, DIT, T_4 and T_3) are released and the active hormones pass out of the cell. Iodine (I^-) is reaccumulated. (From Greer and Haibach, 1974.)

of Ca^{2+} into the cell which results in hormone release (Fig. 4.16a). The activation of the enzyme adenyl cyclase and the formation of cyclic AMP (see p. 158) also appear to be involved in the release of insulin (as well as several other hormones, see below) where it may promote the accumulation of Ca^{2+} or act more directly in the excitation–release coupling process.

Thyroid hormones are stored extracellularly, combined with thyro-globulin in the colloid of the thyroid gland follicles. Stimulation by TSH results in an activation of adenyl cyclase in the thyroid cells and the forma-tion of cyclic AMP. In a manner that is not yet understood, this nucleotide is thought to stimulate pinocytosis during which fragments of follicular colloid are broken off and taken up by the thyroid cells; this process is thus called endocytosis. Lysosomes in the cell cytoplasm (Fig. 4.16b), that contain digestive protease enzymes, combine with these ingested pieces of colloid to form phagolysosomes which move towards the basal regions of the cell. During this migration, the colloid is digested and its components are separated so that thyroxine and tri-iodothyronine are freed to pass out of the cell. The iodotyrosines present are deiodinated and the iodine is either retained by the cell or, if also extruded, is subsequently reaccumulated.

The precise mechanism by which steroid hormones are released from the cell is not clear. The amounts stored in lipid droplets within the cell are relatively small compared to the large amounts of those hormones that are accumulated in granules. The synthesis and release of steroid hormones is promoted by various substances, including trophic hormones. Cortico-trophin and LH activate adenyl cyclase so that cyclic AMP is formed which plays a vital part in the initiation of the hormone synthesis and release: these two processes have not, however, been satisfactorily separated. It is possible that an increased hormone synthesis inevitably leads to its release by a type of 'overflow mechanism' but this is usually considered to be an unsatisfactory explanation.

The concentration of hormones in the blood

Hormones are normally present in the peripheral plasma at concentrations as low as 10^{-12} M and in-vitro observations suggest that these levels may, in some instances, be similar to those necessary to stimulate the effector. However, for several reasons it is difficult to predict with certainty what the latter concentration will be. As we have seen, endocrines discharge their secretions in response to a great variety of stimuli; in some instances they may be sudden, or acute, occurrences while at other times they are of a more sustained, or chronic, nature. In the first case the hormones may appear as a single 'spurt' of activity in the blood, such as seems to occur with the release of oxytocin in response to suckling. Thus high concentrations of hormones may exist locally, which briefly stimulate the effector and which are subsequently dissipated so that the remaining concentrations in the peripheral blood are not effective ones. It is difficult to obtain precise information about the temporal pattern of a hormone's release but undoubtedly many hormones, like those from the thyroid and gonads, are released in such a manner that fairly uniform levels appear and persist in the circulation. This may occur continually as a series of small spurts such as seen with corticosteroids (Fig. 4.11). A sustained release of a hormone may be necessitated by a continual demand by many tissues in the body.

Differences in the patterns of release combined with seasonal and diurnal rhythms make it difficult to generalize as to the concentrations of hormones that are normally present in the blood. The antidiuretic hormone is normally present in the peripheral plasma of mammals at concentrations of 10^{-12} to 10^{-11} M. It is higher in small animals, like the mouse, 2×10^{-11} M, than large ones such as man, where it is 4×10^{-12} M (see Table 4.2). In the

TABLE 4.2. *The concentrations of vasopressin or vasotocin in the plasma of vertebrates. The animals were normally hydrated except where indicated.* (From Heller, 1966: Heller and Štulc, 1960; Robinson and MacFarlane, 1957; Bentley, 1969)

	Hormone concentration (moles/l)
Mammals	*Vasopressin*
Mouse	2×10^{-11}
Rat (i) normal hydration	1×10^{-11}
(ii) dehydrated	2.5×10^{-10}
Guinea-pig	8×10^{-12}
Cat	2×10^{-12}
Rabbit	1.6×10^{-12}
Dog	2×10^{-12}
Man	4×10^{-12}
Bandicoot, wallaby and possum (Marsupials) dehydrated	1.8×10^{-10}
Amphibians	*Vasotocin*
(Anura) dehydrated	4 to 8×10^{-10}

laboratory rat, ADH concentration increases about 25-fold during dehydration: from 10^{-11} M to 2.5×10^{-10} M. Some dehydrated amphibians (frogs and toads) have similar circulating levels of vasotocin: about 6×10^{-10} M (Bentley, 1969). In others such as the mudpuppy and a teleost fish, the eel *Anguilla rostrata*, the hormone is not detectable (less than 10^{-11} M) under such conditions. Other polypeptide hormones like glucagon and insulin are normally present in the blood of mammals at a concentration of about 10^{-9} M.

The steroid hormones (see Idler, 1972) attain much higher concentrations in the blood than the polypeptides, thus plasma cortisol and corticosterone levels in teleostean fish are about 10^{-7} M though in the holostean and chondrostean fishes it is 10 times less than this. In chondrichthyeans 1α-hydroxycorticosterone usually has a concentration of 10^{-8} M in plasma. Aldosterone, compared to other corticosteroids, is present at much lower concentrations, about 10^{-10} M in mammals and 10^{-9} M in birds, though it is 10^{-8} M in amphibians. The steroid sex hormones are, due to the cyclical nature of their release, present in widely differing concentrations in the peripheral plasma: 10^{-7} M to 10^{-10} M.

Hormones act when present in extremely dilute solution. The levels vary somewhat with different species. The reasons for variation appear to be related to the animal's size, metabolic rate and the hormone-binding capacity of the plasma. It seems likely that differences in concentration may also be influenced by the body temperatures at which the animals usually function, because they affect the rate of the hormones' reactions with the effectors. Differences in the potencies of homologous hormones, as reflected in their chemical structure, may also contribute to variations in plasma levels. A mutation that initiates the formation of a less active hormone analogue could be compensated for physiologically by its release in greater quantity. An example of such an adaptation may be seen in mammals (Suiformes) that have lysine- instead of arginine-vasopressin.

Transport of hormones in the blood

Following their release from storage sites in the endocrine glands the hormones are carried, in the blood, to their various effector sites. This may be for very short distances, such as from the median eminence to the adenohypophysis, or involve much longer journeys, like from the neurohypophysis to the kidney. Some hormones have a relatively long life in the circulation and may recirculate many times before they are finally destroyed or excreted.

The blood plasma is an aqueous solution that contains high concentrations of proteins made up of several distinct components, or fractions. The hormone molecules may be dissolved in this solution or a substantial proportion can be bound to some of the proteins that are present. The binding of hormones in the blood is particularly important in the instances of the thyroid and steroid hormones.

The chemical nature of the binding of such hormones to plasma proteins is such that an equilibrium exists between the molecules that are bound and those that remain in solution. In other words, the binding is a reversible phenomenon and thus involves relatively weak chemical forces such as hydrogen and ionic bonds and Van der Waal's forces. The relative strength of these bonds, however, may vary considerably (high and low affinity) depending on the particular hormone and the nature of the binding protein. As we shall see, in some instances special proteins are present that appear to be able specifically to bind certain hormones.

There are several consequences of a hormone's binding to a protein in the plasma.

(1) Hormones are usually assumed to be unable to initiate their effects when so bound. The receptor is envisaged as interacting with hormone molecules present in the aqueous phase. Their removal from solution shifts the equilibrium and may thus result in a dissociation into solution of some of the hormone that is bound to the proteins. It is implicit that the receptor has an even stronger ability than the plasma proteins to bind the hormone and it is even possible that more direct exchanges may occur between binding proteins and the receptors.

(2) The process of the hormone's inactivation, such as can occur in the liver, and its excretion, mainly in the urine and bile, is delayed while it is in the bound form.

(3) The bound hormone may constitute a circulating pool of the excitant that can extend or moderate the hormone's action. For instance, in pregnant women and guinea-pigs, a plasma protein is formed that can bind testosterone and so may protect the mother from the effects of this steroid.

(4) The distribution of the hormone within the body can be influenced by its binding to plasma proteins (Keller, Richardson and Yates, 1969). A hormone–protein complex may have a special propensity to pass through the capillaries in certain vascular beds. It may thus help specifically to determine where the hormone is going.

(5) Protein-binding may contribute to the specificity of a hormone's action in the cell (Funder, Feldman and Edelman, 1973). In the instance of the corticosteroids, two types of response can occur, mineralocorticoid and glucocorticoid (referring to their effects on electrolytes and intermediary metabolism), for which there are two types of receptors in the cell. Although the aldosterone receptors have a higher affinity for aldosterone than corticosterone, the normal excess of the latter hormone (in the plasma of rats) would be sufficient to negate this difference so that corticosterone would be expected, inappropriately, to occupy the aldosterone receptors. Most of the corticosterone in the plasma, unlike aldosterone, is bound to proteins in such a way that it does not obscure the effect of the aldosterone.

High proportions of hormones often exist in a bound form in the plasma. In the instance of testosterone, it is usually greater than 90% of the total present though it is less in other cases. Cortisol binding in the plasma of teleost fishes varies from 30 to 55% of the total (Idler and Truscott, 1972). Thyroxine and tri-iodothyronine also are substantially associated with plasma proteins; thus, in the plasma of kangaroos, more than 95% of the thyroxine and 90% of the tri-iodothyronine is so bound (Davis, Gregerman and Poole, 1969).

The hormones may be bound to different protein components present in the blood plasma. In some instances, specialized proteins are present that have a high affinity (they are strongly bound) for the hormones. These include a globulin that binds cortisol, cortisol-binding globulin or *CBG*, which is also called *transcortin*. This protein is present in all vertebrates though quantitative differences in cortisol-binding capacities of the plasma exist, indicating that interspecific variations occur (Seal and Doe, 1963). Cortisol binding is much less in the plasma of fishes than in most mammals (Idler and Truscott, 1972) and it is unlikely that the proteins are identical in the different species. Transcortin also binds corticosterone and progesterone (but not aldosterone) and in mammals its levels increase during pregnancy as a result of the action of estrogen which promotes its formation in the liver. Another plasma globulin that binds sex hormones, *sex hormone-binding globulin* or *SHBG*, is found in women (where its level also increases in pregnancy) and various other vertebrates including chondrichthyean and teleostean fishes and amphibians (Ozon, 1972). It is unlikely that this protein is chemically identical in all these groups. Many mammals possess an α-globulin that preferentially binds thyroxine, *thyroxine-binding globulin* or *TBG*, but which is absent in non-mammals (Farer *et al.*, 1962; Tanabe, Ishii and Tamaki, 1969). It also binds tri-iodothyronine but only about one-third as strongly as it does thyroxine. Although such plasma proteins strongly bind hormones, they are present in relatively small quantities so that the total amount of hormone that they can associate with is limited (low binding capacity). The plasma albumins on the other hand, have a high binding capacity though their affinity is low. In species that lack such specialized hormone-binding proteins, or if the amount of hormone present exceeds their binding capacity, large amounts are bound to plasma albumin. In primates, and even their phyletically distant relatives, the marsupials, thyroxine is principally bound to a pre-albumin (thyroxine-binding pre-albumin, *TBPA*) while tri-iodothyronine combines with albumin. In some birds, reptiles and fishes, thyroxine is bound principally to albumin while in others it is associated with both pre-albumin and albumin (Tanabe *et al.*, 1969).

Differences in binding proteins exist between species. In some Australian marsupials (kangaroos), 75% of the thyroxine is bound to the pre-albumin while only about 10% is associated with the post-albumin. On the other hand, in an American marsupial, the opossum, about 60% of the thyroxine is bound to the post-albumin (Davis and Jurgelski, 1973). Even among these opossums differences were observed: the post-albumin exists in two different polymorphic forms PtA_1 and PtA_2. Six, out of 177 sera examined, possessed PtA_2, which has different physicochemical properties to PtA_1; for example, PtA_1 binds tri-iodothyronine but PtA_2 does not.

Peripheral activation of hormones

Some hormones are chemically altered, at sites peripheral to the endocrine gland from where they originated, in a manner that enhances their biological activity. This process is sometimes referred to as 'activation' and may occur in the plasma or at tissue sites.

Angiotensin I is thus converted to its active octa-peptide form, angiotensin II, by the action of converting enzymes. Not all the tri-iodothyronine present in the circulation is directly released from the thyroid gland. Some is formed from thyroxine in the liver and kidneys (Sterling, Brenner and Saldanha, 1973). An integral step in the action of testosterone involves its conversion at its target site by 5α-reductase, to 5α-dihydroxytestosterone. Cortisone can also be converted peripherally to the more active steroid cortisol. Cholecalciferol (vitamin D_3) undergoes several transformations before it can exert its effects. These changes involve the formation of 25-hydroxycholecalciferol in the liver and a further hydroxylation to 1,25-dihydroxycholecalciferol in the kidney (Lawson *et al.*, 1971). The latter has twice the biological activity of its immediate parent compound. It is also possible that some of the large protein hormones are fragmented peripherally into smaller pieces prior to their action. There is evidence that this may occur in the instance of parathormone while large amounts of pro-insulin are normally present in the plasma which, possibly, may also be converted into the active hormone.

Termination of the actions of hormones

The durations of action of hormones vary: they may persist for many hours and even days, or only have a short-lived, transitory, effect. This is usually in keeping with the nature of the homeostatic processes that they mediate. Even a prolonged effect, however, will necessitate the renewed release of hormones because of the metabolic and excretory processes that inevitably result in their inactivation and elimination from the body.

Hormones persist in the circulation for different periods of time. The half-life of human vasopressin is about 15 minutes, that of cortisol is about one hour, while that of thyroxine is nearly a week. This reflects the speed of their degradation in the body, the rate that they may be eliminated in the urine and bile and the protection afforded them as a result of binding to proteins. The effects of the latter are well illustrated by comparing the rates of removal (clearance) of corticosteroid hormones from the blood. In man, about 1600 liters of plasma are normally completely cleared of aldosterone each day. In the instance of cortisol, however, only about 180 liters are so purged in this time. The difference principally reflects the strong binding of cortisol to cortisol-binding globulin (transcortin) in the plasma. Considerable interspecific differences are apparent in the half-lives of hormones: arginine-vasopressin (ADH) has a half-life of about one minute in laboratory rats, compared to 15 to 20 minutes in man. This probably reflects the effects of size: small animals destroy and eliminate hormones more rapidly than large ones. The precise chemical structure of the hormone may also be important; lysine-vasopressin has a half-life that is nearly twice as long in the rat as arginine-vasopressin (Ginsburg, 1968). The considerable differences that have been observed in the potency of human, porcine and salmon calcitonin substantially reflect the differences in their degradation rates (Habener *et al.*, 1971; DeLuise *et al.*, 1972). In rats, porcine calcitonin is destroyed by the liver while that from man and the salmon are degraded by the kidney. Salmon calcitonin is much more resistant to inactivation (in rats, dogs and man) than the mammalian hormones. Body temperature will also be expected to have an effect on the rate of inactivation of hormones so that, in cold-blooded vertebrates, the inactivation and excretion process will be modified accordingly. In the toad *Bufo marinus*, at 26 °C arginine-vasotocin has a half-life of 33 minutes while in the domestic fowl (at 43 °C) it is only 18 minutes (Hasan and Heller, 1968). The effects of differences in size, species and temperature on hormone metabolism have not yet, however, been thoroughly evaluated.

The action of a hormone may be terminated in several ways.

(1) In order to act it must attain a certain critical concentration in the neighborhood of its receptor site. If a hormone is released in a short burst as, for instance, usually occurs with oxytocin, the receptor will respond to a local high concentration that is sequestered like a small packet in the plasma. The response will then be terminated simply as a result of the subsequent dilution and redistribution of the hormone in the body fluids. A hormone may also be removed if it is bound or has accumulated at tissue sites. Adrenaline is readily taken up by the adrenergic nerve terminals. Less-specific binding to tissues, such as that of neurohypophysial peptides to skeletal muscle, may also contribute to the removal of hormones from the circulation.

(2) Small amounts of hormones may be eliminated unchanged in the urine and bile but this usually amounts to less than 5 % of the total released. The activity of hormones is generally reduced or destroyed as a result of their metabolism by enzymes in the tissues, particularly in the liver and kidneys. They are transformed in various ways and the by-products are usually then excreted in the urine and bile. The hormone's chemical structure may be altered in several ways. The catecholamines can be methylated by catechol-*O*-methyl transferase (COMT) or, to a lesser extent, they may be deaminated as a result of the action of monoamine oxidase (MAO). The action of the thyroid hormones is largely destroyed by the removal of iodine from the molecule by a deiodinase enzyme. Protein hormones are broken up by proteolytic enzymes. Steroid hormones (and to some extent thyroxine) are combined chemically with glucuronic and sulfuric acids, in a process called conjugation that results in increased solubility which enhances their chances for excretion in the urine and bile. Prior to such a conjugation considerable changes may be wrought in the chemical architecture of the steroids. Despite such chemical alterations, some hormones may still retain some of their biological activity. This is especially apparent in the gonadotrophins that appear in large quantities in the urine of mammals during pregnancy. The activity of such urine in promoting ovulation and spermiation in various animals is used as a basis for pregnancy tests. Despite the retention of such biological effects, the products in the urine exhibit a different chemical behavior to that of the gonadotrophins present in the pituitary gland. Urinary FSH and LH apparently undergo some changes prior to their excretion, possibly involving a fragmentation of their molecules.

Mechanisms of hormone action

Hormones exert actions on every major group of tissues in the body. The nature of their effects are numerous and include changes in the intermediary metabolism of fats, proteins and carbohydrates, growth and development of the tissues, changes in the permeability of membranes and the contraction or relaxation of muscles. The precise manner in which they effect such processes is only incompletely understood.

It is generally considered that a hormone, in order to act, must eventually influence an enzyme activity. It may do this by a process of activation, promoting enzyme formation (induction) or possibly acting as, or providing, a cofactor in the chemical reaction that it promotes. As described earlier, this process is thought to be initiated as a consequence of the hormone's combination with a precise chemical moiety in the cell which is called its receptor. The exact role of this hormone–receptor unit in triggering the response is uncertain. It may provide a carrier for the hormone and so facilitate its transfer to a vital site in the cell or, conversely, the hormone may facilitate transfer of the hormone receptor to an essential site. It is also possible that the receptor–hormone complex itself could act as an enzyme activator (or inhibitor) or cofactor. The receptor could exist as a part of an enzyme so that an activating effect of a hormone could be direct, but this is uncertain.

Hormonally mediated responses to hormones may take place in successive steps involving numerous enzymes. Conceivably, by acting at a single rate-limiting stage in such a process and altering the supply of an essential metabolite, a hormone could influence a complex series of metabolic events in the cell. It is also possible that a hormone may directly influence more than a single process in such a chain of events.

Hormones often initiate responses in several different tissues in the body. In such instances, the basic effects may be similar or even differ from one another. For instance, aldosterone initiates a change in membrane permeability to sodium in the kidney, salivary glands, sweat glands and, in amphibians, the skin and urinary bladder. Alternatively, adrenaline mobilizes fatty acids from fat cells but glucose from muscle cells. Despite the diverse tissue sites and the nature of the processes involved, the underlying initiating mechanism is generally the same. For instance, the effects of adrenaline on fat and muscle cells are both mediated by the action of cyclic AMP (see later).

At the present time the mechanisms of action of hormones are thought to be affected through either of two major groups of processes in the cell (others may also exist). The nucleotide adenosine-3′,5′-monophosphate, or cyclic AMP, is a vital link in many endocrine responses and its formation is promoted, or inhibited, by numerous hormones. Other hormones, by regulating genetic transcription in the cell nucleus, can control the formation of proteins and enzymes that are essential for a response. Not all the actions of all the hormones are even partially understood. It is likely that some which have multiple actions may utilize several different mechanisms. Some hormones, notably insulin, do not easily fit into either of the above classifications.

The role of adenosine-3′,5′-monophosphate (cyclic AMP)

Our knowledge of the part played by adenosine-3′,5′-monophosphate, or cyclic AMP, in the action of many hormones is primarily due to the work of Earl Sutherland (see Robison, Butcher and Sutherland, 1971; Sutherland, 1972). This adenine nucleotide is chemically related to ATP from which it is formed. Its chemical structure is shown in Fig. 4.17.

The discovery of the endocrine role of cyclic AMP was made during an investigation of the mechanism of action by which adrenaline and glucagon convert glycogen to glucose in the liver. This reaction is dependent on several enzymes but the presence of a phosphorylase is rate limiting. This enzyme exists in two forms, a relatively inactive phosphorylase *b*, which on the incorporation of phosphate is converted to the much more active form, phosphorylase *a*. This activation takes place not only in intact cells but also in broken-cell preparations upon the addition of adrenaline or glucagon to them. The phosphorylase is a soluble enzyme which can be separated in

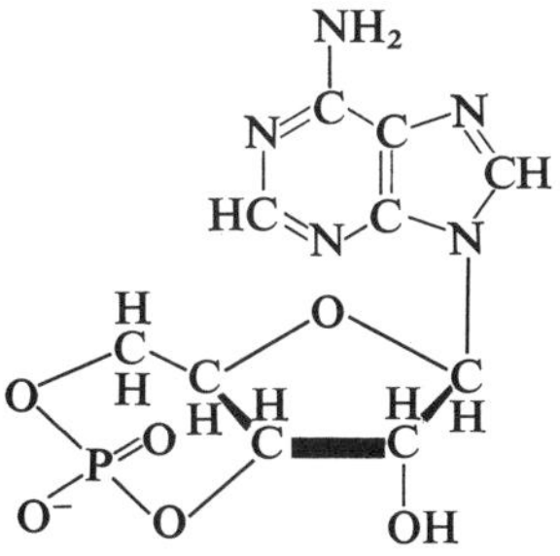

Fig. 4.17. The structural formula of adenosine-3′,5′-monophosate (cyclic AMP).

the supernatant fraction of the broken cells. This enzyme, however, cannot be activated by the hormones when alone in this solution; the presence of the particulate cell material is a necessary condition. When the latter fraction is exposed to adrenaline or glucagon a substance can subsequently be washed from the cell particles which, when added to the supernatant, activates the phosphorylase. This activating chemical was found to be cyclic AMP and it is formed from ATP as a result of the action of an enzyme, *adenyl cyclase* (more correctly adenylate cyclase), that is part of the cell membrane. In the liver, this enzyme is activated by adrenaline or glucagon and in skeletal muscle preparations only by adrenaline.

Adenyl cyclase has been found in many animals: mammals, birds, amphibians and fishes, as well as many invertebrates. This enzyme is also present in bacteria, though apparently not in higher plants. Most tissues in the body contain adenyl cyclase. Apart from liver, they include muscle, kidney, heart, brain, adipose tissue, bone and many endocrine glands (Table 4.3).

The role of adenyl cyclase and cyclic AMP in mediating the effect of hormones was first described for the actions of glucagon or adrenaline on glycolysis in the liver or skeletal muscle. A description of this process can thus be used as a prototype for its role. It should, however, be remembered that although the initiating reactions may be similar for many tissues and various hormones, the responses of the final effector (or responding system) will often differ. As summarized in Fig. 4.18, the initiating event is the interaction of the hormone and its receptor, in the cell membrane, which results in the activation of (or sometimes the inhibition) adenyl cyclase. The nature of the relationship between the receptor and the enzyme is uncertain; it could involve a series of chemical reactions or the two components may constitute part of a single structural complex. The simplest interpretation is that the receptor and enzyme are considered to constitute two subunits, an outward-facing receptor portion and an inner enzyme part which has access to the cellular ATP. The activated adenyl cyclase reacts with the latter to form the cyclic AMP. This does not directly

TABLE 4.3. *Distribution and hormonal sensitivity of mammalian adenyl cyclase.* (From Robison *et al.*, 1971)

Tissue	Hormone
Liver	Glucagon and adrenaline
Skeletal muscle	Adrenaline
Cardiac muscle	Catecholamines
	Glucagon
	Tri-iodothyronine
Kidney	Vasopressin
	Parathyroid
Bone	Parathyroid
	Calcitonin
Brain	Catecholamines
Adrenal	ACTH
Corpus luteum	LH and prostaglandins
Ovary	LH
Testes	LH and FSH
Thyroid	TSH
	Prostaglandins
Parotid	Catecholamines
Pineal	Catecholamines
Lung	Adrenaline
Spleen	Adrenaline
Adipose	Adrenaline
Brown adipose	Catecholamines
Platelets	Prostaglandins
Leucocytes	Catecholamines and prostaglandins
Erythrocytes	None demonstrated
Uterus	Catecholamines
Pancreas	None demonstrated
Anterior pituitary	Several
Vascular smooth muscle	None demonstrated

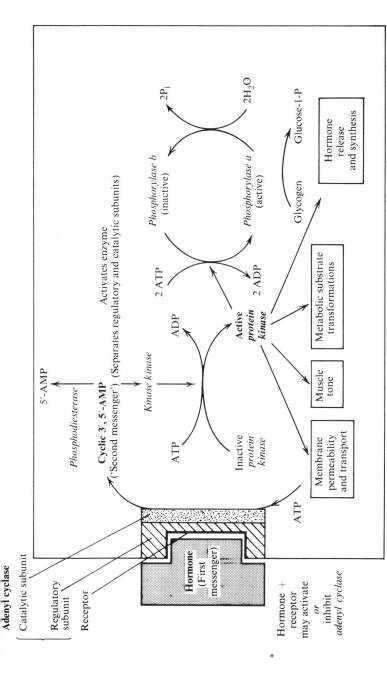

Fig. 4.18. A diagrammatic representation of the role of cyclic AMP as a 'second messenger' in the mechanism of a hormones action.

A variety of protein kinases are thought to be present in different cells which, when activated by cyclic AMP, can mediate the phosphorylation or dephosphorylation of certain proteins. This may result in diverse responses.

activate the phosphorylase *b* enzyme but initiates a sequence of chemical events that has this effect. Initially, the cyclic AMP activates a protein kinase kinase (also called kinase II or phosphorylase *b* kinase kinase) enzyme in the cell. The inactive form of the enzyme is thought to exist as a combination of a regulatory and a catalytic subunit. The nucleotide combines with the former upon which the subunit separates so that the catalytic part can exert its action. This kinase kinase activity, with the aid of ATP, activates phosphorylase *b* kinase which, in turn, converts phosphorylase *b* to phosphorylase *a*. The latter enzyme initiates the breakdown of glucose to glucose-1-phosphate.

Cyclic AMP can mimic the actions of many hormones on a variety of tissues. The formation of this nucleotide, as a result of hormonal activation of adenyl cyclase, appears to mediate the actions of such hormones as (apart from adrenaline and glucagon) melanocyte-stimulating hormone, vasopressin, corticotrophin, thyrotrophic hormone, luteinizing hormone and parathormone (see Table 4.4). The ultimate responses are very diverse and

TABLE 4.4. *Relations between the metabolic actions of some hormones and cyclic AMP*

Hormone	Actions shared by cyclic AMP
Adrenaline	Glycolysis (liver and muscle), lipolysis (fat cell)
	Heart muscle contraction
ACTH	Steroid synthesis in adrenal cortex
LH	Steroid synthesis by corpus luteum
TSH	Production and release of thyroid hormones
Growth hormone	Lipolysis (?)
Vasopressin	Increased water movement in renal tubule
MSH	Melanophore, dispersion of melanin
Glucagon	Glycolysis (liver), lipolysis (fat cell)
Parathormone	Mobilization of bone calcium

include glycolysis, lipolysis, changes in the permeability of membranes to water, sodium and calcium, the dispersion of melanin in the melanophores and the formation and release of many other hormones including thyroxine, corticosteroids and sex steroids.

The nature of the effector response to hormones differs widely. It has, however, been suggested that in those processes that involve the action of cyclic AMP as a 'second messenger', this is always the result of the phosphorylation, or dephosphorylation, of a protein and involves a protein kinase. In the instance of the glycolytic and lipolytic effects of adrenaline and glucagon, this results from enzyme activation. The formation of

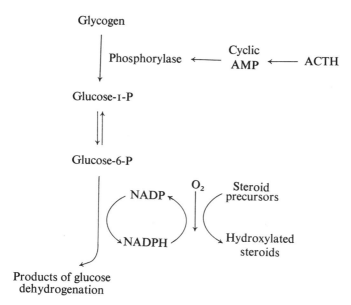

Fig. 4.19. The mechanisms of action of corticotrophin (ACTH) and cyclic AMP in stimulating the synthesis of corticosteroids by the adrenal cortex. This is the Haynes hypothesis.

steroids in the adrenal cortex and gonads may proceed as a result of an increased supply of NADPH from glucose-6-phosphate that also results from the activation of a phosphorylase and glycolysis (Fig. 4.19). In addition, an increased supply of cholesterol due to an increased activity of cholesterol esterase may also contribute to steroid synthesis. The mechanism by which membrane permeability is changed in response to neurohypophysial hormones is poorly understood but it may involve conformational alteration of constituent proteins as a result of their dephosphorylation by a protein phosphatase. The responses are stopped as a result of the replacement of phosphate at its effector sites by a phosphorylase enzyme.

Apart from initiating increases in levels of cyclic AMP in cells it is possible that some hormones may act by bringing about a decrease in the nucleotide's concentration. As described earlier, the catecholamine hormones exhibit two types of actions called α-adrenergic and β-adrenergic effects. The latter are associated with activation of adenyl cyclase and increases in cyclic AMP. On the other hand, α-adrenergic responses involve reduced levels of the nucleotide; possibly mediated by an inhibition of adenyl cyclase. Such inhibitory effects of catecholamines have been described in processes involving the release of insulin from B-cells, increase of the osmotic permeability of membranes by neurohypophysial peptides and the dispersion of melanin in melanophores by MSH. It has been suggested that some hormone receptors may contain a moiety which, when combined with α-adrenergic agents, initiates this inhibition. The anti-lipolytic effects of insulin in fat cells are accompanied by a decline in cyclic AMP levels. Prostaglandins are a group of excitants which have been found in extracts from many tissues and which have diverse effects. These substances, which possibly function as locally acting hormones, can also decrease (though sometimes they increase) cyclic AMP levels in tissues and this could mediate some of their effects. Processes may thus exist in cells which alter the concentrations of cyclic AMP in either direction and so determine the nature of the ultimate response.

The inactivation of cyclic $3',5'$-AMP to $5'$-AMP in cells is normally due to the action of the enzyme *phosphodiesterase*. This enzyme could provide a potential site for the action of hormones but such effects have not been described. Phosphodiesterase can be inhibited by certain drugs, notably the methylxanthines which include caffeine.

The activation of adenyl cyclase can be accomplished by several hormones, in various tissues, and can result in diverse responses. The actions of hormones are, however, relatively specific, so how, when several utilize cyclic AMP, can a separation of their roles and effects be accomplished? In some instances more than one hormone is thought to act on a single adenyl cyclase. This seems to be so in the effects of adrenaline, glucagon, ACTH and TSH on fat cells *in vitro*. The effects of the two latter hormones may not be physiological but reflect the relative non-specific reactivity of the receptors, and the adenyl cyclase, that are present. The adenyl cyclase that mediates the effects of ADH on the kidney is much more specific in its reaction with hormones. This selectivity is probably dictated by the nature of the hormone receptors. In some instances, a tissue exhibits more than one adenyl cyclase-mediated response to a single hormone. Neurohypophysial hormones increase both water and sodium transfer across amphibian skin and urinary bladder and these effects are mediated by different receptors and effectors. In the amphibian membranes, it seems likely that two adenyl cyclases and/or two 'pools' of cyclic AMP exist in the tissue, each mediating a distinct response.

Differences in the ability of various hormones to activate adenyl cyclase from different tissues or species could reflect a polymorphism in the enzyme's structure. If the receptor is considered to be an integral part of the enzyme, then this polymorphism undoubtedly occurs, but it is not known if such an association is present. Cell-free preparations of adenyl cyclase exhibit different quantitative responses to hormone analogues. Adenyl cyclase preparations from rat, mouse, rabbit and ox kidney are more readily activated by arginine-vasopressin than by lysine-vasopressin (and much less by oxytocin). On the other hand, the enzyme prepared from the kidneys of pigs is most responsive to lysine-vasopressin (Dousa *et al.*, 1971). This relative sensitivity corresponds to that of the homologous hormones present in each species, while oxytocin does not normally act on the kidney anyway. These differences, however, as discussed above, may only reflect the response of a receptor that need not necessarily be a part of the enzyme itself. The properties of renal adenyl cyclase from a number of vertebrates has been compared (see Table 4.5). Mammal, bird, reptile and amphibian enzymes were all strongly activated in the presence of fluoride. The mammalian enzymes were also activated by neurohypophysial

TABLE 4.5. *Stimulation of renal adenyl cyclase in various vertebrate species.* (From Dousa *et al.*, 1972)

Species	Neurohypophysial hormones	Parathyroid hormone	Fluoride
Rat	++++	++++	++++
Mouse	++++	++++	++++
Pigeon	o	+++	++++
Alligator	o	++	++++
Toad	±	o	++++
Bullfrog	+	o	++++

+ Indicates strength of response; o, no response; ±, rudimentary response. Homologous neurohypophysial hormones were tested.

hormones but this response was much less prominent, or even undetectable, in the non-mammals. This lack of response could, however, merely reflect the presence of only small amounts of the appropriate enzyme in the kidneys of the non-mammals or its lack of importance in mediating the renal responses in those species. It is notable that adenyl cyclase obtained from amphibian urinary bladder is strongly stimulated by neurohypophysial hormones. The renal adenyl cyclase that is activated by parathormone (a distinct entity from that responding to the neurohypophysial hormones) was stimulated in enzyme preparations made from the kidneys of mammals, birds and reptiles but not in those made from amphibian kidneys. This may reflect an absence of a renal effect for parathormone in the Amphibia. Except in bacteria, adenyl cylase has not been prepared as a soluble enzyme so that the precise studies that may confirm its polymorphism are not feasible. This limitation and the uncertainties as to its relationship to the hormone's receptor limit speculation as to the possible evolution of this enzyme.

Hormonal effects mediated by changes in the transcription of DNA in the cell nucleus

The synthesis of proteins is primarily controlled by the cell nucleus through the coding DNA that is contained in the chromosomes. Our knowledge of the manner by which this genetic material is utilized to regulate the synthesis of specific proteins in the cell is largely the result of the work of Jacques Monod and his collaborators (Monod, 1966).

For an understanding of how hormones may act at this level it is necessary to recapitulate, briefly, the nuclear processes thought to control protein synthesis. It should be noted however, that these have been principally worked out using bacteria. The basic unit carrying the genetic code for the synthesis of proteins is the *structural gene* whose actions are regulated in a complex manner. Adjacent to the structural gene is a special genetic segment called the *operator*.[1] The structural gene and the operator together form a unit called the *operon*. The operator is, in turn, controlled negatively by a *regulatory gene* that can produce a molecule, a *repressor*, that may specifically combine with, and so inhibit, the operator (Fig. 4.20). This regulatory compound has an ability to bind with other materials that can either result in its activation, which will result in decreased protein synthesis or its inactivation that will produce an increase. There are thus several possible places where an external controlling substance, such as a hormone, can act. The position of the site of action is still largely open to speculation but it could involve a repression of the regulatory gene, the inactivation of the repressor, or the masking of the operon from repressor action.

The effects of the steroid hormones, as well as the thyroid hormones and growth hormone, are most obviously manifested as increases in growth, development and differentiation of tissues. These changes are especially clear when one observes the effects of estradiol-17β on the uterus. Such growth is associated with a marked increase in the rate of protein synthesis. More subtle changes may accompany other responses to such hormones. For instance, aldosterone, which increases the rate of sodium transfer across some membranes, may induce the formation of a permease, or some other enzyme, that increases active sodium transport. It is not always possible to identify the specific proteins formed, but progesterone is known to increase the formation of avidin in the chick oviduct, while cortisol

---

[1] In the *lac* operon a 'promoter' is present which is a binding site for RNA polymerase. This enzyme is released when the repressor is absent from its site on the operator.

enhances the production of Na–K activated ATPase in several tissues including the kidney and, in some fish, the gills and intestine.

The evidence about the effects of hormones on the synthesis of proteins can be derived from direct measurements of their concentrations or from changes in the rate of their incorporation of radioactively labelled amino acids. Protein synthesis is a cytoplasmic process taking place in association with the ribosomes which follow the translation pattern provided by the messenger RNA which is derived from the chromosomal DNA.

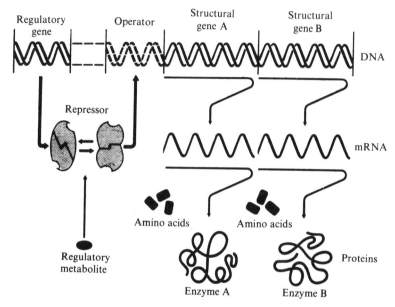

Fig. 4.20. A diagrammatic representation of the function and regulation of the operon. (From Monod, 1966.)

While it is possible that hormones exert some direct effects on the ribosomal protein formation this does not usually appear to be so. One of the earliest distinguishable effects of the action of hormones that influence protein synthesis is the formation of RNA by the nucleus and the incorporation of tritiated uridine into it. This process is accompanied by an induction or release of RNA polymerase, that catalyzes the separation of the messenger-RNA from DNA and which can be inhibited by actinomycin D. The effects of steroid hormones can be prevented by this antibiotic. The protein synthesis by the ribosomes can also be blocked by puromycin, which acts on the cytoplasmic translational process but does

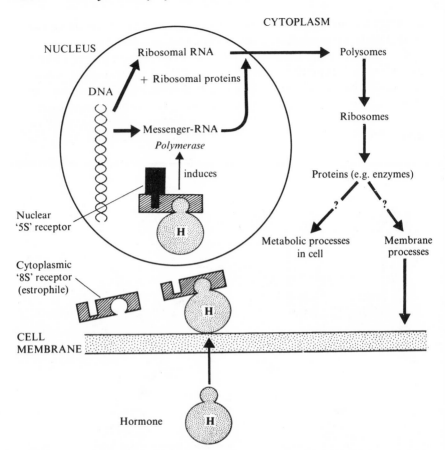

Fig. 4.21. A diagrammatic representation of the mechanism of action of a steroid hormone that is acting by influencing genetic transcription. This scheme is not completely understood and will undoubtedly be modified as new information becomes available.

not stop the early formation of the messenger-RNA. A further indication as to the nuclear site of action of hormones is their autoradiographic identification in the chromatin.

The processes by which hormones act to initiate protein synthesis, via increases in nuclear transcription of the genome, have been most completely described for the action of estradiol-17β on the uterus (see Jensen and DeSombre, 1972, 1973; O'Malley and Means, 1974) and have been summarized in Fig. 4.21. Comparable mechanisms also appear to exist for progesterone, the androgens, cortisol, aldosterone and also probably thyroid hormone and growth hormone.

The estrogenic hormone molecule, upon crossing the plasma membrane and entering the cell's cytoplasm, is bound to a protein which has been called its cytoplasmic receptor. These proteinaceous units are also called *estrophiles*. The hormone–receptor combinations can be isolated *in vitro*. They pass intact through a gel filtration column and can be separated from other cell constituents by ultracentrifugation in a sucrose gradient where they have a sedimentation coefficient of about 8 S. In the presence of salt solutions this changes to 4 S, indicating that the receptor–hormone unit can be broken into subunits, and it is possible that these may have functional significance in mediating the subsequent effects. There are about 100 000 estrophiles in each uterine cell, which is a much higher number than is present in non-target tissues. The binding for a particular hormone is relatively specific and can be prevented by substances that block their effects *in vivo*.

At a temperature of 2 °C, 75 % of the hormone is present in the 8 S complex in the cytoplasm, but if the tissue is warmed to 37 °C most of it moves across into the nucleus. The estrophilic complex may be acting as a carrier for the hormone or the hormone may facilitate the transfer of one of the subunits of 8 S that could initiate the formation of messenger-RNA. The estradiol-17β in the nucleus is bound in another proteinaceous form that can be separated in KCl solution and which has a sedimentation coefficient of 5 S. This complex is clearly distinguishable from the 8 S one in the cytoplasm. The 5 S nuclear complex can be freed from the chromatin by the action of DNAase, indicating that it is closely associated with DNA. Other evidence suggests an interaction with acidic proteins in the chromatin. The 5 S nuclear receptor has not been found in the nucleus prior to its exposure to the 8 S complex, suggesting that the latter may first be converted to 5 S which has an acceptor site in the nucleus. The free hormone does not readily bind to the chromatin. The precise site of the combination with the chromatin is not known nor is it clear how a hormone–nuclear receptor complex influences the metabolism of messenger-RNA. It could for instance promote its synthesis on the chromatin by combining with a repressor gene, or prevent the activation of a repressor substance, or by association with the operon be masking it from the latter's action. Alternatively, an activation of RNA polymerase would also have such an effect. Other mechanisms, however, could be involved, such as a reduced rate of destruction of RNA and an increase in the rate of its transfer from the nucleus to the cytoplasm.

At the present time not all the actions of hormones can be accounted for by the two preceding types of mechanism. Hormones have multiple effects and not all the effects of a single hormone can always be accounted for by a single mechanism. Thyroxine, growth hormone and prolactin can stimulate protein synthesis in the cell and, while these actions can be prevented by inhibition of RNA polymerase (with actinomycin D), just how primary, or universal, this effect is in these hormones' actions is not clear. Insulin influences protein synthesis not by direct involvement of genetic translation but by post-translational events. Insulin has profound effects in facilitating uptake of glucose and amino acids across the cell membrane and many of its effects can be accounted for by this transport process which involves a mechanism which is at present unknown. Other effects of insulin, however, such as the formation of glycogen in the liver and muscle, cannot be accounted for completely by this transport mechanism. Changes in the activity of glycogen synthetase are involved but just how this is brought about is not clear. Growth hormone also increases protein synthesis and it is possible that it also acts on the cell membrane.

Some speculations on the evolution of the actions of hormones in cells

The two basic control mechanisms involving cyclic AMP and the regulation of genetic transcription exist in all animals as well as many micro-organisms. We may thus suspect that these basic mechanisms, upon which hormones can impinge their actions, have always been present in animals, even in unicellular ones.

Cyclic AMP exists in bacteria, where it also appears to have a role in regulating cellular activities. Adenyl cyclase in bacteria is an intracellular enzyme and it has been suggested that this represents the primitive condition, though it may be more true to say that it is the situation in unicellular organisms. This internal site would appear to be most suitable for an enzyme that has to respond to changes in the intracellular nutrients and metabolites. In metazoan animals, adenyl cyclase appears to be confined to the cell membrane, a position that may be more apt for its interaction with metabolites and chemicals coming from other cells. These include the hormones. Such a membrane site for adenyl cyclase may thus be more opportune for intercellular cooperation and especially for interactions with molecules like the polypeptide hormones that cross cell membranes with some difficulty. The lack of information about the nature of the relationship of the hormone receptor and adenyl cyclase has already been discussed. It is unnecessary to postulate evolutionary changes in the structure of adenyl cyclase, only in the receptor which may be a subunit of it. The specificity of the receptor for a particular hormone has also been described earlier. The simultaneous evolution of the complementary nature of both the hormone and its receptor is difficult to envisage. One wonders, when considering all the possible differences in structure, how they ever got together and how, at the same time, they acquired their complementary relationships.

The transcription of genetic material plays a basic role in the life of cells. In unicellular organisms, this need only be controlled by internal accumulations of nutrients and metabolites. These may act by combining with structural units, analogous to the subunits of the 8 S cytosol receptors. With the onset of need for intercellular communication these controlling subunits may have contracted an ability to combine with materials originating outside the cell. In other words, they may have acquired, or been transformed so as to incorporate, a hormonal receptor. Steroid hormones are non-polar materials that readily gain access to cells and so would appear to be well suited to such an intracellular reaction. They would, however, because of their water-insoluble, non-ionic, non-polar nature be unsuitable as agents that act allosterically on enzymes. The required mobility of the 8 S-type regulatory units would, in such a simple system, preclude fixed receptor sites on the membrane.

It is not difficult to envisage, using the Monod model, how hormones with a nuclear site of action may have acquired different functions. Genetic rearrangement of the chromosomal material could, for instance (see Monod, 1966), move the control of a structural gene from one operator segment to another. If a hormone were acting through such an operator this would be accompanied by the transfer of its endocrine control to another structural gene. Such a mechanism could be completely inappropriate, but in some instances could provide a basis for the change, and evolution, of the role of a hormone.

Conclusions

In the following chapters we will be examining the roles of hormones in coordinating different physiological processes in the body. In the present chapter we have looked at the manner by which the endocrine system itself works. Although information about non-mammals is rather sparse it appears that the underlying mechanisms of a hormone's synthesis, release, transport, mechanism of action and its destruction are rather similar in all vertebrates. Even when different hormones are involved, the general underlying processes involved are often similar; but major differences are often apparent between the general types of hormones, especially those made from cholesterol (steroids), and those derived from amino acids. There are a number of interspecific differences in the 'life history' of particular hormones in the body and these can be related to the animal's manner of life. The natures of the stimuli that initiate a particular hormone's release are especially variable among the vertebrates and are dictated by the different physiological roles that a hormone may have assumed. In addition, quantitative differences may arise with respect to a hormone's rates of synthesis, destruction, the quantities that are stored in the gland and the concentrations that appear in the blood. Such differences can arise at distinct stages of the life cycle of an animal but they are also observed between various species where they can be related to such characteristics as size, rates of metabolism, and environmental factors like temperature and the availability of different nutrients, salts, and water.

5. Hormones and nutrition

Animals require a continual supply of food in order to sustain life. Such nutrients, in the first instance, are obtained from the external environment. These materials are used as an energy supply, as building blocks for growth and reproduction, and also as a source of certain essential chemicals necessary to the adequate functioning of the metabolic machinery in the body. The processes involved are thus basic to life and are regulated to a considerable extent by hormones.

Animal cells, including tissues isolated from metazoan species, can survive *in vitro*, in the absence of hormones, for extended periods of time. Except, however, for cancer cells, their life and continual perpetuation cannot go on indefinitely. Even the more limited survival of normal tissues *in vitro* depends on an adequate supply of special nutrients that are chosen for their ability to be utilized by that particular tissue. One cannot run a gasoline engine on diesel fuel and in the same way cells can only metabolize certain forms of nutrients.

The foods that animals obtain from the environments where they live are usually chemically far more complex than can be used by their cells. The original nutrients are transformed in the body into compounds that may sometimes be immediately metabolized by the cells, or may be converted into substances that can be stored for subsequent transformation into such compounds.

Hormones play an important role in regulating the interconversions of nutrients to metabolic substrates and their stored forms. The endocrine secretions may help to regulate the levels of nutrients by contributing to the control of their absorption from the gut, their levels in the blood, the nature and rate of their storage, their release from tissues, and their assembly into the structural elements of the body.

Animals lead diverse lives in a plethora of environmental conditions. The definitive metabolic processes are basically similar in all animals and lead to the utilization of ATP, for the supply of energy, and the building of cells. Nevertheless the physiological processes leading to these accomplishments may differ considerably. Such processes are dictated by numerous circumstances and events.

The chemical nature of the foodstuffs that animals obtain from their environments may differ greatly. In their feeding habits, animals may be carnivorous, herbivorous or omnivorous. Even within these categories considerable differences exist in the types of food animals eat. Some animals may feed principally on invertebrates such as insects, molluscs and worms that live in terrestrial, freshwater or marine environments. Other animals feed on vertebrates. Plants from equally diverse situations are also used for food. The possibilities for gastronomic experiments thus appear to be endless but only a limited number can furnish a particular species with its needs.

Animals have different patterns of feeding. Some eat almost continually, such as cattle and sheep that nibble plants for hour after hour. Large predatory carnivores, like lions, snakes and crocodiles, may only feed intermittently with days or even weeks separating their meal-times. Circumstances, such as an unexpected drought, may inadvertently result in enforced fasting or even starvation. A dependence on body stores of nutrients for prolonged periods of time may be a fairly predictable part of an animal's life cycle, such as dictated by hibernation during winter, estivation during hot, dry summers and migrations to more equitable regions for food and in order to breed.

The nutritive requirements of animals may differ considerably. The normal rate of metabolism of different species can differ by more than 100-fold. Warm-blooded homeothermic animals usually have a higher metabolic rate than cold-blooded poikilotherms. Even among homeotherms the basal metabolic rate differs considerably; for instance it is about 35 times greater in the shrew than in the elephant. Factors such as size, patterns of activity and the environmental temperatures experienced, contribute to the differences in metabolic requirements of animals. Young, growing animals have special nutrient requirements, while breeding and care of the young alter the needs of adults.

Dominating all these differences, and dictating many of them, is the phylogeny of the species. The genetic constitution of a species determines the pattern of its nutrition and the mechanisms involved in the regulation of it. These physiological processes are presumably the result of a prolonged evolution and adaptation to environmental conditions. This is reflected in the diversity of the endocrine mechanisms that control the metabolism of animals.

Endocrines and digestion

Apart from catching or collecting and then eating food, the first physiological event in the nutritional process is digestion. Food is usually broken down to simpler chemical compounds prior to its absorption from the gut. This process involves the actions of acids, alkalis and enzymes secreted

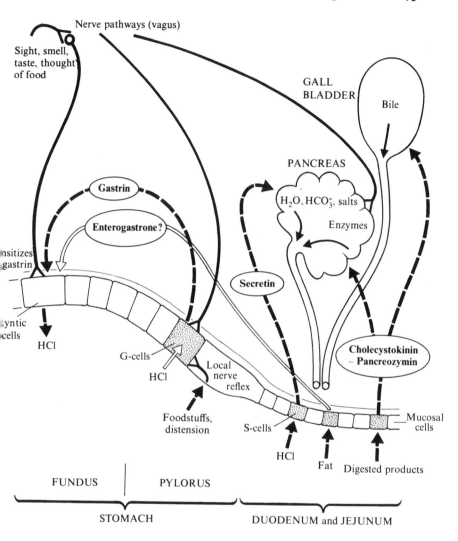

Fig. 5.1. The role of hormones in controlling gastric acid secretion, pancreatic secretion of salts and enzymes, and the contraction of the gall bladder. *Gastrin*, from the pylorus, initiates secretion of hydrochloric acid by the oxyntic cells in the fundus.

The duodenal–jejunal hormones, *secretin* and *cholecystokinin–pancreozymin*, initiate the secretion of, respectively, pancreatic juice and enzymes. *Enterogastrone*, from the duodenum–jejunum, inhibits gastric acid secretion (the existence of enterogastrone is in doubt). The double open arrows indicate an inhibitory effect; the dashed ones a stimulation.

It should be noted that neural pathways can initiate and modify these effects.

by glands in the wall of the stomach and intestines as well as the exocrine, or acinar, cells of the pancreas. The orderly flow of these juices is controlled by hormones as well as by nerves (see Fig. 5.1). Indeed the discovery of the role of *secretin* in stimulating the secretion of the exocrine pancreas into the intestine, was the first unequivocal demonstration of the role of a hormone in the body and it was in connection with this discovery that the term 'hormone' was initially used. Bayliss and Starling performed the crucial experiment in 1902. A loop of the jejunum of an anesthetized dog was tied at both ends and was denervated. When acid (0.4% HCl) was introduced into this sac (or the duodenum) secretion of pancreatic juice was stimulated. An extract of the jejunal mucosa was made by rubbing it with sand in the presence of the HCl solution. When (after filtering) this was injected into the jugular vein of the dog, pancreatic secretion was also stimulated. Starling remarked 'then it must be a chemical reflex'. This experiment was performed at University College, London on the afternoon of 16 January, 1902 and has been summarized thus (Sir Charles Martin, see Gregory, 1962): 'it was a great afternoon'.

Secretin that is released as a result of the action of acid on the duodenal and jejunal mucosa stimulates the formation of a voluminous pancreatic juice, rich in bicarbonate and salt but poor in enzymes. The secretion of proteolytic, amylolytic and lipolytic enzymes (but not water and salt) can be stimulated by the vagus nerve as a result of feeding. Another hormone, however, also assists this process. This is *cholecystokinin–pancreozymin*, the role of which was not established until 40 years after that of secretin (Harper and Raper, 1943). This endocrine secretion is also formed in the upper parts of the intestine from which it is released in response to the presence of the digestive products.

Gastric secretion is also controlled by nerves and hormones. Stimulation by the vagus initiates the formation of acid by the oxyntic cells and enzymes from the chief cells. The action of the vagus on acid secretion is mediated by the release of *gastrin*, a hormone formed in the pyloric region of the stomach, which stimulates the oxyntic cells to secrete acid. The vagus also sensitizes the oxyntic cells to the action of gastrin. The complete reflex arc, which is initiated by feeding, initially involves nerve stimulation along cholinergic nerves to the pyloric cells that release gastrin. This hormone then closes the reflex arc and acts on the oxyntic cells. Secretion of gastrin also results from the initiation of a local nerve reflex due to the presence of food in the stomach. Gastric acid secretion can be inhibited when fat or oils pass into the upper parts of the intestine. This stimulus apparently initiates the release of another hormone, *enterogastrone*, which inhibits secretion from the oxyntic cells. As described previously, the distinct hormonal characteristics of this material remain in some doubt.

The presence of fats in the intestine initiates the release of bile from the gall bladder, and this also involves an endocrine reflex due to the release of *cholecystokinin–pancreozymin* that contracts the gall bladder and relaxes the sphincter of Oddi (Ivy and Oldberg, 1928). This hormone has several roles in the body and, apart from its actions on the gall bladder and the endocrine and exocrine pancreas, can elicit a sensation of satiety. Injected cholecystokinin–pancreozymin thus reduces feeding in rats (Gibbs, Young and Smith, 1973).

Secretin, gastrin and cholecystokinin–pancreozymin can also initiate the release of insulin or glucagon. The physiological significance of this stimulation is uncertain but it has been suggested that such effects may contribute to the homeostasis during feeding. Such signals could mobilize insulin and glucagon in anticipation of the absorption of digested nutrients.

The presence of these humoral reflex arcs influencing digestion have been shown in several mammals but direct evidence as to their presence in other vertebrates is lacking. It seems likely that, from the sporadic evidence available, they exist. Indeed, further experiments by Bayliss and Starling in 1903 indicated that this is so with respect to the effect of secretin on the pancreas. They performed experiments on a variety of mammals, including monkey, dog, cat, rabbit, and a bird (a goose, 'in the process of fattening for Christmas') and confirmed the wider phyletic distribution of this humoral reflex. Secretin-like activity was also shown to be present in the duodenum of man, ox, sheep, pig, squirrel, pigeon, domestic fowl, tortoise frog, salmon, dogfish and skate. This interesting paper by Bayliss and Starling (1903) is entitled 'On the uniformity of the pancreatic mechanism in vertebrates' and must be one of the earliest contributions to comparative endocrinology.

The transformation of metabolic substrates: the role of hormones

The diversity of intermediary metabolism in vertebrates

Nutrients that are utilized for the production of energy can be classified into three major groups: carbohydrates, fats and proteins. These compounds, especially the latter, are also incorporated into the cell structure and so are essential for growth and reproduction.

The nature of the nutrients upon which the animal's metabolism is based depends, in the first instance, on its diet. In carnivores, this consists mainly of protein and fat. The carbohydrates obtained by herbivores may consist of materials such as starches and sugars that can be broken down into simpler sugars by the digestive enzymes. The major organic constituent of most plants is cellulose which is fermented by micro-organisms present

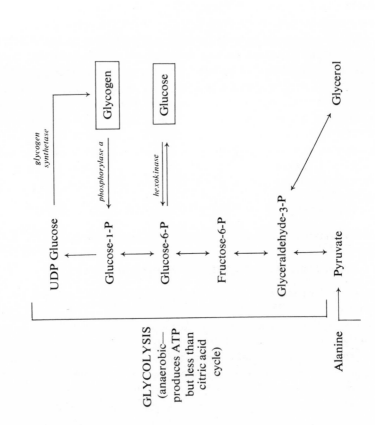

UDP Glucose

glycogen synthetase

Glucose-1-P

phosphorylase a

Glycogen

Glucose-6-P

hexokinase

Glucose

Fructose-6-P

Glyceraldehyde-3-P

Glycerol

Pyruvate

GLYCOLYSIS
(anaerobic—
produces ATP
but less than
citric acid
cycle)

Alanine

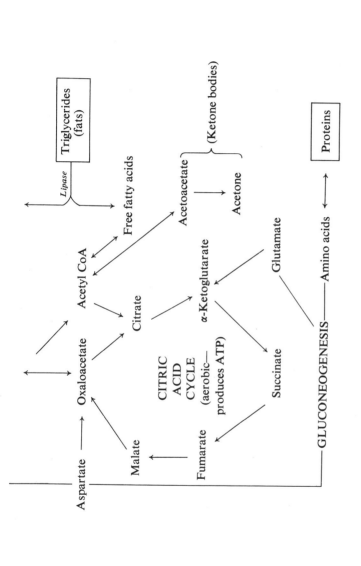

Fig. 5.2. A diagrammatic summary of the processes, enzymes and major compounds involved in the metabolic transformations of glucose, triglycerides and proteins.

Hormones may either increase or decrease certain of these reactions, as indicated in the text and subsequent Figures.

in various compartments of the gut (the cecum, the colon, or the rumen) and which produces short-chain fatty acids. These are mainly propionate, butyrate, and acetate. Such fermentation by symbiotic micro-organisms is widespread in herbivorous mammals, especially in the sacculated rumen of ruminants (cattle and sheep) as well as the colon of horses and the cecum of lagomorphs, like rabbits, as well as the sacculated stomach of some marsupials. In non-mammals the situation is less clear but micro-organisms undoubtedly aid digestion in these animals also.

Different species of animals thus show differing dietary dependencies on proteins, fats and carbohydrates. Proteins can be broken down to their constituent amino acids, fats to fatty acids and glycerol, and carbohydrates to simple sugars, like glucose, or, with the aid of micro-organisms, can give rise to fatty acids. The resulting basic subunits can be utilized directly for the production of energy or they may undergo transformations (see Fig. 5.2) into forms that can be stored and provide a readily accessible reserve. Apart from their reassembly into more complex units, inter-conversions of one such type of chemical compound into another may also take place in the body. Glucose can thus be readily converted in liver, adipose tissue and the mammary glands into triglycerides (fats). Some amino acids are transformed by the process of gluconeogenesis, in the liver, to glucose while others are changed into fatty acids. The transformation of fatty acids to sugars is not as common, though in ruminants propionate, which is formed in large amounts, is converted to glucose. Glycerol can also be transformed into glucose.

The reserves of protein that are maintained in the body are small and in any case it is not a very suitable substrate for the storage of energy. During starvation, protein may, nevertheless, make an important contribution to an animal's energy requirements. Substantial amounts of glucose are stored as glycogen in the liver and muscles but these reserves are inadequate to maintain an animal for prolonged periods of time. Triglycerides provide the most economic and convenient storage-form for energy. One gram of fat furnishes 9500 calories while the same amount of carbohydrate and protein, respectively, only supply 4200 and 4300 calories.

Stored fat may be dispersed widely among the tissues in the body but it usually predominates at certain sites. Adipose tissues thus exist at sub-cutaneous, mesenteric, perirenal and periepididymal sites in mammals. The large fat-bodies near the gonads in the abdominal cavity of Amphibia are familiar to student dissectors. The tail of urodeles and lacertilians is also a common site for fat storage. Large quantities of fats may also be stored in the liver of poikilotherms and this is especially important in chondrichthyean fishes, though it is also seen in other vertebrates. In many fishes, fats are stored in close proximity to the muscle fibers which directly utilize their fatty acids. A characteristic type of adipose tissue called *brown fat* is present in embryonic mammals and also some adults such as rats and hibernating species like hedgehogs, moles, bats and squirrels. Brown fat is capable of undergoing very rapid metabolism, with considerable production of heat, particularly during arousal of hibernating animals.

The transformation, storage and utilization of fats, proteins and carbohydrates is regulated to a considerable extent by hormones. The relative differences in the availability and importance of such substrates in different species, not surprisingly, may be reflected in the animal's particular response to hormones. Quantitative, or even qualitative, differences may be observed. Some such variations in the responsiveness of different species of reptiles and amphibians to injections of insulin and glucagon, as well as pancreatectomy, are shown in Fig. 5.3. It can be seen that changes in the blood glucose levels following these treatments show considerable interspecific variability in the speed of onset, and the magnitude and the duration of the responses.

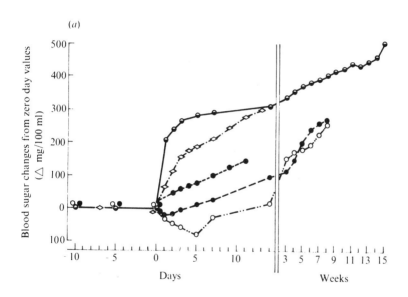

Fig. 5.3. Effects of various experimental treatments on blood glucose concentrations in amphibians and reptiles.

(*a*) Changes in the blood glucose concentrations following pancreatectomy in various reptiles and amphibians. It can be seen that the elevation in the glucose occurred relatively promptly in the alligators and toads but was considerably delayed in the lizards and snakes.

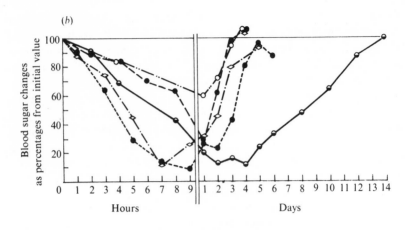

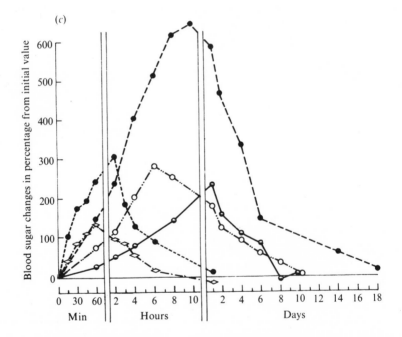

(*b*) Effects of injected insulin (1 unit/100 g body weight). The amphibians responded more rapidly than the reptiles but the response was usually more prolonged in the latter.

(*c*) Effects of injected glucagon (10 μg/100 g body weight). The reptiles responded more slowly than the amphibians but the response in the former lasted for a longer time. ●—●, Alligators; ◇—·—◇, toads; ●---●, snakes; ○—··—○, lizards; ●---●, frogs. (From Penhos and Ramey, 1973.)

Hormones that influence intermediary metabolism

When considering hormones that influence the transformation, deposition, mobilization and utilization of fats, carbohydrates and proteins in the body, we should be careful to distinguish between those that function physiologically and those actions that probably do not normally occur in the animals (pharmacological effects). For instance, lipolytic activity can be exhibited by at least seven pituitary hormones and also several secretions from other endocrine glands. It is possible, however, that a pharmacological action of a hormone in one species, or tissue preparation, may reflect a physiological role in some phyletically distant species.

Insulin plays a central role in intermediary metabolism and this may be associated, especially during fasting, with the actions of the *corticosteroids*. *Glucagon* and *adrenaline* also contribute to the control-system. *Growth hormone* and *thyroid hormones* modulate the processes involved in many

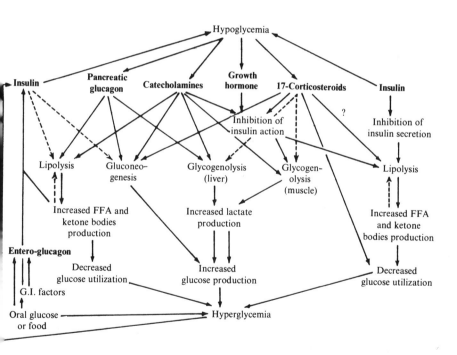

Fig. 5.4. A summary of the role of hormones in stimulating, or inhibiting, metabolic processes and transformations that control the glucose concentrations in the blood. Solid line, stimulation or increase; dotted line, inhibition or decrease. (Modified from Foà, 1972.)

chemical pathways. *Prolactin* (Bern and Nicoll, 1969) stimulates the formation of milk proteins in the mammary glands and in the pigeon crop-sac. Prolactin also, when injected, may have a hyperglycemic action and promote the deposition of fat in teleost fish, amphibians, reptiles and birds. Several hormones usually interact with each other in such metabolic processes, none can function normally in complete isolation from all the others (Fig. 5.4). As so well stated by Tepperman and Tepperman (1970), 'It is virtually impossible to separate out one set of signals which control lipogenesis, another which controls gluconeogenesis and a third which controls ketogenesis. All these processes share so much of the metabolic machinery . . .'. With this reservation in mind we can consider the effects of various hormones on the processes of intermediary metabolism.

Insulin

This hormone decreases plasma glucose concentrations and, in mammals at least, also reduces free fatty acid levels. In the absence of insulin, muscle wasting occurs due to the excessive mobilization of proteins. These effects are the result of several actions, at various sites; principally the liver, muscle and adipose tissue. *The processes mediated by insulin* are basically (see Fritz, 1972) as follows:

(i) Increase in the rate of uptake of glucose across the cell membranes of skeletal and cardiac muscle, adipose tissue, and mammary gland. Organs like liver, kidney or brain do not respond in this way to insulin.

(ii) Insulin facilitates glycogen formation from glucose. This is partly the result of the more rapid accumulation of glucose but is also due to an increased activity of glycogen synthetase in liver, muscle and possibly even adipose tissue.

(iii) An inhibition of the mobilization of glycogen to form glucose.

(iv) Inhibition of gluconeogenesis from amino acids.

(v) Insulin increases the accumulation of fatty acids across the cell membranes of adipose cells.

(vi) An inhibition of lipolysis of triglycerides to form fatty acids.

(vii) Insulin increases lipogenesis, from glucose, in adipose tissue.

(viii) There is an increased uptake of amino acids by muscle and liver cells.

(ix) An inhibition, by insulin, of the mobilization of amino acids from protein [related to (iv)].

The effects of insulin can thus be broadly divided into its actions on the accumulation of nutrients across cell membranes and its facilitation, or inhibition, of metabolic synthesis in cells.

Variations in the responsiveness to insulin among vertebrates

The actions of insulin on intermediary metabolism have been principally studied (*in vitro* and *in vivo*) in mammals, usually in laboratory rats. The hypoglycemic action of insulin nevertheless appears to be widespread in vertebrates (see Bentley and Follett, 1965; Falkmer and Patent, 1972); it occurs among animals ranging from cyclostomes to mammals. Differences in sensitivity nevertheless are apparent, reptiles and especially birds being relatively insensitive to injected mammalian insulin. The pancreas of birds contains little insulin and this is only released 'sluggishly' in response to hyperglycemia (Hazelwood, 1973) so that this hormone may not play such an important role in these as in other groups of vertebrates. Some urodeles fail to respond to injected insulin; yet others do respond (McMillian and Wilkinson, 1972). Among mammals, ruminants are less sensitive than carnivores. Such variations appear to depend on the source of the exogenous insulin, the metabolic rate and condition of the animals, and the presence of compensatory mechanisms, such as release of glucagon and adrenaline, as well as the normal diet of the animals. The hypoglycemic effect of insulin appears to be a universal one among vertebrates. Differences have been observed, however, in its ability to promote deposition of glycogen in liver or muscle or both. For instance, in lamprey liver, but not muscle, glycogen is increased by insulin injections. Muscle glycogen is increased in the skate (Chondrichthyes) and scorpion fish (Teleostei) (Leibson and Plisetskaya, 1968) where the effects on the liver are less pronounced. Insulin has unpredictable effects on plasma fatty acids; it decreases them in mammals but may have no effect or even increase their concentration in birds.

The sensitivity of different species to insulin may be related to their dietary habits. It has been noted that herbivorous mammals withstand an absence of insulin far more readily than carnivorous ones (Gorbman and Bern, 1962; Fritz, 1972). Carnivores only eat periodically so that they may have a sudden large intake of nutrients which must be stored for utilization during the fast between meals. Herbivores, on the other hand, graze for long periods of the day and are continually absorbing the products, mainly fatty acids, from the large stores of digesting food in their guts. Coordination of the storage and mobilization of nutrients in the body are thus expected to be more important for periodic eaters like carnivores, than herbivores (which have aptly been called 'nibblers').

Catecholamines

Adrenaline can increase the concentration of both glucose and fatty acids in the plasma. These effects are mediated in the liver and muscles, as a result of activation of phosphorylase, and in adipose tissue by activation of lipase. Both effects are due to the formation of cyclic AMP. The hyperglycemic effect of adrenaline is seen in species from all the main groups of vertebrates though the site of its action may differ. For instance, in lampreys, adrenaline mobilizes glycogen in liver but not muscle, while in the Chondrichthyes, glycogen from both sites is depleted, as in mammals (Bentley and Follett, 1965; DeRoos and DeRoos, 1972). The hyperlipidemic response to adrenaline has not been studied on such a broad phyletic scale and even mammalian adipose tissue from all species studied is not uniformly responsive. Thus mobilization of fatty acids in response to adrenaline has been shown in the rat, dog, goose and owl but not the duck, chicken, rabbit or pig (Prigge and Grande, 1971; Table 5.1). Adrenaline increases plasma fatty acids in the eel *Anguilla anguilla* (Larsson, 1973) but this is not seen in all fish (Minick and Chavin, 1973) and may reflect an indirect effect by inhibiting release of insulin.

There is some doubt, at least in mammals, as to the efficacy of the circulating concentrations of adrenaline in stimulating glycogenolysis. The concentration usually observed in the plasma appears to be too low to act physiologically on the liver. Adrenergic effects on mobilization of glucose in the liver may thus be mediated by stimulation of the sympathetic nerves. Whether this applies to all tissues, and other species, is unknown.

Glucagon

This pancreatic hormone exerts a hyperglycemic action in mammals as a result of the mobilization of liver glycogen following activation of phosphorylase. It does not have this action on skeletal muscle. Gluconeogenesis is increased. Plasma fatty acid levels are elevated by glucagon in birds and mammals but at somewhat higher concentrations than affect glucose. Lipolysis is promoted in adipose tissue from rat, rabbit, goose, duck and fowl, but not in the dog (Prigge and Grande, 1971). Glucagon has widespread hyperglycemic effects in mammals, birds and reptiles but is relatively ineffective on cyclostomes, chondrichthyeans and some teleosts though it is very effective in the eel (Patent, 1970; Larsson and Lewander, 1972). In contrast to mammals and birds, eel plasma fatty acids are unaffected by glucagon (Larsson and Lewander, 1972).

Other peptide hormones

Other hormones can also influence glucose and fatty acid metabolism. The neurohypophysial peptides (when injected), vasopressin, oxytocin and vasotocin, exhibit hyperglycemic effects that can be demonstrated in

TABLE 5.1. *Species sensitivity to hormonal stimulation of free fatty acid release from adipose tissue.* (Based on Shafrir and Wertheimer, 1965)

	Rat	Mouse	Rabbit	Hamster	Guinea-pig	Cat	Dog	Pig	Chicken	Man
Adrenaline or noradrenaline	++	++	o	++	o	+	++	o	o	+
Corticotrophin	++	++	+	++	+	+	o	o	+	++
Cortisol	+		++		++				o	
Glucagon	+			+	+				+	
Thyrotrophin	+ or ++		o	o	++	o	+	+		
MSH (α or β)	o		++	o	+		+	o		
Vasopressin	o		++	o	++		o	o		

++, Strong response; +, moderate or weak response; o, no response.

cyclostome fishes, amphibians, birds and mammals (see Bentley, 1966) and have also been more recently described in reptiles (LaPointe and Jacobson, 1974).

Vasotocin also increases plasma fatty acid levels in birds (John and George, 1973). The pituitary hormones MSH, ACTH, TSH and growth hormone may all, depending on the species and the adipose tissue preparation used, facilitate mobilization of fatty acids from mammalian adipose tissue (Table 5.2; Mirsky, 1965). These actions are probably not normal physiological ones and it is unknown whether they are widespread in non-mammals.

TABLE 5.2. *Hormonal effects on fat mobilization.* (Based on Fritz and Lee, 1972)

	Fat mobilization from adipose tissue	
Hormones	*In vitro*	*In vivo* (FFA) plasma
Glucagon	↑	↑
Adrenaline	↑	↑
Noradrenaline	↑	↑
ACTH	↑	↑
TSH	↑	
Growth hormone	↑	↑
Insulin	↓	↓
Cyclic 3′,5′,-AMP	↑	

ACTH: adrenocorticotrophic hormone; TSH: thyroid-stimulating hormone; ↑: increase; ↓: decrease, in rate of metabolic process indicated in column by hormone designated in row; (FFA) plasma: plasma concentration of free fatty acids.

Adrenocorticosteroids

The steroid hormones have profound effects on intermediary metabolism (especially in fasting animals), reproductive processes, growth and lactation. The corticosteroids increased blood glucose concentrations and promote gluconeogenesis and the deposition of glycogen in the liver. In excess, corticosteroids promote muscle wasting and a negative nitrogen balance, while in young animals they inhibit growth. These two effects reflect their action on protein catabolism. *The actions of corticosteroids on intermediary metabolism* can be summarized thus:

(i) Increase in gluconeogenesis in the liver following a mobilization of proteins from skeletal muscle and the deamination of the amino acids that are released. This action is most important, especially during fasting.

(ii) Glycogen is deposited in the liver because of an increase of the glycogen synthetase reaction.

(iii) Corticosteroids inhibit glycogenolysis.

(iv) There is a reduction in peripheral oxidation and utilization of glucose.

(v) Corticosteroids inhibit the conversion of amino acids to proteins, and fatty acids to triglycerides.

Such effects appear to be widespread in the vertebrates although the information available in non-mammals is sporadic (see Chester Jones *et al.*, 1972). Hyperglycemia in response to the injection of corticosteroids has been shown in vertebrates that range phyletically from cyclostomes to mammals. This response is associated with gluconeogenesis and elevation of tissue glycogen levels. The facilitation of gluconeogenesis is associated with increased levels of liver transaminase enzymes of teleost fishes, amphibians, birds and mammals (though it had once been thought that this increase only occurred in the latter two groups of vertebrates) (see Janssens, 1967; Freeman and Idler, 1973). An inhibition of growth or loss in body weight has been shown in the domestic fowl and the amphibian *Xenopus laevis*, as well as two species of teleosts, *Salmo gairdneri* and *Salvelinus fontinalis* (Bellamy and Leonard, 1965; Freeman and Idler, 1973; Janssens, 1967), as well as in mammals. The actions of corticosteroids on growth and metabolism appear to be basically the same in all vertebrates.

Steroid sex hormones

Estrogens and androgens have widespread metabolic effects on the growth and differentiation of tissues, especially the reproductive organs. They may, however, also influence other tissues in the body, principally by promoting the formation of proteins. Androgens when administered have anabolic effects on skeletal muscle (myotrophic action), promoting the formation of proteins. The magnitude of this effect depends on the species, age and the hormonal status of the animal. It is greatest in young male animals with deficient circulating androgens; it is also apparent in females but has little effect in older animals. Estrogens increase plasma lipid levels in mammals and also have an anabolic effect but this is principally confined to the mammary glands and reproductive organs. In oviparous species, estrogens promote the formation of lipoproteins in the liver which are incorporated into the yolk of the egg. Progesterone increases the formation of avidin by the oviduct of the chicken and this is also incorporated into the egg.

Thyroid hormones, growth hormone and prolactin

These hormones also contribute to metabolic regulation. The consumption of oxygen by homeotherms is depressed in the absence of thyroid hormones and they are also necessary for adequate growth and differentiation. The actions of several other hormones are not as pronounced in the absence of the thyroid secretion. Such reduced responses are seen for the catecholamines and corticosteroids. The thyroid gland seems to modulate the levels and activity of metabolic enzymes in cells; in the instance of adrenaline this action may involve an increase in adenyl cyclase (Krishna, Hynie and Brodie, 1968). Growth hormone promotes growth and stimulates the formation of protein in cells. At least some of its actions are mediated by *somatomedin* (see later), a protein whose formation it promotes (or into which it is transformed) in the liver. Growth hormone facilitates uptake of amino acids by liver and muscle cells but inhibits the action of insulin on glucose uptake and thus may have a diabetogenic effect. It has a lipolytic action on adipose tissue. Growth hormone has been shown to influence growth in mammals, birds, reptiles, amphibians and teleost fish (see Chapter 3) so that its actions have a wide phyletic distribution. Somatomedin has been identified in several mammals where its action appears to be less specific to any one single species than growth hormone itself (see Tanner, 1972). Prolactin influences the intermediary metabolism of the mammary glands (present in mammals) and the crop-sac of pigeons. As we shall see, prolactin may also promote fat deposition in a number of vertebrates.

Conclusions

Hormones can thus be seen to exhibit widespread actions on intermediary metabolism. In some instances, several secretions can exert similar effects though, in the animal, these may not all be physiologically equivalent. In other cases, hormones may exert opposing effects, either by acting on different processes or by a more direct inhibition. Hormones can also (see Chapter 4) directly influence one another's release and so mimic or oppose the actions of other hormones. Intermediary metabolism, while being extremely complex and involving several tissues, many chemical reactions and numerous metabolites, is a well-integrated process. This is largely the result of the actions of hormones at different types of sites (some of which are summarized in Fig. 5.5), both within the same cell and in different kinds of cells, as well as their ability to act in harmony with each other.

The underlying roles and actions of hormones in regulating intermediary metabolism appear to be basically similar in all vertebrates. The relative

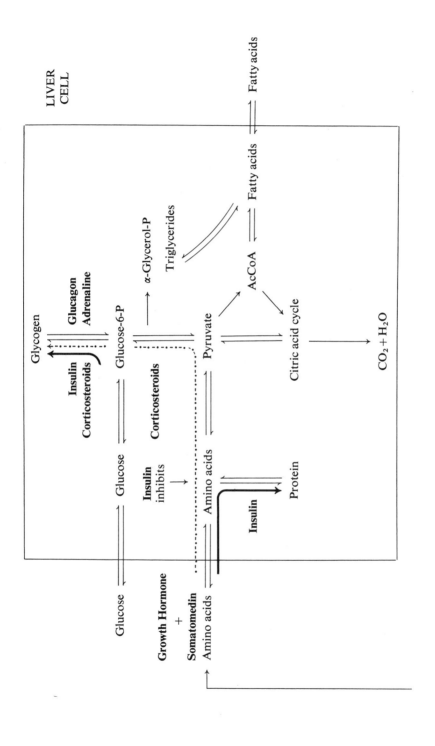

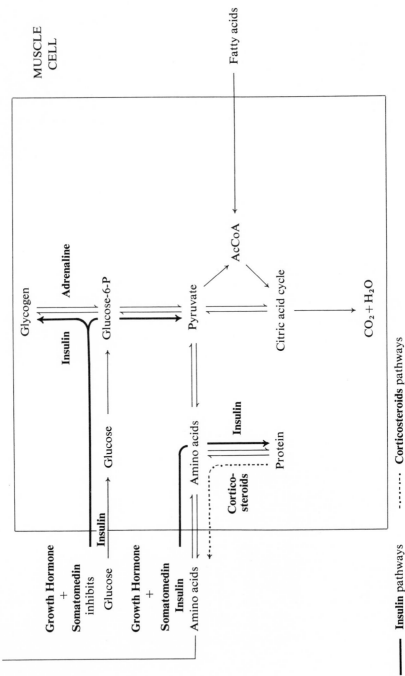

MUSCLE CELL

Growth Hormone + Somatomedin inhibits

Insulin

Growth Hormone + Somatomedin

Insulin

Glucose

Amino acids

Glycogen

Adrenaline

Insulin

Glucose-6-P

Glucose

Amino acids

Pyruvate

AcCoA

Fatty acids

Citric acid cycle

$CO_2 + H_2O$

Cortico-steroids

Insulin

Protein

———— **Insulin** pathways ········· **Corticosteroids** pathways

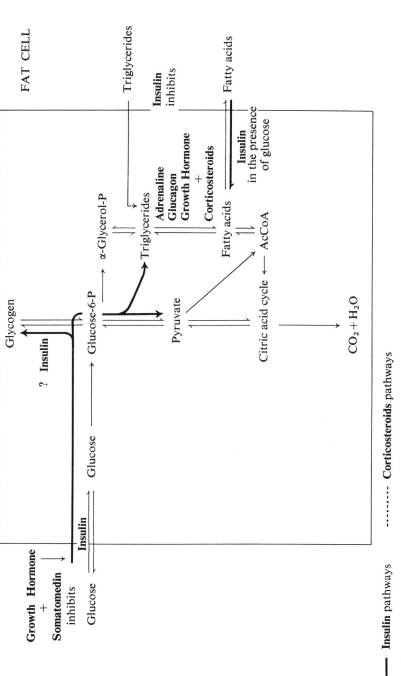

Fig. 5.5. A diagrammatic summary of the effects of hormones on the transfer of metabolites across the cell membranes and the metabolic transformation within liver cells, muscle cells and fat cells.

The principal metabolic transformations that are influenced by insulin are indicated by the heavy solid lines and those by corticosteroids by the broken lines.

—— **Insulin** pathways   ········ **Corticosteroids** pathways importance of each action of the hormone may, however, differ, depending on the animal's usual life-style and the particular stage of its life cycle. Events like fasting, associated with hibernation, estivation and migration, and reproduction (including the formation of eggs, pregnancy, lactation and the growth, maturation and metamorphosis of the young) are all associated with special metabolic needs and hormonally mediated effects. Some of these will be described in more detail.

Hormones and calorigenesis

The production of heat in the body

The chemical reactions that are continually taking place in cells are associated with the production of heat. This can be quantitatively expressed in terms of calories; hence the term calorigenesis. This heat represents:

(i) A by-product resulting from the energy requirements of the body. This heat may result from mechanical activity of muscle contractions or from the more ubiquitous metabolic transformations of chemical substrates.

(ii) In homeotherms, the production of heat energy *per se* may be necessary to maintain the body temperature of the animal.

All the tissues of the body contribute to the release of heat in the body and this predominates in skeletal muscle and liver. In some species, the brown fat may also be an important site of calorigenesis, especially in newborn animals and in hibernators during awakening or arousal. In the cold-adapted laboratory rat, the skeletal muscles are thought to contribute about 50 % of the heat produced, the liver 25 % and brown fat only 10 % (this value is greater for newborns). During arousal from hibernation, brown fat may contribute as much as 40 % of the body heat of hamsters, and 60 % in ground squirrels (Janský, 1973).

In adult homeotherms, added heat requirements for maintaining the body temperature in cold environments are largely met by a skeletal muscle activity called shivering. The rest of the heat is produced from other metabolic activities that are continually proceeding in the body. The amount of heat produced can also be increased, in response to homeothermic need, in a manner that is independent of shivering. This is called non-shivering thermogenesis (or NST) and is greater in young animals and small species than in adult or large animals. A hamster can increase its metabolism four-fold in this manner, while in man the change is negligible (see Janský, 1973).

Thyroid hormones

Thyroxine and tri-iodothyronine have an important role in maintaining the production of heat (and the associated oxygen consumption) in mammals and birds. In the absence of the thyroid gland, the oxygen consumption of mammals declines by as much as 50% while an increase in the rate of thyroid secretion can stimulate metabolism by a similar amount. Thyroxine administration can nearly double the basal rate of metabolism in the golden hamster (Janský, 1973). The thyroid is thought to play a permissive role in cell metabolism rather than contribute to acute changes in metabolism. Thyroid hormones act slowly, thyroxine taking several days to exert its maximal effect. They are usually considered to contribute to thermogenesis by maintaining the metabolic machinery in a manner that allows it to function optimally. Nevertheless, in mammals, increased thyroid activity is associated with exposure to low temperatures (thyroidectomized rats have a limited survival in the cold) and it decreases with high temperatures. These changes are mediated by the response of the hypothalamic centers to alteration in the temperature of the blood that perfuses them (Collins and Weiner, 1968).

There are some reports that thyroid hormones increase calorigenesis in poikilotherms though it is not always possible to demonstrate this effect (indeed a decrease has sometimes been observed). The evidence is considered to be equivocal. The injection of thyroxine into the lizard *Eumeces fasciatus* results in a 30% increase in oxygen consumption at 30 °C but there is no effect at 20 °C (Maher, 1965). A similar increase in oxygen consumption has been seen in the leopard frog *Rana pipiens* (McNabb, 1969) at different temperatures. Thyroid activity also varies with the body temperature of poikilotherms; it is greater at higher temperatures, which is the reverse of what happens in homeotherms. The latter's response, however, presumably reflects a homeostatic role of the thyroid in calorigenesis which would not appear to be present in cold-blooded vertebrates.

Catecholamines

Adrenaline and noradrenaline may play an important role in regulating the production of heat in the body. In homeotherms, a sudden exposure to cold conditions results in an increased production of heat. If mice are transferred from a temperature of 25 °C to 0 °C they normally adjust their metabolism and nearly all of them survive. When, however, they are pretreated with drugs that inhibit the β-adrenergic effects of catecholamines (for example propranolol) they all die within three hours (Estler and Ammon, 1969) because of an inability to increase heat production. Normally under these conditions, this thermal response is mediated, in mice, by catecholamines, principally noradrenaline which is released from the nerve endings of the sympathetic nervous system. Adrenaline, which is secreted by the adrenal medulla, is not usually effective (though it is so in rats) but a suggestion has been made that it also may have such an effect in some circumstances and constitutes a second line of defense.

Such an adaptational effect of catecholamines in calorigenesis can be seen in the adjustment of hypothyroid rats to a cold environment (Sellers, Flattery and Steiner, 1974). When rats are exposed to a temperature of 4 °C their urinary excretion of catecholamines increases, which probably reflects their role in increasing heat production. If such rats are thyroidectomized, they cannot then increase their production of heat but the injection of small amounts of thyroxine allows this increase to occur so that they survive. In the hypothyroid rats, the urinary excretion of catecholamines increases even more than in the normal rats exposed to cold. It has been suggested that an increased activity of the sympathetic nervous system, including the adrenal medulla, facilitates the survival of these hypothyroid rats. It should be emphasized, however, that some thyroid hormones are essential as in the absence of the excitants (see p. 198) rats do not exhibit a calorigenesis in response to catecholamines. The two types of hormones, however, appear to act in conjunction with one another in adapting the rats to cold.

Apart from laboratory rodents such as rats and mice there is little information about the role of the thyroid and catecholamines in cold-adaptation of other mammals.

It is notable that adrenaline may, in addition to being calorigenic, help promote heat loss in some mammals. Sweating during exercise in monkeys is dependent on circulating adrenaline and is considerably reduced following denervation of the adrenal medulla (Robertshaw, Taylor and Mazzia, 1973). This poses an interesting conflict of interest in the physiological role of adrenaline, its potential calorigenic effect presumably being minimal in circumstances where it also contributes to the dissipation of heat. The role of adrenaline in temperature regulation is even more diverse when one recalls that adrenergic nerves, by a peripheral vasoconstrictor action, can also promote heat conservation. This affords another interesting example of the evolution of a hormone's role.

There is little information about the effects of catecholamines on calorigenesis and oxygen consumption in non-mammals (see Harri and Hedenstam, 1972). This effect apparently cannot be demonstrated in birds (pigeon, titmouse and gull) nor in fishes. There seem to be no reports about such an action in reptiles either. In European frogs, *Rana temporaria*, injections of noradrenaline and adrenaline increase oxygen consumption by 25 to 35%. The response depended somewhat on the particular frogs used; it was seen in cold- and warm-adapted summer frogs but not warm-adapted winter frogs. The physiological role of catecholamines in calorigenesis of non-mammals remains in doubt.

Mechanisms of actions of hormones on calorigenesis

Calorigenesis is stimulated in tissues as a result of their mechanical activity, as in muscles, and also accompanies more general metabolic transformations. It is on the latter ubiquitous processes that the catecholamines and thyroid hormones act to stimulate thermogenesis. Their precise sites of action in the cells are uncertain. Heat may be produced as a result of an increased turnover of ATP or by an accelerated rate of mitochondrial respiration mediated by an increase in protein synthesis which, it has been suggested, may be influenced by thyroid hormones. The formation of cyclic AMP, from ATP, as a result of the action of hormones probably

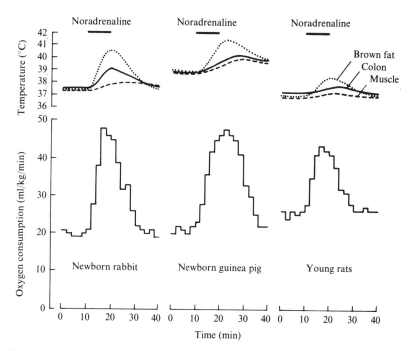

Fig. 5.6. The effects of noradrenaline (intravenous infusion) on the subcutaneous temperatures over *brown fat*, *colon* and *muscle* and the oxygen consumption of newborn rabbits, newborn guinea-pigs and young rats.

The calorigenic effects of noradrenaline are especially prominent in the brown fat while the basal rate of oxygen consumption in each species increased more than two-fold. (From Cockburn, Hull and Walton, 1968.)

results in substantial heat production (Robison *et al.*, 1972). Hormones involved in this process, or released by it, increase the supply of available glucose and fatty acids that can function as substrates for the production of energy in cells. The effects of noradrenaline on oxygen consumption and the associated subcutaneous temperature of brown fat in rabbit, rat and guinea-pig is shown in Fig. 5.6. The calorigenic effects of catechol-amines on brown fat can be seen dramatically *in vitro* where oxygen consumption can be increased four-fold (Table 5.3). This is the result of

TABLE 5.3. *Stimulation of oxygen consumption of slices of rat brown adipose tissue by hormones added* in vitro. (Based on Joel, 1965)

Hormone	Hormone concentration (μg/ml)	Oxygen consumption (μmoles/100 mg fresh tissue per 2 h incubation)	
		Control	Plus hormone
Noradrenaline	0.46	14.9	53.8
Adrenaline	0.50	13.9	64.7
Glucagon	1.0	13.7	35.3
ACTH	10	14.3	28.9
TSH	30	17.7	23.4

mobilization of fatty acids following the activation of lipase by cyclic AMP. It can be seen (Table 5.3) that several other hormones, glucagon, ACTH and TSH mimic the effects of the catecholamines on brown fat but it is unknown whether or not they contribute physiologically to the dramatic increases in metabolism that are observed in this tissue.

The effects of catecholamines on calorigenesis are facilitated (at least in the rat) by the action of thyroid hormones (Lutherer, Fregly and Anton, 1969) at several sites including adipose tissue. In thyroidectomized rats, catecholamines have little effect on oxygen consumption but if thyroxine is injected their effects are increased. This can be related to an increase in the amounts of adenyl cyclase present in the fat cells (Krishna *et al.*, 1968).

Conclusions

The calorigenic actions of hormones are most important in homeothermic vertebrates where they assist in the process of temperature regulation. The production of heat by the cells in such animals is a basic underlying characteristic, the levels of which may, within limits, be modified. Nutrients, especially proteins, obtained from the diet have apparently a direct effect on these processes, termed the 'specific dynamic action of foods'. The hormones act in concert with this effect. Other physiological effects mediated by nerves and hormones indirectly influence thermoregulation by facilitating or diminishing losses of heat from the body. This involves evaporation which occurs from the respiratory tract and the skin where the control of sweat gland activity (referred to above) is important. The blood supply to the peripheral parts of the body, which is controlled by sympathetic nerves that release catecholamines and promote piloerection, influences heat exchanges that occur by conduction. It is unlikely that circulating adrenaline significantly alters peripheral blood flow in such circumstances. Hormones thus do not predominate but do contribute to the regulation of body temperature in mammals and birds. This effect represents an evolution of a physiological role which is not apparent in their phyletic forbears.

The storage of nutrients and their utilization during fasting

The diversity of feeding–fasting patterns in vertebrates

As described earlier, sufficient food may only be available to animals at irregular intervals of time separated by several hours, many months or even years. In some instances, when climatic conditions are unfavorable, animals may sequester themselves in protected havens where metabolic activity is minimal. Hibernation during the winter months is well known in many small mammals (especially among the Insectivora, Rodentia and Chiroptera) that seek refuge in burrows and allow their body temperatures to decline to levels similar to the ambient one. Other animals become inactive during hot, dry periods of drought which result in a limited food supply and a shortage of water that may produce severe osmotic problems. In the latter instance, called 'estivation', the body temperature is also similar to the ambient one, but as this is usually relatively high, metabolism would be expected to be greater than in those animals that are hibernating. Animals may survive for many months or possibly even years under these conditions. A recent report from Russia (*Nature*, **242**, 1973) described a live newt that was found in a piece of ice in Siberia which, according to carbon-dating, had been entombed for nearly 100 years. African lungfish can survive for two to three years in a state of estivation, though more usually it is for four to six months between seasonal rains. Such periods of estivation are also common among amphibians that live in hot, dry deserts.

While hibernation and estivation are associated with minimal activity and metabolic needs, other situations associated with fasting require a high expenditure of energy. Such an occasion is most dramatically seen during seasonal and breeding migrations. Birds may fly many hundreds, or even thousands, of miles from temperate regions, at the beginning of winter, to warmer tropical climes and then return again in the spring. Fishes, such as lampreys and salmon, when they become mature, migrate from the rivers, where they grew up, into the sea from which they later return in order to breed. Eels make the opposite migration from breeding grounds in the sea to rivers and then later the young return to the sea to breed. Other sea-going creatures, such as turtles and whales, also make long journeys. On many of these occasions, the animals do not feed or only do so infrequently. Reserves of nutrients are amassed in the body in preparation for the migrations during which they are expended. The endocrine glands undoubtedly have a role to play in such storage and the subsequent utilization of nutrients but the available information is only fragmentary. Further clues can be obtained from the voluminous studies of endocrine function during normal feeding and fasting.

Endocrines and feeding

The release of hormones

Feeding, like so many other physiological processes, is a process that involves the secretion of several hormones. This begins with the release of hormones that are associated with digestion: gastrin, secretin and cholecystokinin–pancreozymin. These excitants, as well as entero-glucagon from the intestinal tract, promote the early release of insulin and glucagon from the Islets of Langerhans. The absorbed nutrients also influence the secretion of these hormones; glucose increases insulin release, while amino acids initiate the discharge of both hormones. Nerve stimulation, via the vagus, can also stimulate insulin release. The endocrine response differs somewhat with the diet, depending on the relative amounts of protein and carbohydrate present (Table 5.4). A high carbohydrate diet is associated with high insulin and low glucagon levels while a predominance of protein elevates the concentrations of both of these hormones. In ruminants, the fatty acids that are absorbed from the rumen may also stimulate the release of insulin (Manns, Boda and Willes, 1967).

The disposal of absorbed nutrients

The nutrients that are absorbed from the digestive tract can be disposed of in the body in several ways. They may be used immediately as a source of energy. This process may be relatively direct, such as the oxidation of

TABLE 5.4. *Interrelationship of metabolism with the nutritional state of mammals.* (From Cahill, Aoki and Marliss, 1972)

Nutritional states	Hormonal states		Liver				Muscle			Adipose		
	Insulin	Glucagon	Glycolysis	Lipo-genesis	Gluconeo-genesis	Keto-genesis	Glucose uptake	Protein synthesis	Proteo-lysis	Glucose uptake	Lipo-genesis	Lipolysis
Carbohydrate-fed	+++	±	++	++	0	0	++	±	0	++	++	0
Protein-fed	++	+++	0	+	++	0	+	++	0	+	+	0
Carbohydrate- and protein-fed	++++	±	+	+++	0	0	++	++	0	+++	0	0
Fasting (low insulin)	+	++	0	0	++	++	0	0	+	0	0	++
Diabetes (absent insulin)	0	++++	0	0	++++	++++	0	0	++	0	0	++++

+ to 0, either concentration of the hormone or rate of function described.

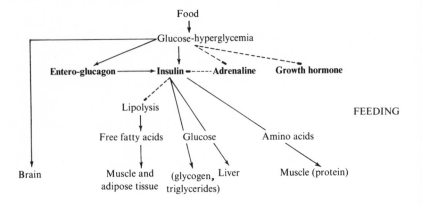

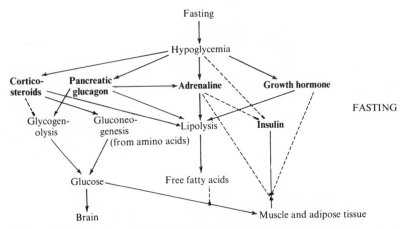

Fig. 5.7. The role of hormones in intermediary metabolism during *feeding* and *fasting*.

Solid line, stimulation or increase; dashed line, inhibition or decrease. (Modified from Foà, 1972.)

glucose and fatty acids. Amino acids can be transformed to glucose in the liver. The large amounts of propionate absorbed by ruminants can also be converted to glucose. Certain amino acids can also be changed into fatty acids. These products can be more readily transformed into energy. Gluconeogenesis is stimulated by glucagon and corticosteroids. This process is especially important in ruminants that must dispose of large amounts of fatty acids. As we shall see, gluconeogenesis is also fundamental for homeostasis during fasting in other species. The over-all processes of oxidation of nutrients are, at least in mammals, chronically influenced by thyroid hormones but more acutely by the levels of the energy substrates themselves, as well as certain other hormones (see section on calorigenesis).

Nutrients that are not required for the immediate production of energy by the animal are stored (usually following their metabolic transformation) and can be subsequently utilized during fasting. Elevated insulin levels play a central role in this process (Fig. 5.7; see also Table 5.4). Insulin facilitates the conversion of glucose to glycogen and triglycerides, of amino acids to protein, and of fatty acids to triglycerides. Simultaneously, insulin inhibits further gluconeogenesis and lipolysis. Other hormones may impinge their influence on these processes but usually in a negative manner. For instance, a decreased secretion of glucagon and adrenaline are favorable to lipogenesis. Growth hormone favors incorporation of amino acids into proteins. Prolactin may (see later), on certain occasions, facilitate the deposition of fat in some species.

The principal energy store utilized by animals during periods of prolonged fasting is fat, though proteins can also undoubtedly be used. Obese animals fare better and survive longer. The immense fat reserves of migrating birds, fish and whales are well known. The continual food collecting and feeding activities of small mammals prior to winter hibernation is almost legendary. The total fat content of the body is normally controlled (by processes that are not understood) within limits that can be considered modest, both physiologically and esthetically. In man, it has recently been suggested that obesity is associated with a decreased secretion of glucagon (Wise, Hendler and Felig, 1972) but this is, almost certainly, only one facet of the endocrine involvement in this process.

Preparations for migration

Animals periodically show dramatic departures from their normal limits of fat content which may be associated with their potential requirements during a period of fasting. The maximum accumulations of fat in several species of migratory birds may be equivalent to about 50% of the total body weight (Table 5.5). The normal fat levels in such birds are 3 to 10% of their body weight. This fat is contained in various parts of the body but especially in cutaneous and subcutaneous sites and can be deposited very rapidly in one to two weeks. The timing of this activity appears to be the result of changes in the length of the day and so is under photoperiodic control. Depending on the season the stimulus may be an increase (as in spring) or a decrease (as in autumn) in the hours of daylight (King and Farner, 1965). Such changes in day-length are associated, in birds, with a nocturnal restlessness and activity (*Zugunruhe*), frantic feeding, development of the gonads and changes in the pituitary gland. It is thus reasonably

TABLE 5.5. *Maximum lipid deposition as indicated by fattest individual so far extracted in samples of 20 or more birds of each of seven species of seven families that undertake long overseas migratory flights in autumn.* (From Odum, 1965)

Species and family	Sex	Total wet wt (g)	Total extracted lipids (g)	Fat-free wet wt (g)	Non-fat dry wt (g)	Fat index (g fat/g non-fat dry wt)
Ruby-throated hummingbird (Trochilidae)	Female	5.65	2.59	3.06	0.74	3.50
Blackpoll warbler (Parulidae)	Male	24.08	12.29	11.79	3.59	3.42
Red-eyed virco (Virconidae)	Male	25.37	11.02	14.35	4.36	2.53
Summer tanager (Thraupidae)	Female	37.82	21.77	16.05	7.08	3.07
Swainson thrush (Turdidae)	Male	53.11	25.94	27.17	9.02	2.88
Bobolink (Ictridae)	Male	50.26	24.67	25.59	9.04	2.73
Yellow-billed cuckoo (Cuculidae)	Female	93.50	46.54	46.96	15.48	3.01

suspected that the deposition of fat in these circumstances is associated with endocrine signals but the evidence is difficult to interpret. The problem is probably largely the result of trying to define the role of single hormones

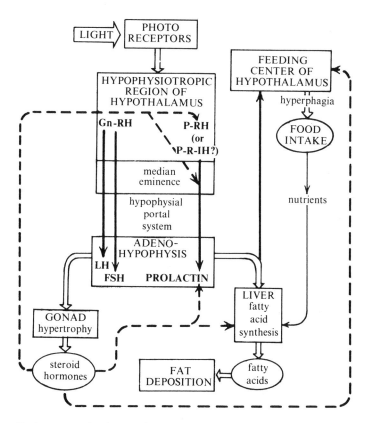

Fig. 5.8. A proposed scheme whereby nerves and hormones may mediate the changes in avian gonadal growth and deposition of fat that occur seasonally in response to photoperiodic stimulation. This scheme is based on observations in white-crowned sparrows (*Zonotrichia l. gambelii*).

The dashed lines represent possible actions of the gonadal steroid hormones and the heavy solid lines neurotransmitter roles of the hormones (that of prolactin on the feeding-center is uncertain). (Modified from Stetson and Erickson, 1972.)

when the interactions of several are involved. The effects of photoperiodic stimulation are (possibly apart from those in the pineal gland) conveyed to the endocrine system through the pituitary gland, which is suspected of playing an early role in the preparations for migration.

Prolactin, when injected, can accelerate deposition of fat in birds (white-crowned sparrows, *Zonotrichia leucophrys gambelii*) whether they have been subjected to photostimulation or not (Meier and Farner, 1964). Such injections must be given at certain times of the day and they act optimally at times related to the cyclical release of thyroxine and corticosteroids (Meier, 1970; Meier *et al.*, 1971). Such an effect of prolactin has also been observed in teleost fishes, amphibians and reptiles. The temporal relationship of gonadal development and fat deposition in birds led to the early proposal that these may be related (Rowan gonadal hypothesis) and this (despite earlier contradictions) has recently been, at least partly, confirmed (Stetson and Erikson, 1972). When white-crowned sparrows are castrated *prior* to photostimulation the hyperphagia and fat deposition do not occur. A proposed scheme that relates some of these endocrine events is shown in Fig. 5.8. It has also been suggested that reduced levels of glucagon may promote fat deposition in migratory birds (Goodridge, 1964), an observation that is consistent with the more recent suggestions as to the causes of obesity in man.

Knowledge of the endocrine processes controlling deposition of fat is, despite a widespread applicability in animal and human nutrition, not well understood. It seems to involve several hormones but these, and the pattern of their effects may differ depending on the species, the diet and the physiological occasion.

Endocrines and fasting

Hormones and the utilization of stored nutrients

During fasting, the animal is dependent on its endogenous stores of nutrients for energy. These must be converted to substrates that can be metabolized, principally glucose, fatty acids and ketone bodies, the levels of which are increased by a low insulin and a high glucagon level in the plasma (Table 5.4).

During periods of fasting that last only a few hours, or in times of sudden acute need, such as for violent action, mobilization of glycogen stores in the liver and muscles may be sufficient. This transformation is increased by the action of glucagon (on the liver) and, on occasions, also catecholamines (Fig. 5.7).

During longer periods of time, when no food is available, the liver and muscle stores of glycogen are usually maintained. Fatty acids are mobilized from triglycerides in adipose tissue and are used to provide energy. This may be a direct process involving β-oxidation or result in the formation of ketone bodies. These latter substances, acetoacetate, β-hydroxybutyrate and acetone, are produced mainly in the liver but in ruminants they are also formed by the rumen epithelium and the mammary glands. Fatty acid mobilization is favored (see Table 5.4 and Fig. 5.7) by low insulin levels, resulting from hypoglycemia as well as elevated glucagon concentrations. Catecholamines and growth hormone can also promote lipolysis. The common denominator mediating this lipid mobilization may be increased levels of the tissue's cyclic AMP. Glucose is also necessary for adequate utilization of ketone bodies in the citric acid cycle while certain tissues, especially brain, have a specific requirement for glucose. This substrate is obtained during starvation, as a result of gluconeogenesis, from certain amino acids as well as from propionate and glycerol. Gluconeogenesis is stimulated by glucagon and corticosteroids. As a prerequisite, amino acids must be mobilized and proteolysis is favored by low insulin levels and is promoted by corticosteroids.

While fat is the main source of energy during prolonged fasting in vertebrates, considerable protein catabolism also occurs and indeed, in the more terminal stages of starvation, may be inevitable. This protein utilization may be more important in poikilotherms. During estivation in amphibians and lungfishes, very high concentrations of urea accumulate in the body fluids (McClanahan, 1967; Smith, 1930) indicating substantial protein catabolism (Janssens, 1964). These animals tolerate high urea concentrations (as much as 600 mM) in their body fluids; levels that would be fatal to a mammal. Aquatic species that form ammonia instead of urea would be more readily able to excrete the toxic by-products of protein catabolism during fasting. In addition, mammals that hibernate and birds that migrate are subjected to circumstances that are often associated with limited supplies of water and may (apart from the associated muscle wasting) find excessive protein catabolism an added disadvantage, because of the problem of extra nitrogen excretion.

Stored nutrients may be mobilized and oxidized, to produce energy at greatly contrasting speeds. A bird during a non-stop migratory flight of up to 2400 km may use up almost its entire fat reserves in 40 to 60 hours (Odum, 1965). Hibernating and estivating animals, on the other hand, have relatively small energy requirements so that the rate of utilization of stored nutrients is much slower than usual. No definitive information is available about the catabolic role of the endocrines in these circumstances.

Migration of fishes

High levels of corticosteroids are present in the plasma of migrating and spawning salmon and rainbow trout (Robertson *et al.*, 1961) and it seems likely that they facilitate gluconeogenesis. These steroids are also released in response to exercise in the trout, *Salmo gairdneri* (Hill and Fromm, 1968) and stress in the sockeye salmon (Fagerlund, 1967). Such stimuli could be occurring during migration. As described earlier, injected glucagon fails to increase free fatty acid levels in the blood of eels, while catecholamines are ineffective in the goldfish. It is nevertheless possible that they may be effective in migrating fishes. An increase in the activity of the thyroid gland occurs in migrating Atlantic salmon (*Salmo salar*) (Leloup and Fontaine, 1960) and may facilitate fasting metabolism.

Lampreys, *Lampetra fluviatilis* (on histological evidence), appear to have an active thyroid gland at the beginning of their breeding migration from the sea into rivers (Pickering, 1972), although this activity declines as sexual maturity is approached. Normally the digestive tract degenerates during the migration but this can be prevented if the lampreys are gonadectomized (Larsen, 1969). It thus appears that gonadal hormones, either by direct action, or possibly due to a change via a negative-feedback mechanism in the pituitary or hypothalamus can influence tissue catabolism in migrating lampreys. Unlike in salmon, corticosteroids cannot be detected in the plasma of the migrating sea lamprey, *Petromyzon marinus* (Weisbart and Idler, 1970), so that their gluconeogenic role in migrating cyclostomes is doubted.

Bird migration

We can only speculate about the effects of hormones in the metabolism of birds during migratory flights. Glucagon has a potent effect in increasing free fatty acid levels in the plasma of birds so that this hormone, which is present in substantial quantities in the avian Islets of Langerhans, may be important on these occasions. The potent effects of injected vasotocin in elevating free fatty acid concentrations in the plasma of pigeons are also interesting. This hormone can be released in response to dehydration and stress, which are situations that may be expected to occur in migrating birds, so that (as suggested by John and George, 1973) its lipolytic action could be useful during migratory flight.

Hibernation in mammals

It has been suggested that there is decline in the activity of the endocrine glands during hibernation (Hoffman, 1964) but there is little information about the direct role of hormones during this period of dormancy or in its 'onset' and 'awakening'. Hibernation is prevented by an active reproductive system and thyroid gland but the adrenal may play a more active role.

It is suspected that both the catecholamine hormones and the corticosteroids may contribute to the events of hibernation but this is largely hypothetical and based on the known effects that such hormones have on intermediary metabolism and calorigenesis. In a hibernating rodent, the woodchuck *Marmota monax*, the rates of urinary excretion of catecholamines have been shown to be lowest in January, just prior to the onset of hibernation, and they may reach their highest values just before awakening in April (Wenberg and Holland, 1973). Metabolites of steroid hormones are excreted at very low levels early in hibernation but these rise steadily towards the time of arousal in spring. Indeed, it has been found that the injection of a corticosteroid hormone, cortisol, into a monotreme, the echidna *Tachyglossus aculeatus*, can delay or prevent the onset of the hibernation (referred to as 'torpor') that these animals may assume when they are exposed to low environmental temperatures during fasting (Augee and McDonald, 1973). It has, in addition, been observed that echidnas pass into torpor far more readily following adrenalectomy (an operation which they are well able to survive!). The rate of metabolism, blood glucose concentration and body temperature all decline in hibernating echidnas and it appears the gluconeogenic actions of corticosteroids may oppose this. It seems likely, however, that such a process must continue during hibernation or after adrenalectomy, albeit at reduced levels, in order to mobilize nutrients that are necessary to sustain the life of the animal.

Estivation in African lungfish

Estivation may last two to three years in African lungfish. During this phase of its life cycle *Protopterus annectens* only secretes thyroid hormone at about 1/75 of the rate that it does in its normal free-swimming state (Leloup and Fontaine, 1960). The TSH content of the pituitary is similar in both conditions so that the release of TSH may be inhibited by estivation. Godet (1961) has suggested that an inhibition of the adenohypophysis precedes estivation in these lungfish. It was found that when the pituitary was removed after estivation had commenced there was no effect on the subsequent torpor. An intact pituitary gland was, however, indispensable for the fishes' survival on emergence from estivation. Many amphibians also estivate during periods of drought, but their endocrinology does not appear to have been investigated on these occasions.

Hormones and lactation

Lactation, or the secretion of nutrients by the mammary glands, in order to feed the young, is an activity confined to the mammals. Somewhat analogous processes, such as the formation of pigeons' milk by the crop-sac of some birds, can occur in other vertebrates.

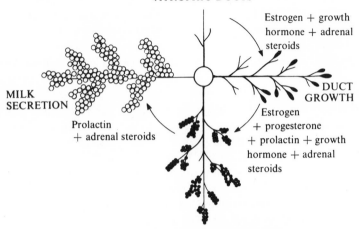

ATROPHIC DUCTS

Estrogen + growth
hormone + adrenal
steroids

MILK
SECRETION

DUCT
GROWTH

Prolactin
+ adrenal steroids

Estrogen
+ progesterone
+ prolactin + growth
hormone + adrenal
steroids

LOBULO-ALVEOLAR GROWTH

Fig. 5.9. The role of hormones in the growth of the mammary gland and the secretion of milk. In this instance the example is the *laboratory rat*, but in other species different combinations of hormones (called the 'lactogenic complex') may be required (see text). (Based on Lyons, 1958, from Cowie, 1972.)

Milk is a nutrient solution, the composition of which differs depending on the species and the duration of the lactation. (For a complete account see Cowie and Tindal, 1971.) It contains fats, carbohydrates and proteins as well as minerals and other essential dietary items. The fat content of milk varies from 0.3% in the rhinoceros to 49% in porpoises; the lactose from 0.3% in whales to 7% in man and the protein from 1.2% in man to 13% in whales. The contained energy is equivalent to 500 kcalories/kg milk in the rhinoceros to 2773 kcalories/kg in the reindeer (Kleiber, 1961). The nutritional requirements of the mother may thus be considerable.

The mammary glands, like adipose tissue, can make triglycerides and also lactose. The proteins present are largely synthesized by the mammary tissue though some, such as serum albumin and immunoglobulins, are transferred directly from the plasma. In order to deliver these nutrients efficiently to the young, the mammary tissue differentiates into a complex system of alveoli and ducts. These processes, morphological differentiation and the secretion of nutrients, are regulated by several hormones so that the mammary glands afford an example of the multiple actions that hormones may have on a tissue.

A diagram of the arrangement of the mammary gland tissues and a summary of the activities of hormones on them is given in Fig. 5.9. The milk is secreted into the alveoli and passes down the duct prior to release from the teat or nipple. Experiments on animals with an intact pituitary, but with the ovaries removed, indicate that development of the alveoli is influenced largely by estrogens and progesterone while the duct system responds principally to estrogens. The neurohypophysial hormone, oxytocin, is released (see Chapter 4) as a result of a suckling-stimulus on the nipple. It contracts the myoepithelial cells that surround the alveoli, and results in milk let-down (the *galactobolic effect*). Initiation of the secretion of milk (*lactogenesis*) by the alveoli is dependent on prolactin which is released during parturition and as a result of suckling. While these hormones are directly essential for the activities mentioned, adequate circulating levels of other hormones, including thyroid hormones, cortico-steroids and growth hormone, are also necessary. The experiments that demonstrate this are complex ones and involve restoration of the growth and function of the mammary glands in animals from which the pituitary and the ovaries have been removed. No single hormone appears to be effective on one function but combinations involving ACTH or cortico-steroids, prolactin, growth hormone, thyroid hormones, estrogens and progesterone are necessary. The metabolic effects of the absence of certain of these hormones appear to inhibit or limit the actions of other hormones that act more directly.

Species differences are common. In the hypophysectomized rabbit, lactogenesis can be initiated by a single hormone, prolactin. In hypophys-ectomized goats, prolactin plus growth hormone plus corticosteroid plus tri-iodothyronine are required, while in rats and mice corticotrophin plus prolactin are needed. Some strains of mice also require growth hormone. When a group of hormones is needed for normal lactation (the more usual situation) this is referred to as a '*lactogenic complex*'.

The maintenance of lactation is called *galactopoiesis* and the hormonal requirements of this are often difficult to distinguish from lactogenesis. Species differences again exist. Generally, the following occurs: removal of the ovaries once lactation has been established does not influence lactation; indeed estrogens, and estrogens plus progesterone, are often used to terminate, or dry-up, the secretion of milk. Removal of either the thyroid, adrenals, parathyroids or the endocrine pancreas depresses lactation. The administration of thyroid hormones has a well-established effect in increasing milk production in dairy cows. At one time this treatment was seriously considered as a method to increase the yield of milk. The role of such hormones in maintaining lactation probably reflects their roles in maintaining the metabolic transformations of energy substrates in cells generally. It is not unexpected that such metabolic processes are necessary, directly and indirectly, for the adequate secretion of milk by the mammary tissue.

At the cellular level three important stages in the process of lactation have been distinguished (Baldwin, 1969; Turkington, 1972). These events (usually based on *in vitro* observations in rats and mice) are as follows:

(i) Mammary cell proliferation. The increase in the rate of division of the mammary tissue cells has been shown *in vitro* to depend on the presence of insulin and is enhanced by estradiol-17β. This cell division is manifested as an increase in DNA synthesis. Progesterone also assists this process and may direct, or organize, it in such a manner as to allow the orderly development of the duct system.

(ii) Differentiation of the mammary cells, which includes the acquisition of the enzymes necessary for the formation of the constituents of the milk. This requires (*in vitro*) the presence of insulin, cortisol and prolactin (placental somatomammotrophin is also effective). Prolactin initiates transcriptional processes and the formation of messenger RNA which appears to mediate the formation of milk proteins and the enzyme lipoprotein lipase (Zinder *et al.*, 1974). The latter regulates the uptake of triglycerides by the mammary tissue.

(iii) The utilization (or expression) of the alveolar cells' capacity to produce milk protein, triglycerides and lactose. Little is known about this process but in order to function optimally an adequate and continuing complement of hormones is necessary.

The mammalian lactational process is a phylogenetically novel one that has arisen in the later stages of vertebrate evolution. Its complexities, not surprisingly, necessitate coordination by the endocrines. The hormones involved probably existed prior to the mammary gland itself so that an evolution of their role in the body has occurred in this case. Certain processes occur in vertebrates which are analogous to mammalian lactation. The formation of pigeon's milk, by the crop-sac of pigeons and doves, with which they feed their young, is an example. This process, which involves the proliferation of the crop-sac epithelium, can be induced by prolactin. In certain teleost fishes, mucous secretion from the skin can be stimulated by prolactin (see Bern and Nicoll, 1968). In one such fish (*Symphysodon*), the newly hatched young have been observed to feed off the surface of the parental fish, suggesting another possible 'lactational' role for prolactin.

Storage of nutrients in the egg

The early development of the young is supported by nutrients stored in the egg. These materials may be sufficient for the complete embryonic development of the young (megalecithal eggs) or only support more limited differentiation (alecithal) that leads to the emergence of a free-swimming larva. An even more limited growth may precede viviparous development in the uterus. Alecithal eggs are usually confined to species that develop in a watery environment. Megalecithal eggs are common in terrestrial species as well as certain fishes including the Agnatha and Chondrichthyes (see Fig. 5.10).

Estrogens and progesterone may play important roles in the growth and maturation of the egg which are summarized in Fig. 9.17.

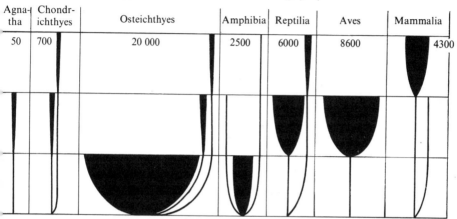

Agna-tha	Chondr-ichthyes	Osteichthyes	Amphibia	Reptilia	Aves	Mammalia
50	700	20 000	2500	6000	8600	4300

Fig. 5.10. A diagrammatic representation of the relative frequency of the occurrence in the vertebrates of megalecithal eggs (with large amounts of stored yolk) and alecithal eggs. Alecithal eggs are confined to aquatic species; the production of eggs with small amounts of yolk in terrestrial vertebrates necessitates the viviparous habit which is also included. It should, however, be noted that megalecithal eggs and viviparity also occur among the aquatic fishes and amphibians. Numbers = no. of species. (From Browning, 1969, based on a drawing by H. A. Bern.)

The 'white' of the megalecithal eggs contains water, salts and proteins, including avidin, which, as we have seen, can be formed by glands in the chicken oviduct as a result of stimulation by progesterone. Progestins may also stimulate the secretion of the jelly from the oviductal glands that surrounds alecithal eggs, as in amphibians.

The yolk of eggs contains most of the nutrients required for the predicted development of the young. In the domestic fowl, about 50% of this is composed of fats and proteins. The number of eggs produced successively in one period of time (the clutch) varies considerably in different species but 20 to 30 are not uncommon in birds and reptiles. Under domestic conditions, a chicken may produce 200 to 300 eggs in a year. Even species that produce alecithal eggs must contribute considerable amounts of nutrients to the eggs, especially as such species often produce many hundreds of such eggs.

The deposition of the yolk is called *vitellogenesis*. Estrogens can promote the appearance in the blood of elevated concentrations of calcium and phosphorus, and certain proteins and lipids. These usually exist as a single calcium-binding phospholipoprotein, though in birds two components, phosphovitin and lipovitellin, can be distinguished. These phospholipoproteins (also called vitellogenin) are formed in the liver in response to stimulation by estrogen. They are then carried in the blood to the ovaries for transfer to, and deposition in, the eggs. This effect of estrogens is very specific (it is not seen when other steroid hormones are injected) and is prominent in oviparous species (Urist and Scheide, 1961; Urist, 1963; Urist *et al.*, 1972). The vitellogenic effect of estrogens is absent in mammals, cyclostomes and chondrichthyeans but is seen in teleosts, lungfishes, amphibians, reptiles and birds. The situation in the Monotremata (oviparous mammals) does not seem to have been explored. Apart from their vitellogenic effects estrogens exert remarkably similar actions in all vertebrates but hormones may acquire special roles of which this effect is a good example for the estrogens.

Hormones and growth and development

The process of growth may place considerable nutritional demands on an animal. The development of the young is associated with rapid increases in the mass of the body and thus has special nutritional requirements. The development of the young *in utero* also contributes indirectly to the mother's nutritional requirements. In addition, growth processes also occur continually in the bodies of adult animals and involve cyclical changes in the reproductive organs as well as the replacement of cells, due to age and general wear and tear. These growth processes principally involve the formation of proteins (and thus a positive nitrogen balance) and phospholipids. The contribution of minerals, especially calcium and phosphorus, should not be forgotten.

Like lactation, successful growth requires the harmonious regulatory activities of several hormones. Those more directly involved in development and maturation of the young are growth hormone and the thyroid hormones. Androgens and estrogens also contribute to the control of the maturation of the reproductive system and, in adults, help maintain it and regulate the cyclical changes in these tissues. Androgens, as we have seen, may also exert anabolic effects and promote protein formation in muscle. A hormonal background of adequate insulin, that promotes conversion of amino acids to protein, and a not over-adequate supply of corticosteroids, that facilitate proteolysis, is also important.

The process of growth consists of (*a*) an increase in the number of cells;

(b) an enlargement in their size; (c) increases in their complexity and differentiation; (d) replacement of cells as a result of loss due to age, wear and tear, injury or as part of morphogenetic differentiation as in (c).

Growth hormone produces a striking increase in cell DNA, indicating an increase in cell number. Cell size is also increased. This effect is manifested in all tissues, including cartilage, and leads to the building of bone. Growth hormone promotes the accumulation of amino acids across the cell membrane and enhances the formation of messenger-RNA which determines the formation of specific proteins, which presumably contribute to the structural elements as well as cell growth and reproduction. Growth hormone also promotes the incorporation of sulfate into cartilage. This observation led to the discovery that growth hormone was not necessarily acting directly on the tissues (see Daughaday, 1971). The incorporation of sulfate into cartilage was found to be due to a protein present in the plasma which was initially called 'sulfation factor'. In addition, this factor also promotes the incorporation of uridine into RNA and thymidine into DNA in the tissue. The formation of this intermediary is promoted in the liver by growth hormone and it has been renamed *somatomedin* (Daughaday *et al.*, 1972). It seems likely that somatomedin also mediates other effects of growth hormone. Somatomedin persists for a longer time in the plasma than growth hormone which thus need not be released continually in order to exert its effects. It is also interesting that somatomedin has a stronger effect in tissues from weanling rats than from fetal and old rats. The last two are known to be unresponsive *in vivo* to such effects of growth hormone (Heins, Garland and Daughaday, 1970). The widespread actions of growth hormone in mammals, birds, reptiles, amphibians and teleost fish have already been described (Chapter 3).

As indicated earlier (Chapter 3) prolactin and growth hormone share a number of activities, a property that may reflect their structural similarities. It is therefore not surprising to find that in some species injected prolactin may influence growth. This effect of injected mammalian prolactin has been observed in the pigeon, the lizard, *Lacerta s. sicula*, and tadpoles of the frogs *Rana temporaria* and *R. pipiens* (Bates, Miller and Garrison, 1962; Licht and Hoyer, 1968; Frye, Brown and Snyder, 1972). Metamorphosis in the larval amphibians is prevented by injected prolactin. It is uncertain if such effects reflect another physiological role of prolactin or are merely due to similarities between its structure and activity to that of the homologous growth hormones.

Apart from their calorigenic effects in homeotherms, thyroid hormones play an important role in the body, contributing to an optimal and orderly growth and development. This is most apparent in mammals that lack sufficient thyroid hormone during postnatal development. The thyroid hormones are especially necessary for adequate development of the brain, skeletal system and reproductive organs. The disease of cretinism that results from insufficient thyroid hormones in man during infancy has been observed in other domestic animals. In newborn rats deprived of thyroid hormone, it has been shown that development of the brain is slowed but the same number of cells (DNA content) appear to develop though they remain smaller (higher concentration of DNA) than in the normal rat. The rate of multiplication of the cells is reduced in thyroid deficiency. The smaller size of the cells is reflected in neural tissue by a reduction in the development of the nerve-endings while the incorporation of amino acids into proteins is diminished (Balazs *et al.*, 1971). When tadpoles are deprived of thyroid hormone, they continue to grow but fail to metamorphose. On the other hand, if tadpoles are fed thyroid hormone they undergo an earlier metamorphosis. This latter transformation is associated with the acquisition of many adult characteristics along with the induction of certain enzymes that mediate reabsorption of the tail. The thyroid is also necessary for the maturation of the gonads in some teleost fish (Sage, 1973).

The morphological effects of thyroid hormones in mammals are also not strictly analogous to their morphogenetic effects in amphibians. Myant (1971) suggests that amphibian metamorphosis may have arisen as a special adaptation in which differentiation is suppressed during a period of rapid growth, and is then stimulated under the influence of thyroxine (see Chapter 9). The morphogenetic effect of thyroid hormones in amphibians has no true analogy in other vertebrates and appears to be a specialization. Thyroid hormones, however, appear to have a basic ability to stimulate protein synthesis in cells of homeotherms and possibly some poikilotherms like tadpoles. Such a basic effect, if exerted at strategic sites in the cell, could support multiple, different actions of the hormones on cells. The thyroid hormones thus may potentially be able to assume various functions and be especially suited to their proposed 'permissive' role in cellular processes; at least in homeotherms and amphibian tadpoles.

Conclusions

The processes of nutrition, whether involving the digestion of food materials or their transformation, storage and utilization, following absorption, or the feeding and growth of the young are very dependent on the actions of hormones. It is especially notable that several hormones are usually involved in such processes ('multihormonal' effects) probably more often than elsewhere, and these excitants have widespread effects both with respect to the types of tissues that they act upon and the nature of the underlying biochemical events that they influence. Such ubiquitous endocrine effects are well integrated with each other by a series of feedback controls that involve the levels of metabolites and the hormones themselves. The latter indeed often directly influence each others' release. The hormonal pathways that function, or predominate, in the control of nutrition of a particular species seem to depend more on its feeding habits and the stage of its life cycle, than on its phylogenetic origins. This probably reflects the common biochemical pathways underlying nutritional processes in all vertebrates. The principal species' differences in responsiveness to hormones are quantitative ones, though when novel processes have evolved, such as lactation in mammals, this has been accompanied by adaptations of the roles of hormones to integrate the new systematic features.

6. Hormones and calcium metabolism

Calcium is vital for animal life. In vertebrates, this divalent ion is present at various sites, including the body fluids, structural parts of the cell (especially the endoplasmic reticulum), and in most species it is also a major component of the endoskeleton. The outer shell of the eggs of birds and many reptiles also consists principally of calcium. The physiological role of calcium appears to be the result of a rather unique set of physicochemical properties. In aqueous solution, calcium can exist in a soluble form, which is important for its mobility in the body, and yet, equally essential to its role, many of its salts, including phosphates and carbonates, have a low solubility so that a physicochemical equilibrium may exist between its solid and aqueous forms. The quantity of calcium free in solution is thus restricted in the presence of certain anions. In addition, such relatively insoluble salts in certain of their crystalline forms, principally calcium phosphate and calcium carbonate, can contribute to the mechanical support and stability of biological structures. Calcium also has a ready propensity to associate with, and combine to, proteins. Such combinations are seen in the body fluids, where a considerable portion of the calcium is bound to serum proteins, and in cells, where it contributes to their structural stability by helping maintain essential ionic bridges at vital points in protein molecules; thus, when tissues are placed in calcium-free solutions they tend to disintegrate and the cells swell and fall apart.

Calcium plays an essential role in coordinating many events in the body that may reflect those general properties described above. Calcium stabilizes membranes and this effect can be seen in the hyperactivity of nerve fibers placed in solutions with low calcium concentrations. Such instability and the repetitive electrical depolarization of the nerve cell membrane result in tetanic contractions of the muscles they supply. Muscle contraction requires the presence of calcium; when released from the sarcoplasmic reticulum within the cell, calcium couples the initiating electrical depolarization of the cell membrane to those processes that initiate changes in the contractile proteins. In a comparable manner, calcium is necessary for the ultimate initiation of many endocrine events such as the release of hormones (see Chapter 4, p. 148) and the responses of their definitive effectors.

The availability of calcium to animals varies considerably, depending on the environment and their diet. The physiological need for calcium may also vary a great deal. Young, growing animals, especially those that are forming large amounts of bony tissue, have much greater requirements than adults. The latter, however, periodically need more calcium for reproductive processes, as during pregnancy and lactation in mammals, and in birds and reptiles, the production of large cleidoic eggs that are covered with a shell of calcium carbonate. Even fishes and amphibians that do not have cleidoic eggs, require substantial amounts of calcium for the production of their eggs. The availability of calcium in the environment varies a great deal. The concentration in sea-water is even higher than that in the body fluids of vertebrates but in fresh water, little calcium is usually present (Table 6.1); thus vertebrates living in the sea or fresh water may be expected to experience very different problems with respect to the availability of this ion. Terrestrial animals obtain most of their calcium from their diet which may include appreciable amounts obtained from

TABLE 6.1. *Calcium and phosphate in serum and in the environment.* (From Copp, 1969)

Environment and representative species	Total Ca (mmole/liter)	Ionic Ca²⁺ (mmole/liter)	Total P (mmole/liter)
Environment			
Pacific Ocean	10.0 ± 0.1	10.0	0.00
Brackish water	2–5	–	–
Lake Huron	0.9 ± 0.1	0.9 ± 0.1	0.003
Marine invertebrate			
Nephrops	11.95	–	–
Cyclostomes			
Hagfish, *Eptatretus stoutii*	5.4 ± 0.1	3.0 ± 0.4	1.5 ± 0.2
Lamprey, *Petromyzon marinus* (from fresh water)	2.60 ± 0.1	1.74 ± 0.2	1.3 ± 0.1
Chimaeroid			
Ratfish, *Hydrolagus colliei*	4.8 ± 0.3	–	2.2 ± 0.4
Elasmobranchs			
Marine shark, *Carcharhinus leucas*	4.50 ± 0.9	3.10 ± 0.4	2.0 ± 0.6
Freshwater shark, *C. leucas nicaraquensis*	3.0 ± 0.2	1.7 ± 0.1	1.6 ± 1.7
Teleost			
Marine, *Paralabrax clathratus*	3.2 ± 0.8	2.0 ± 0.9	2.0 ± 0.2
Freshwater, *Megalops atlanticus*	2.5 ± 0.2	1.8 ± 0.2	1.2 ± 0.4
Mammal			
Man, *Homo sapiens*	2.32 ± 0.08	1.15	–

certain drinking waters. In times of extra need for calcium, vertebrates, especially birds during egg-laying, may consume inorganic calcium-containing minerals, such as the so-called 'grit' fed to domestic fowl. The ultimate acquisition of calcium, whether from the external bathing solutions, drinking water or food, takes place principally from the gut. Absorption across the intestine is regulated in relation to the animal's needs.

The calcium that is not absorbed from the intestine, either due to a lack of physiological need for it, or, if it is present in chemical combination that makes this process impossible, is excreted in the feces. Additional amounts of calcium are also excreted by the kidney in the urine. Urinary losses of calcium should not only be viewed as an excretory mechanism for ridding the body of an excess of this ion but also as part of an unavoidable loss that results from the formation of urine. The calcium that is not bound to plasma proteins is filtered across the renal glomerulus but most of this ion is subsequently reabsorbed by the renal tubules. This conservation is less effective in the presence of a high plasma calcium concentration so that extra amounts of this ion are excreted in the urine. Conversely a hypo-calcemia results in an increased renal reabsorption of calcium which is accompanied by an increased phosphate excretion. The excretion of calcium and phosphate in the urine is thus subject to physiological control.

The bony skeleton possessed by most vertebrates plays a central role in calcium metabolism. Calcium phosphate salts are an integral part of this structure and make the principal contribution to skeletal rigidity. Not all vertebrates, however, have a bony calcareous skeleton; it is absent in the cyclostome and chondrichthyean fishes that have a more elastic cartilaginous skeleton in which little calcium is deposited. In the bony vertebrates, the skeleton is the predominant quantitative site of calcium in the body though this should not be taken to reflect its relative importance, as its presence is equally vital to the soft tissues. The calcium in the bones and that in the body fluids and soft tissues exists in equilibria with each other. Two processes are involved in this: first, there is a relatively static *physico-chemical* equilibrium that reflects the solubility (and insolubility) of the calcium salts in the body and which only results in small exchanges of calcium; second, the major exchanges of calcium between the bones and body fluids are due to a *dynamic*, physiologically mediated equilibrium that results from the activities of cells in the bone; the *osteoblasts, osteocytes* and *osteoclasts*.

Bone, despite its mineral-like appearance, is a living tissue. Some fishes, however, possess acellular bone which, once formed, is much less labile than the cellular-type of bone that is characteristic of the tetrapods. Cellular bone has a microscopic honeycomb-like appearance due to the presence of numerous small chambers, or lacunae, each of which contains an osteocyte bone cell. Numerous fine channels (canaliculi) radiate from the lacunae to the surrounding mineralized tissue and these provide a pathway through which the bone fluids and the dendritic-like extensions of the osteocytes maintain contact with the tissue. The osteocytes act as sentinels controlling local mineral exchanges. Bone also has an extensive network of blood vessels through which the supply of blood can be regulated according to its metabolic needs. Exchanges of bone minerals take place at any of the bone's free surfaces; the canaliculi, at the inner and outer borders of the limb bones (periosteum and endosteum), the channels through which the blood vessels pass and special tunnels, called 'cutting cones', which are excavated in the tissue. Minerals are principally mobilized from the cortical regions of the shafts of the long bones but a labile store also occurs in the medullary regions of the bones of birds where it is an especially important store during the egg-laying cycle.

Bone is formed by the osteoblast bone cells. These cells extrude collagen which is laid down as a matrix into which calcium, phosphate and some carbonate are subsequently deposited (called 'accretion'). It is possible that the osteoblasts also play some direct role in the process of deposition but this is not essential. Minerals are deposited as two phases, initially as an amorphous form of calcium phosphate which is subsequently changed to a structurally stronger, crystallized form resembling the mineral apatite. The osteoblast, after surrounding itself with such tissue, then is transformed into an osteocyte. The mineralized bone is not necessarily a permanent structure but can be modified and remodelled and the calcium and phosphate may be returned to the tissue fluids. This process of 'resorption' occurs more readily from the amorphous than the crystallized phase but the latter can also be broken down. Resorption of minerals from bone is associated with an increase in the activity of the osteoclasts, and the osteocytes are also involved. The process of resorption involves the secretion of enzymes by the bone cells and these facilitate the dissolution of the minerals. This process can be regulated.

Physiological control of calcium levels in vertebrates can thus be affected at three major sites: the intestine, by the control of absorption; the kidney, by the regulation of reabsorption from the glomerular filtrate, and the bones which act as a storage site for calcium phosphate. The very important role of bone in calcium homeostasis cannot be overemphasized but it should be recalled that this tissue is not present in all vertebrates. The coordination of the exchanges of calcium at these three sites is under the control of hormones. These are parathormone, from the parathyroids (which are absent in fishes), and calcitonin from the thyroid 'C' cells and the ultimobranchial bodies. To these hormones can be added another (or others) that are derived from vitamin D_3. These are 25-hydroxycholecalciferol and 1,25-dihydroxycholecalciferol. It is of further interest that removal of the corpuscles of Stannius that are present in certain bony fishes (see Chapter 2) results in an elevation of plasma calcium levels, suggesting that these tissues may also secrete a hormone that regulates calcium metabolism in such fish. Estrogens aid mobilization of medullary bone in birds during egg-laying.

A historical note about the discovery of parathormone and calcitonin

The discovery of the roles of the parathyroids, the mammalian thyroid 'C' cells, and the ultimobranchial bodies in the regulation of calcium metabolism is an interesting endocrine tale. It is not an example of a triumph of 'goal-oriented' research but rather that of serendipity and the persistent following up of a series of unexpected observations. Species differences in the endocrine tissues have played an important part in establishing the role of these glands.

As we have seen (Chapter 2), two types of endocrine tissues are now known to be concerned with regulating calcium metabolism. These tissues are present in close morphological association with the thyroid gland. Early efforts to remove thyroid tissue surgically in man were sometimes seen to result in tetanic muscular contractions such as are associated with low plasma calcium concentrations. Closer examination revealed that in these instances the parathyroid tissues had been removed. Subsequently other experiments in animals confirmed the importance of the parathyroids in maintaining optimal calcium levels in the blood and extracts of these glands were shown to exhibit a hypercalcemic action. Such experiments, to demonstrate the role of the parathyroids, are relatively simple in the rat, where there are two distinct bodies of parathyroid tissue present on the surface of the thyroid gland. In another favored experimental animal, the dog, substantial amounts of parathyroid tissue (as well as 'C' cells) are present, embedded deeply in the thyroid gland, so that this species is not an ideal one for such experiments and has, in the past, contributed to some misinterpretations.

As related by Hirsch and Munson (1969), the rat is an ideal species, and favorite subject, for parathyroidectomy and following this operation provides an excellent preparation for the bioassay of injected parathormone. Two methods have been used to remove the rat parathyroids. A simple surgical removal of the two glands can be performed or they can be destroyed with an electrocautery. In the 1950s, the latter procedure was more popular. When, however, a comparison, partly retrospective, of blood calcium levels following each type of operation was made, the resulting hypocalcemia was observed to be much greater following electrocautery than following surgical excision. This observation probably did not initially gain the serious attention it deserved. In retrospect it has, especially when it was observed that stimulation of the thyroid gland by the electro-cautery at sites removed from the parathyroid tissue also produces a drop in blood calcium concentration. The possibility was then considered that this stimulation resulted in a release of a substance from the thyroid that exhibited a hypocalcemic action. A couple of years prior to the latter observation, D. H. Copp (see Copp *et al.*, 1962) proposed the presence of a hypocalcemic hormone in mammals that he called calcitonin, which at the time he considered to be formed by the parathyroid glands. The proposal of the presence of a new hormone was based on experimental observations involving perfusion of the thyroid–parathyroid complex in dogs. Blood containing abnormally high or low concentrations of calcium was perfused through the arteries supplying these tissues and the resulting venous outflow was then passed back into the dog and the effects on the general, systemic, plasma calcium levels were observed. It was found that, when the perfusing blood had a low calcium concentration, the outflowing blood, when passed into the dogs' general circulation, produced a hypercalcemia such as could be accounted for by the presence of parathormone. In the opposite type of experiment, in which the thyroid–parathyroid complex was perfused with blood having a high calcium concentration, the parathormone level would be expected to decline; as indeed it does. The basic question that was asked was this: Is such a decline in parathormone sufficient to bring about a decline in the calcium concentrations from hypercalcemic to normal calcium levels? In other words, while parathor-mone exerts a positive effect in elevating blood calcium concentration, is the mere absence of this hormone all that is needed to adjust the calcium level in a downward direction? Copp found that this was not so. While the venous perfusate from the thyroid–parathyroid complex that was exposed to high calcium concentrations produced a drop in calcium levels, when infused systemically, this hypocalcemia was much greater than could be produced after removing the glandular complex (Fig. 6.1). The implication was drawn that the response is a positive one involving the action of a hormone that has a hypocalcemic action, which Copp called calcitonin.

It was thought at first that calcitonin came from the parathyroids and, as commented upon above, in view of the intermixture of tissues that occurs in the thyroid region in dogs such an error was not surprising. Hirsch and Munson, on the basis of their experiments on rats, proposed the presence of a hypocalcemic hormone in the thyroid gland itself, which, in order to distinguish it from Copp's hormone, they called thyrocalcitonin. These two hormones are in fact identical and in mammals this hormone originates from the thyroid gland. By the choice of an appropriate species,

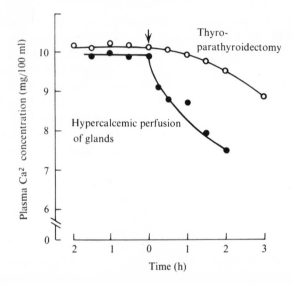

Fig. 6.1. The hypocalcemic response of dogs following perfusion of the thyroid–parathyroid gland complex with hypercalcemic solutions. In intact animals, the decline in plasma calcium concentration (at time zero) occurs promptly, but if the glandular complex is removed, the resulting hypocalcemic response is much slower. This suggests the action of a hypocalcemic hormone in the intact dogs. (From Hirsch and Munson, 1969; from data of Copp *et al.*, 1962.)

this time the pig, in which the thyroid contains no parathyroid tissue, appropriate perfusion experiments of the thyroid alone, similar to those described above, demonstrated the presence of calcitonin in the venous effluent blood.

The question then arose as to what is the site of origin of calcitonin in the thyroid tissue. Is it also produced by the same tissue that forms thyroxine? The answer is no. Calcitonin is formed by the parafollicular or 'C' cells that are present in the mammalian thyroid and this tissue is quite distinct from that which secretes thyroxine. The presence of these secretory cells was first described by E. C. Baber in 1876 but their function was unknown until recently. Radioimmunofluorescent studies, using fluorescent antibodies to calcitonin, show quite clearly that the 'C' cells form this hormone.

The 'C' cells are present in the mammalian thyroid but not that of non-mammals. Embryologically they are derived from the ultimobranchial bodies that are present in all non-mammals except the cyclostomes. An early clue to the function of this tissue was described by Rasquin and Rosenbloom in 1954, several years before the discovery of calcitonin. It was found that Mexican cave fish *Astyanax mexicanus*, when kept for prolonged periods of time in complete darkness, suffered skeletal deformities associated with an hyperplasia of the ultimobranchial bodies. Rasquin and Rosenbloom suggested that the ultimobranchials contained a parathyroid-like hormone. Extracts of the ultimobranchial bodies obtained from chickens and subsequently from many other species, including even dogfish, showed that, when injected into rats, they exhibited a hypocalcemic effect caused by the presence of a calcitonin-like hormone which was not present in the thyroid glands of these species (Copp, Cockcroft and Keuk, 1967*a*).

Vitamin D and 1,25-dihydroxycholecalciferol

Vitamin D has been classified as a vitamin as it is principally acquired in the diet. It exists in two forms, a synthetic one, ergocalciferol (D_2), and the natural material, cholecalciferol (D_3), which is a precursor or pro-hormone (see Fig. 3.1). Cholecalciferol is converted in the liver into 25-hydroxycholecalciferol (25-OHD$_3$). This steroid is further hydroxylated in the kidney into the active hormonal substance 1,25-dihydroxycholecalciferol or 1,25-$(OH)_2D_3$. Production of this latter hormone, in common to other endocrine secretions, is subject to feedback control. The production of 1,25$(OH)_2D_3$ by the kidney is reduced during hypercalcemia and is promoted by hypocalcemia or hypophosphatemia (see Tanaka, Frank and DeLuca, 1973). An increased production of 1,25-$(OH)_2D_3$ can be stimulated by parathormone, which is secreted in response to hypocalcemia, probably by decreasing the inorganic phosphorus levels in the renal cells. Low inorganic phosphorus concentrations, however, can stimulate formation of 1,25-$(OH)_2D_3$ even in conditions when there is a hypercalcemia and parathyroid secretion is, presumably, depressed. Such an effect is seen in rats on low phosphorus diets and suggests that phosphorus levels in the renal cells are of basic importance in controlling the formation of 1,25-$(OH)_2D_3$.

Mechanisms and interactions of parathormone, calcitonin and vitamin D on calcium metabolism

Regulation of the calcium levels in the body depends on the interactions of three effectors: bone, intestine and kidney. These respond to various combinations of parathormone, calcitonin and hormonal metabolites of vitamin D.

The initial accumulation of calcium in the body depends on its absorption across the wall of the intestine and this process is controlled by vitamin D_3. The metabolites of this steroid stimulate the active transport of calcium from the intestinal lumen to the blood. It is notable that the effect of vitamin D_3 on the intestinal calcium transport is not prompt but is delayed for several hours following its administration (see DeLuca, 1971). This delay is due to several important events that concern the vitamin's mechanism of action. Vitamin D_3, as described above, must first be converted to 1,25-dihydroxycholecalciferol, which is the active hormone. Following this activation of vitamin D_3 there is a further delay in the response that reflects a process involving the formation of RNA and protein synthesis. Actinomycin D, which prevents the formation of RNA, inhibits the effect of vitamin D_3 on the intestine. The events that occur in the intestinal mucosal cells in response to vitamin D_3 are summarized in Fig. 6.2. 1,25-Dihydroxycholecalciferol may act at a genetic locus in the cell nucleus to stimulate the

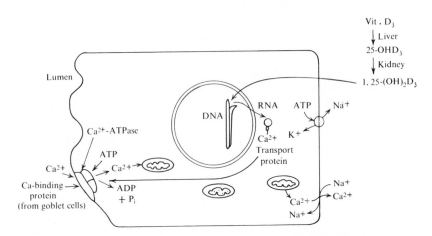

Fig. 6.2. Diagrammatic representation of the action of vitamin D in increasing calcium absorption from the intestine. 25-OHD$_3$ = 25-hydroxycholecalciferol; 1,25-(OH)$_2$D$_3$ = 1,25-dihydroxycholcalciferol.
The calcium is actively absorbed with the aid of a Ca-activated ATPase at the mucosal (luminal) side of the intestine, carried across the cell in the mitochondria and is then extruded from the serosal side of the cell in a process that is Na-dependent. (Modified from DeLuca, 1971.)

formation of a calcium-transport protein. This protein may be the calcium-binding protein in the microvilli or calcium-dependent ATPase that is present at the luminal boundary of the cell. This enzyme mediates the accumulation of calcium (phosphate follows this ion) which is then taken up by the mitochondria from which it is subsequently released and extruded, with the aid of a Na–K ATPase, across the opposite, serosal, side of the cell. It is considered unlikely that parathormone or calcitonin influence calcium transport across the intestine but there is some evidence to suggest that they might.[1] 1,25-Dihydroxycholecalciferol, however, certainly exerts a prominent action on calcium transport at this site and parathormone is probably acting indirectly by increasing the formation of this steroid (Garabedian *et al.*, 1974). The calcium that is absorbed into the blood may be either deposited in the bones or excreted by the kidney.

Regulation of the urinary excretion of calcium appears to be principally due to the action of parathormone. This hormone promotes phosphate excretion by inhibiting its reabsorption from the proximal renal tubule and it also reduces urinary calcium loss by promoting its reabsorption from the distal tubule (Agus *et al.*, 1973). These processes are the result of an activation of adenyl cyclase and the formation of cyclic AMP in the kidney cortex. Calcitonin and vitamin D do not appear to have direct effects on the renal control of calcium or phosphate excretion but this possibility has not been completely excluded.

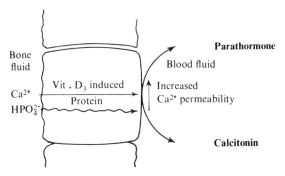

Fig. 6.3. A model for the endocrine control of calcium exchange in bone. The process of calcium resorption from bone is dependent on the presence of vitamin D_3 which apparently induces the formation of a protein that acts as a 'Ca-carrier' that transfers calcium from the bone fluids to the vicinity of the blood plasma. The final 'jump' across the cell membrane depends on its permeability to calcium and this is increased by parathormone and reduced by calcitonin. (Modified from DeLuca, Morii and Melancon, 1968.)

The bones are the sites where vitamin D_3, parathormone and calcitonin interact with each other to promote, or inhibit the resorption, into the blood, of calcium and phosphate. Parathormone stimulates resorption from the bone by an action on both the osteocytes and, principally, the osteoclasts. In vitamin D-deficient animals this effect is, however, considerably reduced or even absent. Vitamin D_3 appears to stimulate the formation of a substrate that functions as a carrier for the transfer of

---

[1] Swaminathan, Ker and Care (1974) have found that the prolonged administration of calcitonin to sheep and parathyroidectomized pigs decreases the absorption of calcium from the intestine.

calcium from the bone fluids into the osteoclasts. This accumulated calcium may then be released from the opposite side of these cells and pass into the blood. Parathormone increases the permeability of the outer membrane to calcium, an effect which, like in the kidney, is mediated by the formation of cyclic AMP. Calcitonin opposes the resorption process in bone, and is thought to oppose directly the action of parathormone on the outer cell membrane. These events are summarized in Fig. 6.3, and the over-all integrated process in bone, intestine and kidney is shown in Fig. 6.4.

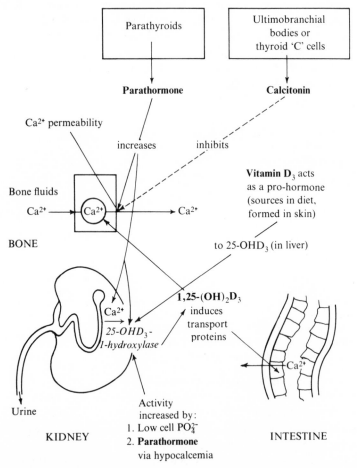

Fig. 6.4. Summary of the effects of parathormone, calcitonin and vitamin D_3 on calcium metabolism in mammals. There are three main sites of action; the kidneys, intestine and bone. The possibility that calcitonin acts on the kidney and parathormone and/or calcitonin changes absorption of calcium across the intestine has not been completely excluded.

The hormonal regulation of calcium metabolism outlined above is based principally on experiments using mammals and, in a few instances, birds. Calcium metabolism in non-mammalian vertebrates, however, may exhibit some interesting differences. Such variations are to be expected on quantitative grounds, as in the hen during its egg-laying when relatively vast amounts of calcium are rapidly utilized. Qualitatively predictable differences arise as the parathyroids are absent in fishes while the cyclostomes appear to have neither these glands nor ultimobranchial bodies. The absence of a calcified bony skeleton in the chondrichthyeans and cyclostomes may also be expected to be a matter of some physiological consequence in calcium metabolism.

Phyletic differences in the role of hormones in calcium metabolism

Mammals

Mammalian calcium metabolism has been described in some detail in the preceding section along with some of the anatomical variations in the distribution of the parathyroid tissues. A variety of mammals including man, dog, rat, pig, sheep and goat have been examined, and the regulation of calcium levels in the blood is related to the activities of the parathyroids and the thyroid 'C' cells. The former have a vital role to play, as mammals deprived of these tissues suffer tetanic seizures because of a hypocalcemia. The rachitic effects of vitamin D-deficiency on bone are also well known. It is, however, not clear how physiologically essential the thyroid 'C' cells normally are in mammals. During calcium stress, when large increases in blood calcium levels occur, calcitonin undoubtedly facilitates the homeostatic adjustment of the concentration of this ion; however, the role of calcitonin in the regulation of the smaller and more usual changes in calcium concentration is uncertain.

This question of what is the normal physiological role of calcitonin in mammals, has been examined in young pigs (Swaminathan, Bates and Care, 1972). Removal of the thyroid results in a rapid rise (in one to two hours) of the plasma calcium concentration. This hypercalcemia is presumed to be the result of a lack of calcitonin, as it can be corrected by infusing small amounts of this hormone into the animal (Fig. 6.5). It is notable, however, that when these thyroidectomized pigs were allowed to recover, without injections of calcitonin, blood calcium levels returned to normal after 24 to 48 hours. This recovery is probably the result of an adjustment in the rate of secretion of parathormone. It has also been found that calcitonin aids, but is not essential, in regulating calcium metabolism in young, growing rats. In adult rats on a normal diet, the evidence to

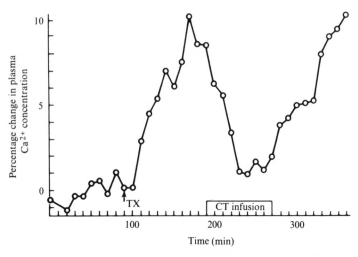

Fig. 6.5. The observed increase in plasma calcium concentration in young pigs following removal (TX) of the thyroid (and its contained calcitonin-secreting 'C' cells). Infusion of calcitonin (CT infusion) restored the calcium concentration to normal, but as soon as this ceased the levels climbed again. After about a day, the calcium levels returned to normal, probably as a result of other physiological adjustments including a decline in secretion of parathormone. (From Swaminathan *et al.*, 1972.)

indicate that calcitonin has a physiological role is contradictory (Kumar and Sturtridge, 1973; Harper and Toverud, 1973).

Thus, while calcitonin seems to contribute to the regulation of calcium levels in mammals, its role may only be vital when there is a large increase in blood calcium. In this respect, it is difficult to see how it could be essential to life, but then neither are some other hormones, including those from the neurohypophysis.

Birds

Birds, especially during their egg-laying cycle, have a very high rate of calcium turnover (Copp, 1972). A domestic hen may then utilize an amount of calcium equivalent to 10% of that in its body each day. Most of this calcium is derived directly from the food. When compared on a unit body-weight basis the domestic hen during egg-laying absorbs calcium across its intestine 100 times more rapidly than a man. Vitamin D, as in mammals, increases the rate of calcium absorption from the intestine of the domestic fowl (Fig. 6.6). At other times, such as during the nocturnal fast, calcium is also mobilized from the bones, principally from medullary bone rather than cortical bone. This distinction is of some endocrine importance as

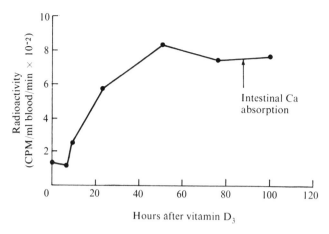

Fig. 6.6. The effect, in young chicks, of an oral dose of vitamin D_3 on calcium absorption from the intestine. The chicks were on a low-calcium diet. (Modified from Harmeyer and DeLuca, 1969.)

estrogens can influence the turnover of calcium in medullary bone while parathormone only acts on cortical bone.

As described in Chapter 5, estrogens facilitate the formation of a particular phospholipoprotein (vitellin) in the liver of female birds that appears in the blood and is incorporated into the yolk of the developing egg. This plasma protein binds calcium and so its presence is associated with elevated blood calcium concentrations. This interesting hypercalcemic effect of estrogens is prominent in birds as well as other oviparous 'bony' vertebrates, but is absent in mammals which may be related to their viviparity. Such a response is especially appropriate to the needs of vertebrates that produce large megalecithal eggs that contain a lot of calcium.

Both the parathyroids and the ultimobranchial bodies hypertrophy in egg-laying hens, suggesting that they are involved in regulating the calcium turnover in such birds. However, the precise contribution of each of these glands to avian calcium metabolism is still not clear.

Removal of the parathyroids results in hypocalcemia in birds and this is particularly dramatic in young, growing chicks. Injections of parathormone elevate blood calcium concentrations through an effect on cortical bone and, possibly, also by increasing intestinal absorption of this ion. The latter effect is suggested (Copp, 1972) as the response to parathormone is much less in domestic fowl on a low than on a higher calcium diet (see for instance Fig. 6.7). Injected parathormone has an extremely rapid hypercalcemic action in the laying hen (Fig. 6.7); it acts six to eight times more rapidly than in the dog (Mueller *et al.*, 1973). This rapid initial phase of the

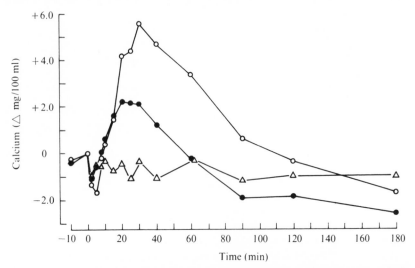

Fig. 6.7. The hypercalcemic effects of parathormone injections into female domestic fowl. The response was enhanced in those birds receiving a higher supplement of dietary calcium, suggesting that this hormone may be increasing intestinal calcium-absorption or the diet decreases basal PTH levels: o, 31 units PTH/kg, 5.00% dietary calcium; •, 31 units PTH/kg, 2.26% dietary calcium; △, control. (From Mueller *et al.*, 1973.)

response in these birds cannot readily be related to an increased activity of the osteoclasts, suggesting that a dual mechanism of action of parathormone may exist. Parathormone preparations when injected have been shown to increase the blood supply to bone and this effect could be important for the rapid and massive mobilization of calcium that occurs during egg-laying in birds. Alternatively an inhibition of the rate of accretion of calcium into bone may be involved (Kenny and Dacke, 1974).

Calcitonin has been identified in the blood of several species of birds including: pigeon, goose, duck, domestic fowl and Japanese quail (Kenny, 1971; Boelkins and Kenny, 1973). The plasma concentrations of calcitonin are much higher in these birds than normally observed in mammals and they can be increased by injecting calcium solutions (Fig. 6.8). In immature Japanese quail, the increase can be as great as 15-fold. It is, therefore, rather surprising that it has not been possible to demonstrate a physiological role of the ultimobranchial bodies in normal calcium metabolism in birds. Injections of calcitonin usually fail to elicit a hypocalcemia but it is possible that this reflects a high basal rate of calcium turnover and a rapid compensatory release of the endogenous parathormone. It thus seems possible that calcitonin could be especially important during egg-laying when it could reduce excessive oscillations in blood calcium levels and

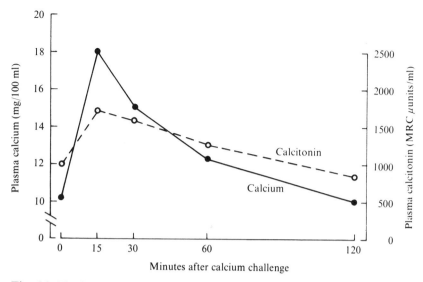

Fig. 6.8. The increase in calcitonin level in the plasma of Japanese quail given intraperitoneal injections of calcium (50 mg/kg). (Modified from Boelkins and Kenny, 1973.)

possibly even influence calcium deposition in the egg. The blood calcitonin levels, however, do not change during the egg-laying cycle of the domestic hen, nor does ultimobranchialectomy significantly influence calcium metabolism in these birds (Speers, Perey and Brown, 1970). In addition, this operation does not significantly influence the development of the skeleton in growing chickens; thus, although some people believe that the ultimobranchial bodies may have a role in regulating calcium metabolism in birds this has not been established. Possibly calcitonin has some other physiological role.

Reptiles

Parathyroidectomy (see Clark, 1972) results in a hypocalcemia and tetanic muscular contractions in several species of snakes and lizards. Such effects are, however, difficult to demonstrate in turtles though a small decline in plasma calcium concentration has been demonstrated following removal of the parathyroids of the Japanese turtle *Geoclemys reevesii* (Oguro and Tomisawa, 1972). It has also been shown, histologically, that in three species of young growing turtles parathormone increases osteocytic calcium mobilization from bones and this effect is inhibited by calcitonin (Bélanger, Dimond and Copp, 1973). It thus seems unlikely that basic differences of a phyletic origin exist between this response in saurian and chelonian reptiles. Rather the variation in the response may reflect the relative ease with which calcium can be normally mobilized and it has been suggested that the bony carapace of turtles provides a calcium store that may facilitate adjustments of the calcium in the blood.

The sites of action of parathormone in reptiles are uncertain. While bone seems to respond, the kidney may not do so. Parathormone stimulates phosphate excretion in the urine of snakes (genus *Natrix*) but neither the injection of this hormone nor parathyroidectomy could be shown to alter urinary calcium loss (Clark and Dantzler, 1972). The latter observation is in contrast to the effect of parathormone in mammals but the experiments should be extended to other reptilian groups before any generalizations as to broad phyletic differences are made.

Despite repeated attempts (on lizards, turtles and snakes), injections of calcitonin have not been shown to exhibit a hypocalcemic effect in reptiles. Although these observations are somewhat unexpected they conform with those in birds which, we are told, are descended from reptiles. Reptiles, like birds, exhibit a hypercalcemic response to injected estrogens (see Clark, 1967).

Reptiles have common phyletic affinities to mammals as well as birds so that their endocrine function is of rather special interest. The reptiles share with birds the problems associated with the production of large cleidoic eggs that in many species are covered with a calcareous shell. It should thus not be surprising to observe similarities in their calcium metabolisms. Reptiles are, however, poikilotherms and this may influence the relative importance of processes that are involved in the metabolic coordination of calcium levels as the speed of the adjustments need not be as great.

Amphibians

The amphibians have a special position with regard to our understanding of vertebrate calcium metabolism, as it is in this group that the parathyroid glands first appear on the phyletic scale. On the other hand, the ultimobranchial bodies have persisted from their piscine ancestors. The information that is available is relatively sparse and is not consistent with the special interest that these animals deserve.

Parathyroidectomy reduces plasma calcium levels in anuran amphibians and this effect seems to be the result of a decreased rate of calcium resorption from bone (see Cortelyou, 1967). Injections of parathormone have a hypercalcemic action. The effects of parathormone on renal calcium excretion are not clear as parathormone, as well as parathyroidectomy, increase urinary calcium loss. It has been suggested that the response to parathormone could be complicated because of the presence of different thresholds in the sensitivity of the kidney and bone, but further observations seem to be needed to clarify this paradox.

Parathyroidectomy has inconsistent effects in urodele amphibians (Oguro, 1973). Hypocalcemia, accompanied by tetanic convulsions, has been observed following removal of these glands in the newt, *Cyanops pyrrhogaster*. The Japanese giant salamander, *Megalobatrachus davidianus*, is, however, unresponsive to parathyroidectomy and this is reminiscent of early observations in three other species of urodeles in which this operation failed to induce tetany. The giant salamander appears to have parathormone present in its parathyroids, as extracts of these tissues have a hypercalcemic effect when injected into parathyroidectomized *C. pyrrhogaster*. It has thus been suggested that this salamander lacks a target-organ system for parathormone and this may also be so in some other urodeles.

Anuran amphibians (*Rana pipiens* has been principally studied) possess 'lime-sacs' that are novel sites for the storage of calcium in the body (see Robertson, 1969*a*, *b*), These organs are extensions of the lymph sacs and they extend caudally along the vertebral canal and emerge between the vertebrae. They contain calcium carbonate, instead of calcium phosphate as in bone, and this exhibits a mobility that includes an added storage following the administration of calcium chloride or vitamin D. Such treatment, that produces a hypercalcemia, also results in the hypertrophy of the ultimobranchial bodies (Robertson, 1968).

Removal of the ultimobranchial bodies in *Rana pipiens* results in an initial elevation in plasma calcium concentration. After about six weeks, however, a hypocalcemia occurs (Robertson, 1969*a*, *b*). It is thought that these responses are the result of an excessive mobilization of calcium from the lime sacs (as well as the bones) which subsequently become depleted. Calcitonin may promote the laying down of calcium at these two sites and it may also reduce the excretion of this ion in the urine.

The lime-sacs are also present in tadpoles where they may contribute to the release of calcium in answer to the needs of metamorphosis. Ultimobranchialectomy in young tadpoles limits their ability to accumulate calcium in the lime sacs whether they are kept in a solution with either a high or a low calcium concentration (Table 6.2). At metamorphosis, the tadpoles in the high-calcium medium can accumulate calcium in their lime-sacs despite the absence of the ultimobranchial bodies. Ultimobranchialectomized tadpoles in the low-calcium medium, however, fail to do this and as a result the bones of the adults are poorly ossified.

Although the number of species studied is limited the Amphibia appear to utilize parathormone, calcitonin and vitamin D_3 for the regulation of calcium metabolism. The sites of their effects, however, are not always clear but in the Anura include an interesting and novel effector, the paravertebral lime-sacs.

TABLE 6.2. *Subjective analysis of amount of calcium carbonate in the paravertebral lime sacs of tadpoles* (Rana pipiens) *tabulated as estimated from X-rays.** (From Robertson, 1971)

Tadpole	Tap water		High calcium	
development stage	Normal	UBX	Normal	UBX
I–V	+	±	+	±
VI–X	++	±	++	±
XI–XVII	++	±	++	±
XVII–XXV	++	±	++	+++

* Concentration of calcium in tap water was 3 meq/l. High-calcium water was 15 meq/l with treatment extended for 6 weeks in all groups except in limb bud stages (I–V), which was for 2 weeks. UBX, ultimobranchialectomized; +, degree of response.

The fishes

A considerable diversity in the mechanisms for the regulation of calcium metabolism is not unexpected in the fishes. This may result from their great phyletic diversity, their ability to live in aqueous environments containing high (sea-water) or low (most freshwater) levels of calcium, the possession of either a bony or a cartilaginous skeleton and the presence or absence of the ultimobranchial bodies and the corpuscles of Stannius.

Paleontological evidence suggests that the ostracoderms, which were the jawless ancestors of modern fishes, had a bony dermal exoskeleton and, sometimes, also a bony endoskeleton that had the appearance of a tissue that functions as a store for calcium (see Copp, 1969). These fishes lived in fresh water where the calcium concentration was, presumably, low so that calcium storage in the bones may have been physiologically important. When such ancestral fishes returned to the sea, where there was an unlimited quantity of calcium, they lost their bony skeleton, and this situation persists in present day cyclostomes and chondrichthyeans. We can extend this speculation (and that is all it is!) and consider whether or not such vertebrates possessed an endocrine system for controlling calcium concentration in the body. In extant species of cyclostomes and chondrichthyeans, there is as yet no evidence for such a control mechanism. In the ancestral freshwater ostracoderms such a control could have been more important and may possibly have occurred. It is even possible that the chondrichthyean ultimobranchial bodies represent a survival from those times. The ancestral freshwater ostracoderms need not, however, have possessed such a system for controlling calcium metabolism as there is, for instance, no evidence for the endocrine control of such processes in some contemporary freshwater fishes like lampreys and the lungfishes (Urist, 1963; Urist et al., 1972).

Many cyclostome fishes live in the sea but others spend their entire life in fresh water. Lampreys migrate from the sea into fresh water where they survive for several months without feeding and produce large numbers of eggs. Cyclostomes are considered to regulate calcium levels by utilizing a so-called 'open system' whereby calcium transfer takes place across the membranes of the body; such as the intestine, gills and kidney. The absence of the endocrine tissues known to influence calcium metabolism in other vertebrates, however, should not be taken to indicate a lack of such control mechanisms. Further investigation of calcium metabolism of these very interesting fish is clearly to be desired.

While vitamin D_3 has been shown to have a role in regulating calcium metabolism in many tetrapods, there is little information to indicate that it also has such a function in fishes. Vitamin D_3 is stored in the liver of teleosts (usually more in marine than freshwater fish) but little or none of this steroid has been identified in the livers of chondrichthyeans or cyclostomes (Urist, 1963). Neither administration of vitamin D_3 to sharks and rays nor vitamin D_3 and 25-hydroxycholecalciferol to the South American lungfish had any effect on their blood calcium levels (Urist, 1962; Urist et al., 1972). There is thus, at present, little to indicate that vitamin D_3 has an endocrine role in regulating calcium levels in fishes, but we should await more extensive experiments before drawing any phyletic conclusions as to the possible evolution of its role as a hormone.

The chondrichthyeans possess ultimobranchial bodies that contain a calcitonin that has a hypocalcemic action, when injected, in mammals. It is unknown what effects, if any, injected calcitonin has in chondrichthyeans but as they lack a bony skeleton, which is this hormone's principal effector in other vertebrates, a response, if present, would be expected to be somewhat different. Possibly calcitonin exerts other actions in chondrichthyeans that may even be unrelated to calcium metabolism. The injection of estrogens also fails to elicit a hypercalcemia in sharks (*Triakis semifasciata* and *Heterodontus francisci*) though a very small (11 %) increase in calcium concentration has been observed in the dogfish *Scyliorhinus caniculus* (Urist and Scheide, 1961; Woodhead, 1969). A hypercalcemic response to the estrogens is also absent in cyclostomes and may not have developed until the emergence of the bony fishes where they produce a 15-fold increase in protein-bound calcium in the plasma in teleosts and lungfishes (Urist et al., 1972).

The bony fishes are a very diverse group and information about the regulation of calcium metabolism is almost entirely limited to teleosts. Within this group few species have been studied but considerable variability seems to exist. One of the reasons for these differences is that although all teleosteans possess a calcified endoskeleton, in some of these fishes, especially marine species, the bone is acellular so that the calcium deposits are relatively immobile. In other teleosts, notably those that can live in fresh water, the bone is cellular and can participate in calcium regulation. Teleosts like the eel also have large stores of calcium present in their muscles; these may be five times as great as those in tetrapods and could provide an additional site for the regulation of calcium exchange.

The injection of calcitonin into teleost fishes may or may not have a hypocalcemic action (see Chan, 1972). In the killifish *Fundulus heteroclitus*, which has acellular bone, calcitonin has no effect on plasma calcium concentration. In the eels *Anguilla anguilla* and *A. japonica*, and the fresh-water catfish *Ictalurus melas*, calcitonin has been reported to decrease the blood calcium concentration but the results are equivocal (see Pang, 1973) (Fig. 6.9). These calcitonin-responsive teleosts possess cellular bone that displays osteocytic and osteoclastic activity and these cells probably are the principal site of calcitonin's action.

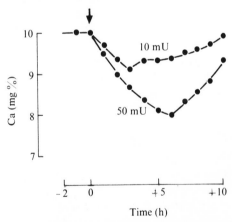

Fig. 6.9. The hypocalcemic effects of injections of two doses (10 and 50 milliunits, mU) of porcine calcitonin in a teleost fish, the eel *Anguilla anguilla*. (From Copp, 1969.)

The classical observations of Rasquin and Rosenbloom in 1954, showing a hypertrophy of the ultimobranchial bodies in Mexican cavefish subjected to continuous darkness, has been described earlier. This change in the activity of the ultimobranchial bodies was associated with osteoporotic degeneration of the bones in these fish.

The corpuscles of Stannius that are present in most bony fishes (see Chapter 2) also may influence calcium metabolism and an interaction between these tissues and the ultimobranchial bodies seems to occur. M. Fontaine, in 1964, found that removal of the corpuscles of Stannius in the European eel *Anguilla anguilla*, results in a marked increase (1.4-fold) in the plasma calcium concentration. This effect of Stanniectomy has been confirmed in other teleosts such as the goldfish *Carassius auratus* and the Asiatic and North American eels *A. japonica* and *A. rostrata* (see Chan, 1972). These hypercalcemic effects are accompanied by a reduced calcium excretion in the urine, and an increased osteoclastic activity which mobilizes calcium from the bone. The hypercalcemic effect of Stanniectomy can be prevented by the transplantation, or the injection of extracts, of the corpuscles of Stannius into the deficient fishes. The hormone has been called 'hypocalcin' which reflects its hypocalcemic action (Pang, Pang and Sawyer, 1974).

The activity (from histological observations) of the corpuscles of Stannius of the killifish *Fundulus heteroclitus*, appears to be greater in normal sea-water than in an artificial sea-water where the calcium concentration is low (see Pang *et al.*, 1973). Stanniectomy in these fish only results in hypercalcemia when they are bathed by solutions with a high calcium concentration; if they are in artificial calcium-poor sea-water or calcium-poor fresh water, blood calcium concentrations are unaffected; thus, the corpuscles of Stannius may play a physiological role in teleosts that live in solutions with a high calcium concentration, such as sea-water, where their secretions exert a hypocalcemic action.

What are the respective roles of the ultimobranchial bodies and the corpuscles of Stannius in teleost fish? Both appear to mediate a hypocalcemic effect! Calcitonin fails to elicit hypocalcemia in eels following Stanniectomy. The absence of its usual effect may reflect a high endogenous release of calcitonin in these fish, in response to the elevated plasma calcium levels, that may mask the action of additional injected calticonin. This possibility is consistent with an observed hypertrophy of the eel's ultimobranchial bodies following Stanniectomy. It is also interesting that ultimobranchialectomy results in an atrophy of the corpuscles of Stannius but it is difficult to interpret this observation which could reflect a direct trophic action on the corpuscles of Stannius, a decrease in their rate of secretion or a release and subsequent exhaustion of this tissue. It is nevertheless noteworthy that an interaction between the ultimobranchial bodies and the corpuscles of Stannius occurs and this is consistent with the possibility that they both interact in the regulation of calcium metabolism in teleosts.

When *Fundulus heteroclitus*, kept in artificial sea-water with a low-calcium concentration, are hypophysectomized they undergo tetanic muscular contractions associated with a considerable decline in the plasma calcium levels (Pang, 1973). This response is not seen if the fish are kept in ordinary sea-water with high concentrations of calcium. The effects in the low-calcium solution can be prevented by injecting the hypophys-ectomized fish with extracts of the pituitary or by transplanting this gland under the skin. The precise nature of the hypercalcemic 'hormone(s)' that may be involved is uncertain but they may be prolactin and corticotrophin. These pituitary hormones may have a hypercalcemic effect in teleosts that live in environments with a low-calcium concentration, like fresh water, while the corpuscles of Stannius may mediate the opposite, hypocalcemic, response in fish living in solutions with a high-calcium level, like sea-water.

The plasma calcium levels in the female plains killifish, *Fundulus kansae*, increase three-fold in the summer compared to the winter (Fleming, Stanley and Meier, 1964). This change in plasma calcium is not seen in the male killifish. In the female, it can, however, be imitated in winter by injecting them with estradiol, which suggests that it is a response associated with breeding. Even the male fish increase their rate of calcium uptake from the external solutions in the summer and in some circumstances this process may be stimulated by injected calcitonin (Fleming, Brehe and Hanson, 1973). The mechanism for the increased accumulation of calcium is not understood.

While knowledge about the regulation of calcium metabolism in fishes is incomplete, several intrinsically interesting facts are known. We have seen that endocrine control of this process may be related to the corpuscles of Stannius which are only present among the Osteichthyes. The secretion of the ultimobranchial bodies, calcitonin, has been shown to exert a hypo-calcemic action which is, however, dependent on the presence of cellular bone and this is similar to its effect in other vertebrates. While chondrich-thyeans possess calcitonin its physiological role is unknown. The estrogens can exert a marked hypercalcemic action, associated with the formation of the egg, and this emerges in the teleosts and is also seen in lungfishes and non-mammallian tetrapods. Almost nothing is known about the regulation of calcium metabolism in sharks and rays while lampreys and hagfishes may be dependent on an 'open system' which does not involve the action of hormones. The regulation of calcium metabolism in cyclostomes and chondrichthyeans undoubtedly will provide a very interesting area in which to pursue this subject further.

Conclusions

Calcium is essential for the life of vertebrates but their requirements for this mineral, and its availability in the environment, vary considerably. This need can be related to vertebrate phylogeny because of the systematic presence or absence of a bony calcareous skeleton and the characteristic life of some groups in the sea which is rich in calcium. It is, therefore, not surprising to observe that the role of hormones in the regulation of calcium metabolism appears to have changed considerably during the course of evolution. In marine fishes that lack a bony skeleton, hormonal regulation of calcium does not seem to occur. When such an endocrine control system is present, as in marine bony fishes, it is concerned with limiting concentration of calcium and lowering its levels in the blood (hypocalcemic effects) and the bone itself is an important effector site. This response possibly involves calcitonin, which is present in all vertebrates except cyclostomes, and, probably in teleosts, a secretion, hypocalcin, from the corpuscles of Stannius. Teleost fishes in fresh water, where the calcium levels are low, may utilize a pituitary hormone, possibly prolactin or corticotrophin, to help maintain adequate concentrations of calcium in their body fluids. Tetrapods, on the other hand, have acquired a 'new' hormone, parathormone, whose role is to mediate a hypercalcemia, and again bone is the major site of its action. A physiological role for calcitonin is doubtful in many species of tetrapods where it may not be as important as it was in ancestral marine fishes. Two other hormones contribute to the regulation of calcium metabolism but apparently in distinct groups of vertebrates, vitamin D has an hormonal function in tetrapods where it facilitates the accumulation of calcium in the body but it is uncertain whether it has such a role in fishes. Estrogens have assumed a 'special' endocrine role in many 'bony' vertebrates where they assist the deposition of calcium in the developing egg; apart from chondrichthyeans and cyclostomes this effect is also absent in mammals.

7. Hormones and the integument

The skin and gills of vertebrates constitute the major external interface between the animal and its environment. This integument is physiologically and anatomically a very important tissue which exhibits considerable diversity reflecting the differences which exist in the physicochemical gradients between the vertebrates and their environments. The integument may thus play a role in the animal's osmoregulation, thermoregulation and respiration. In addition, the integument provides signs and signals that can promote social and sexual contact and can help the animal to blend in with its surroundings and so protect it from predators, or help it catch its food. Of primary importance is the skin's role as an integumental skeleton by which it contains the animal in a condition that facilitates its locomotion. The relative importance of these various roles of the integument varies in different species and the structure varies accordingly also.

In fishes and larval amphibians, the gills, which function as organs of respiration, make up a large part of the animal's external surface. Exchanges of oxygen and carbon dioxide readily occur across these highly vascularized tissues which are also the sites of considerable movements of water and salts. Many fishes contain special cells in their gills called 'chloride-secreting cells' (or 'ionophores') that are the site for active extrusion of salts. The endocrine control mechanisms influencing the permeability of the gills are described in Chapter 8.

The skin is the major non-branchial interface between the animal and its environment. In its simplest form, the skin consists of two major layers of tissues, an outer epidermis which has several strata of cells, and an inner dermis. However, such a simple arrangement does not exist in nature, as various other structures are also included in the skin that modify its properties. These structures include scales, hair, feathers, pigment cells, secretory glands and certain sense organs. Such accessories contribute to the particular physiological properties exhibited by the integument of each species.

The skin is thus a complex tissue that has different physiological needs that depend on the species, the environment and the stage of the animal's life cycle. The constitution of the skin is not static but undergoes continual change commensurate with the normal needs of growth and repair. In addition, rapid changes in the physiological properties of the skin also can

occur, such as involve an increased blood supply and the secretion of sweat
in response to the need to dissipate heat in the mammals, and an increased
osmotic permeability to water, as a result of dehydration, in anuran
amphibians. Many cold-blooded vertebrates can rapidly alter the distribution of pigment in the skin so that they blend more closely with the shades
and hues of their surroundings. Seasonal changes commonly occur in the
integument, such as the changes in pigmentation that may be associated
with breeding and alteration of the color, length and density of fur and
feathers in summer and winter. Such changes in the fur and feathers may
alter the insulative properties of the integument and contribute to the
animal's camouflage.

While the skin has a considerable innate ability to regulate its functions
it is also dependent on the nervous and endocrine systems with whose aid
it can coordinate its activities with the rest of the body. The skin has a
plentiful nerve supply that mediates its sensory functions and regulates its
blood supply. The secretions of cutaneous glands are also predominantly
under neural control, though circulating catecholamines exert some effects
on them. Rapid changes in the distribution of pigment may be controlled
by nerves but hormones are also very important. The endocrines help in the
maintenance of the nutritional and anatomical integrity of the skin as well
as such processes as molting, pigmentation and the function of certain
cutaneous glands. Hormones that influence cutaneous function include
several from the pituitary such as prolactin, MSH, vasotocin, ACTH, LH
and TSH, and also thyroxine, the catecholamines, corticosteroids, gonadal
steroids and melatonin. Some of the actions of these hormones are confined
to relatively few species while the effects of a hormone on the skin may be
quite different in one species as compared with another. The variation that
is observed in the cutaneous effects of particular hormones suggest that
considerable evolution has occurred in their special roles. It is also often
difficult to decide whether the actions following an excess or deficiency of a
hormone are the result of its direct action on a specific cutaneous effector
or are due to merely a more diffuse, indirect effect such as may result from
general changes in the animal's physiological and nutritional status.

Hormones and molting

The epidermis is regularly renewed as its outer layers drop off and are
replaced by new cells that are formed from the underlying epithelium. This
may be a more or less continuous process, such as is common in mammals,
or it may take place suddenly at regular intervals varying from a few days,
as in many amphibians, to several months in certain lizards and snakes.
The hair of mammals and feathers of birds are also subject to such periodic

renewal and this may also occur at precise times of the year such as at the onset of winter or spring or just before, or after, the breeding season (pre- and post-nuptial molts). In reptiles the shedding of the epidermis is often called *sloughing* while the shedding of the pelage in mammals, the plumage of birds and the epidermis in amphibians is called *molting*.

It is generally considered that the regular cyclical molting that occurs in fish, reptiles and amphibians reflects an autonomous rhythm in the skin upon which the actions of hormones can impinge in a permissive manner (Ling, 1972). The seasonal molts that occur commonly in birds and mammals are more closely allied to the external stimuli, principally the photo-period, but they are also modified by the external temperature and the nutritional condition of the animal.

The pituitary and the thyroid glands are the principal endocrines that influence molting in vertebrates.

Removal of the pituitary usually prevents, or considerably prolongs, the length of the reptilian and amphibian molting cycles and blocks the seasonal molts observed in many birds and mammals. This effect of hypophysectomy is the result of the absence of several hormones, a lack of TSH, with its trophic effect on the thyroid, is very important but the lack of prolactin and corticotrophin may also contribute to the debility. In lacertilian reptiles, urodele amphibians, birds and mammals the thyroid hormones accelerate the molting process; thyroidectomy has an inhibitory effect. It is interesting that in ophidian reptiles (snakes) removal of the thyroid results in a decrease in the length of the sloughing cycle which is in direct contrast to what is observed in their lacertilian relatives (Chiu and Lynn, 1972; Maderson, Chiu and Phillips, 1970). Differences in the effects of hormones on molting also occur within the Amphibia, for while this process is facilitated by the thyroid gland in *Ambystoma mexicanum* (Urodela) this is not so in *Bufo bufo* (Anura) (Jorgenson, Larsen and Rosenkilde, 1965). In the latter toads, however, corticotrophin and the corticosteroids (corticosterone is most active) are necessary for successful molting. To further complicate any attempts to define a phyletic uniformity, it has been found that cortico-trophin completely *inhibits* sloughing in the lizard *Gekko gecko* (Chiu and Phillips, 1971*a*).

Prolactin has diverse actions in vertebrates and it is especially notable that many of its effects are on the integument or derivatives of it, most notably the mammary glands (which are merely modified sweat glands). It has been shown that the injection of prolactin decreases the length of the sloughing cycle in the lizard *Anolis carolinensis* (Maderson and Licht, 1967). As thyroxine has a similar effect and prolactin has been shown to exhibit a thyrotrophic action, it seemed possible that the action of prolactin was not direct, but was mediated by its release of thyroxine. However, it has been shown that prolactin can reduce the length of the sloughing cycle in thyroidectomized lizards, *Gekko gecko* (Chiu and Phillips, 1971*b*) so that it may have a direct effect on the skin. Prolactin has been found to facilitate the growth and increase the appetite of lizards so that its effect could also reflect their nutritional condition. In this respect, it should also be remembered that hypophysectomized lizards usually do not eat and are not in perfect health. Prolactin injections have also been shown to accelerate molting in a urodele amphibian, the red eft, *Notophthalmus viridescens* (Chadwick and Jackson, 1948), though such an action has not been demonstrated in other amphibians. It is interesting that, in this newt, prolactin promotes the transition (or metamorphosis) from the terrestrial form into the aquatic breeding stage when it returns to water ('water-drive effect' of prolactin) and this is associated with cutaneous changes. Whether the effect on the skin is a primary one is uncertain, as the prolactin could be stimulating some non-cutaneous process concerned more generally with the metamorphic change.

The seasonal changes that occur in the pelage of mammals and the plumage of birds appear to be principally under photoperiodic control. Changes in the hours of light are transmitted via the eyes and hypothalamus to the pituitary. Such photoperiodic changes also control the gonadal cycles so that it may be difficult to separate the two events. Removal of the pituitary prevents the short-tailed weasel (*Mustela erminea*) from growing a brown spring coat (they stay white), even when they are exposed to a photoperiod that induces this growth in intact weasels (Rust and Meyer, 1969). The effects of photoperiod on the pelage and plumage appear to be mediated through the action of the gonadal steroids, corticosteroids and thyroxine which are in turn controlled by the hypothalamus and pituitary (Ebling and Hale, 1970).

The precise manner in which hormones influence molting is not understood. Cyclical changes in molting in poikilotherms are usually characterized by brief periods of cellular activity and rapid cell division interspersed by periods when little activity occurs, which are referred to as the 'resting-phases'. It is thought that, when they are acting, thyroxine and prolactin shorten the resting-phase (except in snakes!) during which the skin's activity is reduced by lower levels of these hormones. In newts, however, prolactin has been shown to promote active cell division in the epithelium and this is reminiscent of its action on the crop-sac epithelium in pigeons. In toads, the absence of the pituitary does not prevent the formation of new layers of epidermis (or sloughs) but prevents the shedding, or casting off, of these cells. This shedding is promoted in toads by corticotrophin and corticosteroids. It is interesting that some frogs which estivate during periods of drought are protected from excessive dehydration by a cocoon which is composed of accumulated layers of epithelial cells (Lee and Mercer, 1967) and it seems possible that this may reflect a decline in their pituitary and adrenal function.

While not strictly an example of molting, balding (loss of the hair on the scalp) in man is known to be dependent on the activity of androgenic hormones. The sex hormones also control the development at puberty of the typical patterns in distribution of hair in men and women. A rather interesting observation (Fig. 7.1) is that the growth of the facial hair in man shows a rapid increase when female company is restored following a period of abstinence. This change probably reflects a surge in the release of LH and the secretion of testosterone.

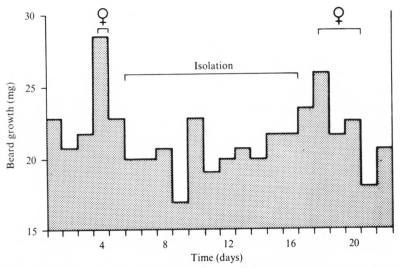

Fig. 7.1. Beard growth in man is less during periods of isolation from women. Restoration of female company results in a sudden 'spurt' in the growth of the facial hair and it has been suggested that this is the result of the release of male sex hormones, probably pituitary LH. (From Anon., 1970.)

Hormones and skin glands

Vertebrates possess several types of secretory glands in their skin which serve a variety of functions (Quay, 1972). These glands can be classified into two major groups: the mucous glands and the proteinaceous glands.

The proteinaceous-type glands have undergone considerable evolutionary modification and include (to name only a few) a variety of venom glands in fishes and amphibians, the uropygial (or preening) gland that is present in many birds and the sebaceous glands, sweat glands and mammary glands of mammals. The sweat glands play an important role in temperature regulation in mammals. The sebaceous glands are associated with hair follicles and the fatty sebum that they secrete serves to protect the hair from wetting. The sebaceous glands and the sweat glands sometimes secrete special odoriferous substances that may play an important role in terri-torial behavior (by defining territorial limits) and also act as sexual attractants. Such scent glands may become enlarged and congregate in distinct areas of the body. Examples of this include the 'side glands' on the heads of shrews, the 'anal glands' and submandibular 'chin glands' in rabbits and the 'ventral gland' in gerbils.

The maturation and function of the sweat glands and sebaceous glands in mammals are influenced by hormones (Strauss and Ebling, 1970). The odoriferous scent glands are commonly observed to be larger in the male than the female while maturation of sebaceous and sweat glands occurs during puberty in man. It thus seems likely that they are influenced by sex hormones, and androgenic steroids are undoubtedly involved. Natural

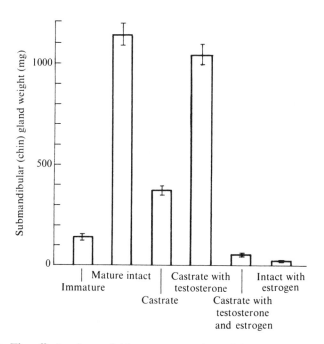

Fig. 7.2. The effects of gonadal hormones on the weight of the submandibular (chin) glands in male rabbits. It can be seen that these glands exhibit considerable increases in weight after sexual maturity, an effect that is prevented by castration. Testosterone injections overcome this effect of castration while injected estrogens bring about an involution of the glands. (From Strauss and Ebling, 1970.)

and experimental differences in the development of the submandibular chin glands in rabbits are shown in Fig. 7.2. These glands are much larger in sexually mature than immature rabbits and they decrease in size when the testes are removed. Injections of testosterone promote the development of the chin glands while estrogens inhibit it. Progesterone also increases their weight but this may be due to the inherent androgenic activity of this steroid hormone. Comparable effects of the gonadal steroids are seen on the sebaceous and apocrine sweat glands of other mammals. Apart from promoting the development of sebaceous glands, androgens also increase the production of sebum.

In primates, especially man, the watery-secretion of the eccrine sweat glands play an important role in providing water that can be evaporated from the body surface and so aid in the dissipation of heat (see Chapter 8). Sweat contains dissolved salts, including sodium and potassium, the concentration of which is influenced by aldosterone; the sodium/potassium ratio declines under its influence. In addition, while the secretion of sweat is primarily under neural control in primates it is also increased, during exercise, by circulating adrenaline.

Mucous glands are present in the integument of fishes and amphibians. Their role is contentious but seems to be related to an aquatic life where it has been suggested that the mucous secretion may have a protective action on the skin as well as serving certain special functions such as the formation of a cocoon in estivating African lungfish and providing food for the young of a cichlid teleost. In the latter fish, this mucous secretion is promoted by the injection of prolactin (Egami and Ishii, 1962). It has also been suggested that prolactin may limit the permeability of teleost fish and urodele amphibians to water and sodium (see Chapter 8) by an action on the mucous glands but the evidence for this is equivocal.

Knowledge about the role of hormones in regulating growth, development and secretion of the integumental glands is incomplete. Clearly, however, some such glands do respond to hormones though the primary importance of neural secretory stimuli should not be forgotten. The functions of such glands vary considerably in different vertebrates and it is interesting that many have attained a responsiveness to certain hormones. The nature of these effects appears to be related to the other basic functions of the hormones in the body; thus gonadal steroids influence skin glands that are involved in sexual activities and aldosterone acts in a manner commensurate with its role as a hormone that conserves sodium in the body.

Information about hormone effects on skin glands in non-mammals is sparse. Adrenaline stimulates chloride secretion from the skin glands of European frogs and this may reflect a direct action of such hormones in the body or a mimicking of a stimulation of the sympathetic nerves.

Adrenaline, as well as vasotocin, has also been shown to stimulate the secretion of a sticky, milky-white material from the proteinaceous skin glands of the South African clawed toad, *Xenopus laevis* (Ireland, 1973). The effect of the catecholamine, but not vasotocin, can be prevented by an α-adrenergic blocking drug.

The control of epidermal skin proliferation – a note on chalones

The control of cell proliferation in the skin has elicited much interest, especially among dermatologists. Apart from the control of the normal replacement of the epithelium, the rapid but controlled processes of repair that accompany wound healing are rather remarkable. These processes, like other changes in the skin, are thought principally to involve an internal control mechanism upon which hormones may impinge their influence.

Chalones are substances that have been isolated from a number of tissues including the mammalian epidermis (see Bullough, 1971). They inhibit the mitoses of epidermal cells including melanocytes and keratinocytes. The chalones are glycoproteins that act both *in vitro* or *in vivo* and show a considerable degree of tissue specificity. It is thought that chalones form an active complex with adrenaline. As cortisol also inhibits mitosis of epidermal cells it may also contribute to this complex. Alternatively, it has been suggested (see Quevedo, 1972) that the effect of adrenaline, in inhibiting cell mitosis, may involve the formation of cyclic AMP and that chalones may be part of the cyclic AMP–adenyl cyclase system. The promotion of rapid epidermal cell growth, such as following a wound, appears to be stimulated as a result of the inactivation of the chalone but it is unknown how this is done. It has been suggested that a local release of prostaglandins may overcome the inhibitory effects of the chalone complex.

Hormones and pigmentation

The integument of most vertebrates contains pigment that makes a major contribution to what is often a very colorful appearance. Pigment may be present within the epidermis or dermis itself or color the integumental appendages, such as scales, hair and feathers. Apart from contributing to man's esthetic delight in contemplating nature, it seems likely that an animal's coloration may be useful to its physiology (Hadley, 1972). Appropriate pigmentation may contribute to the animal's camouflage, protect the internal organs from solar radiation, promote the absorption or reflection of heat and light and so aid in photoreception and contribute to the synthesis of vitamin D in the skin. Integumental colors also provide signs that are important for appropriate dimorphic sexual behavior and reproduction.

Pigments of different colors are usually present in the skin in cells called chromatophores. These cells commonly contain a black or brown pigment called melanin and are called melanocytes or, if the intracellular distribution of pigment can be changed, melanophores. The yellow and red pigments (xanthines and carotenes) that also occur in the skin of vertebrates are contained in, respectively, xanthophores and erythrophores. Some chromatophores also contain pteridine platelets which reflect light, giving an iridiscent appearance, and are thus called iridophores. The complex and beautiful colors of many vertebrates are the result of blending the colors reflected by the various chromatophores.

Many vertebrates can alter their coloration in response to environmental and behavioral needs. Such changes may take place in a relatively slow manner, as when the total amount of pigment in the epidermis, or its appendages, changes. The result is a relatively static coloration that is attained over a period of days or weeks. This process is called *morphological color change* (which may involve melanocytes or melanophores) and is seen when we tan in the sun or when an animal changes the color of its pelage or plumage with the onset of summer or winter or in preparation for the breeding season. In addition, many cold-blooded vertebrates can rapidly change their color, a process that only takes a few minutes or at the most several hours. This relatively rapid response is called *physiological color change*. Both morphological and physiological color change are influenced by the actions of hormones, especially the pituitary melanocyte-stimulating hormone.

The melanophores are cells with long dendritic-like extensions that radiate from a central core. In shape they resemble nerve cells from which they are derived. The melanin is contained within cellular organelles called melanosomes. Darkening and lightening of the skin, as occurs in physiological color change, reflects a migration of the melanosomes in dermal melanophores so that they are widely distributed in the cell (dark color, the melanin is said to be dispersed) or they aggregate in small globs in the center of the cell (light color, the melanophore is said to have a punctate appearance) (Fig. 7.3). The more gross effects of these changes on the color of frog skin is shown in Fig. 7.4. The dispersal of the melanin in the melanophores may depend on a microtubular system in the cell, as certain drugs, for example cytochalasin B, that break such tubules also prevent dispersion of the pigment. The other chromatophores have a rather similar structure to the melanophores but contain different pigments. Iridophores that respond to MSH do so in the opposite manner to that of the melanophores; the platelets of reflecting materials aggregate so that the cell has a punctate appearance. Not all chromatophores exhibit a physiological color change response to MSH and indeed this is not usually seen in the xanthophores and erythrophores (Bagnara, 1969; Taylor and Bagnara, 1972). In epidermal melanocytes (unlike the dermal melanophores), the pigment is relatively fixed in its position so that differences that occur usually reflect

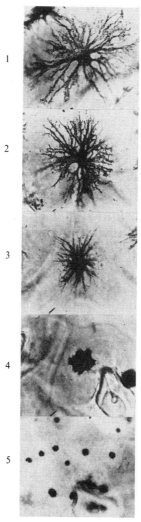

Fig. 7.3. The microscopic appearance of the dermal melanophores of the dogfish *Scyliorhinus canicula*. When the fish is maximally dark, as in 5, the melanosomes are dispersed throughout the cell which can then be seen in outline, while when pale they are aggregated in the central regions (as in 1) so that definition of the cell outline is obscured. The numbers, 1 to 5, correspond to the 'melanophore index'. (From Wilson and Dodd, 1973*a*.)

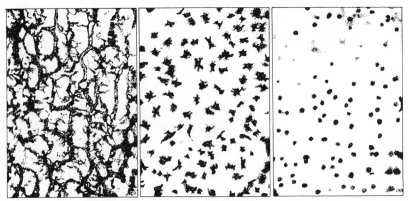

Fig. 7.4. The gross appearance of amphibian melanophores (under a low power microscope) when the animal is (left to right): dark, intermediate and pale in color. (From *The Pigmentary Effector System* by L. Hogben (1924). With permission of the publishers Oliver & Boyd, Edinburgh.)

the total quantities of pigment that are present (morphological color change).

The different types of chromatophores present in colorful animals, including many frogs and lizards, are arranged in layers and changes in the distribution of pigment within these zones alters the transmission of light and the color that is perceived. The innermost layer consists of melanophores and these are overlain by the reflecting iridophores. The xanthophores form a layer closer to the surface of the skin. Changes in the density of the melanophores, that absorb light, and the iridophores that can reflect it, thus can alter the color of a frog from a light to a dark color and influence the display of the colorful pigments in the superficially placed xanthophores. The melanophores, iridophores and xanthophores together make up what is called a 'dermal chromatophore unit'.

The mechanism of release of melanocyte-stimulating hormone

As we shall see in the succeeding sections, the release of MSH plays a most important role in both physiological and morphological color change in vertebrates. The principal stimulus, especially in cold-blooded species, is the receipt of light usually by the lateral eyes but the pineal may also function as a photoreceptor in some species. In addition to directly influencing MSH release, in acute situations, light may also contribute to a cyclical, photoperiodic, release of this hormone in some mammals that seasonally change the color of their pelage. Other stimuli that result in a release of MSH include a deficiency of adrenocorticosteroids and suckling and copulation in mammals, and increases in the osmotic concentration of the plasma in a variety of species. The latter effects may not have any physiological significance but this remains to be further explored. The intimate mechanisms that control the release of MSH from the pars intermedia are only partly understood and appear to be quite complex (see Howe, 1973).

When the pars intermedia is transplanted ectopically, to another part of the body away from the hypothalamus, or if its connections to this part of the brain are severed, MSH is secreted in an apparently uncontrolled manner. The regulation of hormone release thus appears to be under an inhibitory control originating in the hypothalamus.

The pars intermedia has a nerve supply that comes from the base of the brain which contains three types of neurons; aminergic ones that secrete catecholamines, neurosecretory fibers that release peptide hormones and (possibly less important) cholinergic nerves. The vascular supply to the tissue shows considerable species variability but portal vessels coming from the hypothalamus and neural lobe have been described. The controlling stimuli appear to involve the nerves but also possibly portal blood vessels, as in the pars distalis.

Neural stimuli are undoubtedly important in controlling the release of MSH and appear to involve α-adrenergic-type receptors which inhibit the release of MSH. In-vitro studies on rodent and frog pituitaries also indicate that a β-adrenergic-type stimulation can oppose this and prompt MSH release (Bower and Hadley, 1973; Bower, Hadley and Hruby, 1974). As we have seen earlier (Chapter 3), an MSH-release-inhibiting hormone, MSH-R-IH, has been identified in the hypothalamus and this seems to contribute to the control of the hormone's release. MSH-R-IH is a peptide, such as formed by neurosecretory cells, but it is unknown whether it is released directly from such nerve endings in the pars intermedia or is formed in the hypothalamus and carried in portal vessels to the gland. In addition, an MSH-releasing hormone (MRH) may also be present in the hypothalamus, and this could be opposing the action of MSH-R-IH and/or inhibitory neural stimuli. The interactions and possible linkages (for instance do α-adrenergic stimuli act alone or through MSH-R-IH release) between these neural and humoral processes are not yet fully understood. To complicate matters even further, it has been proposed (see p. 268) that melatonin, from the pineal gland, may mediate photoperiodic release of MSH by an action on the hypothalamus. Melatonin injections have been shown to increase release of MSH and a physiological rise in the levels of the former, such as occurs during darkness, could result in the latter's release under normal conditions. Finally, it should be noted that there is evidence to indicate that MSH can exert a negative-feedback inhibition via the hypothalamus which restricts its further release.

Physiological color change

Physiological color changes occur in many cold-blooded vertebrates, from the cyclostomes to reptiles. These changes in the distribution of the pigment in the skin occur in response to a variety of conditions and stimuli. Many vertebrates exhibit a diurnal rhythm in the degree of aggregation and dispersion of melanin in the melanophores. They turn pale at night and dark during the day. This change may reflect the perception of light and be mediated by receptors in the eyes and the pineal, and sometimes can result from a direct stimulation of the melanophores by light. In other instances such as in the lizard *Anolis carolinensis*, a diurnal rhythm can even be seen when the animals are kept in complete darkness. As it is not seen in these lizards after they are hypophysectomized, it probably reflects an inherent diurnal rhythm in the activity of the pituitary gland. Superimposed on such rhythmical changes in skin color are direct, and adaptive, responses to external stimuli. These stimuli include the perception of certain light patterns due to the color and shade of the substrate on which the animal is placed (*background-response*), and to a lesser extent the external temperature and 'excitement'. The latter two effects, which have been observed more commonly in lizards, may override the background-response.

The first recorded observations of physiological color change are more than 2000 years old but our understanding of the mechanism involved is quite recent. An appreciation of the role of hormones in these responses principally resulted from the pioneering studies of Hogben and Winton in the 1920s. An excellent account of the work of the Hogben school in England and that of many others, including Parker in the United States, has been given by Waring (1963). Waring joined the Hogben school in the 1930s and his account of the processes involved in regulating color change is an ideal example of the stringent analytical approach and the application of formal logic that we should all aspire to in scientific investigations. Although physiological color change does not occur in mammals or birds the elucidation of its mechanism has contributed a great deal to our understanding of the role of hormones in physiological coordination. The following account is largely based on Waring's but one should also consult the book by Bagnara and Hadley (1972).

Types of melanophores responses

Non-visual

(*a*) Coordinated; these responses may be abolished by denervation of the skin or the removal of the pituitary or the adrenals.

(*b*) Uncoordinated; where the melanophore (or possibly a skin-receptor close to it) directly responds to a stimulus. This type of response can be seen rather clearly in the horned toad, *Phrynosoma blainvilli.* If these lizards are blinded, hypophysectomized, and the pineal eye is covered, and they are then placed in a black box with no light, they become a pale color. When, however, a thin beam of light is focused on a piece of denervated skin in these lizards this darkens in comparison to the rest of the integument. A localized response to temperature can also be demonstrated in *Phrynosoma*, for when an area of the skin of a maximally dark lizard is exposed to water at 37 °C it pales in that region. Similarly, maximally pale skin will darken locally at a temperature of 1 °C. Chameleons also exhibit dramatic localized changes in skin color; the skin of blinded animals turns dark in light but if a certain area is shaded by an object a lighter colored 'print' or outline of this object can be seen. Such responses do not involve hormones or the ordinary nerve supply and appear to reflect a direct response of the melanophore; however, a local nerve reflex initiated from a nearby cutaneous receptor or a release of a local hormone could be involved.

The visual response

This response is the result of the reception of light by the lateral eyes or in certain species, including some larval amphibians and cyclostomes, and possibly even some lizards, the pineal. The responses may be a generalized lightening (in the dark) or darkening (in the light) of the skin or be influenced by the color of the background: the background-response. When the animal is on a white substrate with overhead illumination, it may turn a pale color and if on a black background (also with overhead illumination) it may turn a dark color. These changes are called the white (or tertiary) and black (or secondary) ocular-background responses. The different effects of light in these two sets of circumstances appear to be due to the stimulation of different parts of the retina; thus, the retina of a frog in a black tank of water (Fig. 7.5) receives light only on the more basal parts of the retina, the 'B' (for black) area, but in a white tank, where the light is reflected into the eye from all sides, the entire retina, including a 'W' (for white) area, is stimulated. Such special receptor areas for light in the retina have also been found in teleost fishes and lizards.

Quantitation of the melanophore response

Early observations of vertebrate color change have been described in general subjective terms such as 'pale', and 'a tint rather dark than pale' that lack adequate precision for a proper scientific analysis and make

(*a*)

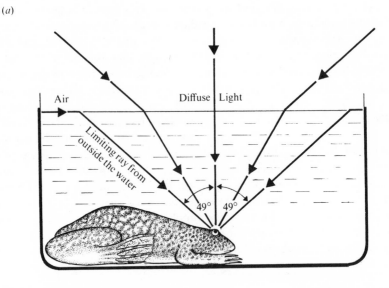

(*b*)   (i)   (ii)

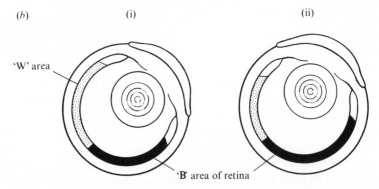

Axis of symmetry 35° to horizontal   Axis of symmetry 65° to horizontal

Fig. 7.5. (*a*) The manner by which the reception of light initiates the dispersion of melanin in the melanophores of a frog (in this instance *Xenopus*) sitting in a tank of water with a black background and overhead illumination.

As the tank has black sides all light will enter the water from above and enter the lens at an angle which is the critical angle for air and water (49°). (From Waring, 1963.) Using this data, as well as the dimensions of the eye and the refractive index of the lens (*Xenopus* has its eyes on the top of its head), Hogben provided a diagram (*b*, i, ii) that shows the area of the retina which receives such light rays. As these conditions result in a darkening of the skin, this has been called the B-area (for black) as opposed to the W-area (for white) which initiates skin lightening in frogs on a white background when light reaches wider areas of the retina. (From Hogben, 1942. Reproduced by permission of the Royal Society.)

comparisons of results from different laboratories almost impossible. Hogben introduced a more stringent quantitative description called the melanophore index (or MI) (Figs. 7.3 and 7.6) which has a gradation of 1, for maximally pale, with the melanin fully aggregated to 5 when the animal is dark and the melanin is fully dispersed. In lizards, this can be translated, as in chameleons, to 1 = yellow, 3 = medium-green and 5 = black. This simple standard of measurement allowed considerable advances to be made in the analysis of the mechanism of color change. Today, electrophoto-receptive devices are also used to quantify the melanophore responses.

Color change in amphibians

The earliest observations on the role of hormones in vertebrate color change were made on European frogs, *Rana temporaria*, and subsequently the

	Appearance of melanophores	Melanophore Index (MI)
Intact Anurans		
In complete darkness		2.5 to 3.0
Light overhead:		
White background		1.5
Dark background		4.5 to 5.0
'Complete' hypophysectomy:		
(light or dark background)		1.0
Denervation of pars intermedia		
(light or dark background)		4.5
Eyeless (light or dark background)		2.5 to 3.0

Chart of Melanophore Index for amphibians:

Fig. 7.6. The melanophore responses, mediated through the eyes, of anurans in relation to the receipt of light when on a white or a dark background. The effects of surgical changes of the pituitary on these responses have been summarized. The melanophore index (MI) in relation to the degree of dispersion, or aggregation, of melanin in the melanophores is given in the lower section. For a description and explanation of these responses the text should be consulted. (Based on Bradshaw and Waring, 1969.)

South African clawed toad, *Xenopus laevis*. With overhead illumination (see Fig. 7.6), on a white background these amphibians are pale (melanophore index about 1.5) and on a black background they are dark (MI about 4.5). When placed in complete darkness, or if they are blinded, they have an intermediate shade (MI = 2.5). In *Xenopus*, this change in melanin distribution in the melanophores is seen as a white or a black coloration but in frogs, which have overlying layers of yellow-green pigment, this appears as a pale green or yellow to a black color. When amphibians are 'completely' hypophysectomized so that no pituitary tissue remains (such remnants commonly *do* remain, as in *Xenopus*) the animals become maximally pale, the MI is 1 and they cannot respond to changes in the background color. If the pars distalis is removed carefully so that the pars intermedia and pars tuberalis remain intact, the background responses are retained. This operation is relatively simple to perform in *Rana* but it is more difficult in *Xenopus* where the pars tuberalis is usually removed together with the pars distalis. This results in an inability of *Xenopus* to display a background response and it becomes permanently dark (MI = 5). The pars intermedia has a nerve supply coming down from the hypothalamus and when this is cut the anurans also have an MI of 5. Removal of the pars tuberalis appears to be associated with an interference of the hypothalamic connections to the pars intermedia and, as these are of an inhibitory nature a sustained release of its secretion, MSH, occurs.

Melanocyte-stimulating hormone (MSH) when released from the pars intermedia is carried in the blood to the melanophores where it promotes a dispersion of melanin so that the animal darkens.

The sequence of events resulting in the black-background response is summarized in the following section (Fig. 7.7). Light from an overhead source, in frogs on a dark background, falls on the 'B' area of the retina where it stimulates receptors that transmit messages along the optic nerve. These messages, travelling along pathways that are as yet unknown, inhibit the normal inhibitory effects of the nerves supplying the pars intermedia and this results in a release of MSH. Whether or not this involves just neural stimuli, a release of MSH-RH and/or an inhibition of release of MSH-R-IH is not clear. When the frogs are on a white background, light also falls on the retina but on the receptors in the 'W' area and this reduces the release of MSH. Normally anurans kept in the dark, as well as blinded animals, have an MI of about 3 which appears to reflect a sustained, but submaximal, release of MSH. Stimulation of the 'W' retinal receptors in some way inhibits this release even further, possibly by increasing the inhibitory nerve impulses or stimulating the release of MSH-R-IH or both.

Pallor of the skin is thus usually thought to result from a decline in the levels of MSH in the blood. In the amphibia, in contrast to some teleosts

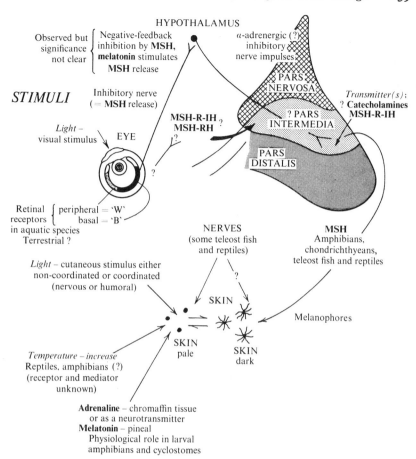

Fig. 7.7. A composite diagram summarizing the humoral and neural control of color change in vertebrates.

and reptiles, nerves are not involved in the aggregation of melanin in the melanophores. The possibility that a melanin-aggregating hormone or MAH may be present in the amphibian pituitary has been considered. This concept was partly based on the observation that an intact pars tuberalis was necessary for the occurrence of a white-background in *Xenopus*. It is now, however, agreed that removal of this tissue results in the severance of the inhibitory hypothalamic connections to the pars intermedia that are necessary for maximum color change to occur (see Bradshaw and Waring, 1969).

The injection of adrenaline into *Rana* produces a skin lightening and an aggregation of melanin in the melanophores which is an α-adrenergic response. On the other hand, in *Xenopus laevis* and in the spade foot toad, *Scaphiopus couchi*, *in vitro*, catecholamines disperse melanin which is due to an increased formation of cyclic AMP following β-adrenergic stimulation (Goldman and Hadley, 1969; Abe *et al.*, 1969). The normal physiological importance of such actions *in vivo* is unknown.

When dried bovine pineal glands are fed to anuran tadpoles, the melanin in the melanophores on the body, but not the tail, aggregates and the animal's body pales (see Bagnara and Hadley, 1970). Under these conditions, the tadpole's internal organs can be seen clearly. This effect is due to the action of melatonin that is formed in the pineal gland. Normally, tadpoles, such as those of *Xenopus laevis*, pale at night and darken during the day and this can be prevented if the pineal, but not the lateral eyes, are removed. Formation and release of melatonin occurs in darkness and this appears to mediate the diurnal rhythm of color change in these tadpoles. The effect of melatonin on the melanophores is a direct one and is not mediated through the pituitary as has sometimes been suggested, as the effect of melatonin is not prevented by hypophysectomy. This physiological effect of melatonin is confined to tadpoles and does not contribute to skin lightening in adult amphibians.

Color change in the chondrichthyes

Many sharks and rays exhibit dramatic changes in color depending on the shade of the background; with overhead illumination they become dark on a black background and pale on a white one. Two dogfish (*Scyliorhinus canicula*) in their dark and pale phases are shown in Fig. 7.8. Waring, in 1936, found that when he transplanted a dogfish pituitary into another dogfish, that was pale in color, it turned dark due to a dispersion of melanin in the melanophores. The release of the MSH is due to the absence of the hypothalamic neural inhibitory control mechanism which is present in these fish (Wilson and Dodd, 1973a). An analysis of the color change in these fish, *Squalus*, *Scyliorhinus* and *Raja*, shows that they exhibit white- and black-background responses that are mediated humorally by MSH, just as in amphibians. Direct neural control of the melanophores does not appear to occur in the Chondrichthyes.

The background response is not seen in blinded dogfish though the fish show a slight paling in darkness which suggests the presence of a non-visual response (Wilson and Dodd, 1973a). When kept in total darkness, the pallor exhibited by these dogfish is not seen if the pineal is removed and they become darker. The pineal may thus contribute to non-visual color change as observed in tadpoles and cyclostomes (see later).

Fig. 7.8. Two dogfish, *Scyliorhinus canicula*, in their dark and pale color phases (×0.33). (J. F. Wilson, personal communication.)

Color change in teleosts

While the chondrichthyean fishes and amphibians that have been examined all have humoral control of their color change the teleosts, which lie phyletically between these two groups, may also possess a neural coordinating mechanism. The teleosts, as has become apparent in the comparison of their other biological systems, exhibit considerable inter-specific differences in the control of color change that presumably reflect the systematic diversity within this large group of fishes.

Stimulation of nerves controlling melanophores usually results in an aggregation of the melanin and a paling of the skin color in teleosts. A dispersion of melanin, in response to neural stimuli, however, may also occur in some species but the evidence for this is equivocal. I have been unable to ascertain with certainty whether the autonomic nerve fibers involved in such color changes are cholinergic or adrenergic, though the aggregating effects of injected adrenaline would tend to favor the latter; however, species differences may exist and it has been suggested that in the eel *Anguilla anguilla*, melanin dispersion is an adrenergic response while the aggregation is a cholinergic one. Alternatively, melanin aggregation may be an α-adrenergic response and dispersion a β-adrenergic one.

Responses of melanophores to neural stimuli are very rapid and may take place within several minutes. The humoral effects of MSH are, in contrast, slow and usually take one to two hours for their completion. This tardiness reflects the gradual build up or removal of MSH from the circulation rather than any lethargy on the part of the melanophores themselves.

One can readily foresee the prospective biological advantages of an ability to change color rapidly as this may help protect the animal from a predator or assist it to catch food. Rapid color change may be especially important in animals that live in places where the background colors are variegated and across which the animals constantly travel in their search for food and sexual companionship. Many teleost fishes that roam about gaily colored reefs may find such rapid color change an especial advantage.

In the Teleostei, color change can be mediated by three types of mechanisms: (*a*) a humoral one, (*b*) a neural one or (*c*) a combination of both neural and humoral processes.

Anguilla (*the eel*)

Eels exhibit black- and white-background responses but only change their color slowly, like amphibians. Hypophysectomy abolishes the full expression of these responses and as the pituitary contains a material that, when injected, disperses melanin in the melanophores, the response is considered to be predominantly a humorally mediated one.

Following hypophysectomy, the eel is not maximally pale but has a melanophore index of 1 to 2 and this has contributed to speculation that a melanocyte-aggregating hormone is also present in the pituitary, as, if only MSH were involved, one would expect a MI of about 1. It has been suggested that, alternatively, melanophore-dispersing nerve fibers are present.

In contrast to amphibians, hypophysectomized eels continue to exhibit a small background response; the MI is 3.5 on a black background and 1.8 on a white one. This response is abolished by the severance of cutaneous nerves that are known to innervate the melanophores; thus, although the predominant mechanism mediating color change in eels is humoral there is an underlying neural control that only becomes apparent after the pituitary is removed.

Adrenaline, when injected, readily produces an aggregation of melanin in the melanophores of eels. This hormone is produced by widely distributed chromaffin cells in teleosts but it is unknown if it has a normal role in mediating their color changes.

Fundulus heteroclitus (*the killifish*)

Killifish exhibit the usual black- and white-background responses but these are *not* abolished following hypophysectomy. *Complete* darkening, or dispersal of melanin, does not occur following this operation, indicating that pituitary MSH may be necessary for the full expression of the black-background response.

The overriding control is, nevertheless, a neural one. Injection of MSH into pale fish does not disperse melanin and electrical stimulation of cutaneous nerves in dark fish evokes pallor. The injection of MSH into pale fish that have had parts of their skin denervated results in a melanin dispersion in these localized areas. Extracts of *Fundulus* pituitaries can evoke such dispersion of melanin.

In *Fundulus*, color change thus occurs in response to neural stimuli to the melanophores; there are melanin-aggregating nerve fibers, and possibly even 'dispersing fibers'. Underlying this mechanism, but generally over-ridden by it, is an ability to respond to MSH and this hormone is necessary for a maximal darkening of the fish.

Phoxinus phoxinus (*the European minnow*)

In minnows, there is little evidence for a role of endogenous hormones in the dispersion of melanin. Black and white-background responses are not prevented by hypophysectomy, though it has been observed that such fish cannot sustain a black coloration as readily as intact fish. Denervation of the skin abolishes the background-responses and the melanin fully disperses. Nerve stimulation evokes an aggregation of melanin and aggregating nerve fibers undoubtedly exist. There is also some evidence that suggests the presence of melanin-dispersing fibers.

The injection of extracts of the pituitaries from *Phoxinus* does not disperse melanin in either the intact or denervated skin of these fish. MSH from anurans will, however, darken the denervated skin of *Phoxinus*. There is no evidence for a melanin-dispersing hormone in *Phoxinus* and indeed pituitary extracts have an opposite, aggregating, effect.

Color change, thus, in the teleost fish may be influenced by three types of mechanisms:

1. A predominantly humoral one that overrides a neural mechanism, as in *Anguilla*.

2. Predominantly neural coordination that overrides a humoral process but which is still important, such as in *Fundulus*.

3. A neural coordinating mechanism with no evidence for an effect of endogenous MSH, as in *Phoxinus*.

Color change in reptiles

The reptiles have either humoral or neural mechanisms coordinating their color changes but as yet no transitional arrangements, which involve both, have been described. Most observations, certainly the most detailed ones, have been made on lacertilians that often display very dramatic changes in color as epitomized by the chameleons.

Snakes and crocodilians also possess chromatophores, and a chelonian, *Chelodina longicolis*, has also been shown to exhibit background-responses that are mediated by MSH (Woolley, 1957).

The lizard, *Anolis carolinensis*, exhibits both visual and non-visual background-responses that are abolished following hypophysectomy. The non-visual response, which can be overridden by the visual one, may be the result of photostimulation of the pineal.

Excitement, such as results from electrical stimulation of the mouth or cloaca, results in a mottling of the skin color patterns in *A. carolinensis* because of a dispersion of melanin in some melanophores and an aggregation in others. This effect can be mimicked by the injection of adrenaline which may normally mediate the response. The melanophores are not innervated and cutting the general nerve supply to the skin does not influence color change.

Chameleons, *Chamaeleo pumila*, and *Lophosaura pumila*, exhibit rapid and dramatic changes in color that are either visual responses or are due to photoreceptors that are apparently present in the skin. The observations that have been made on these responses suggest that there is a neural control of the melanophores; nerve stimulation results in an aggregation of melanin. It is also possible that melanin dispersing nerve fibers are present. It thus seems that the control of color change in chameleons is a neurally coordinated process though, as no experiments seem to have been done on hypophysectomized animals, a subsidiary role of MSH cannot be completely excluded.

Color change in cyclostomes

The control of color change in the lampreys (Petromyzontoidea) and hag-fishes (Myxinoidea) is intrinsically very interesting because of their lowly phyletic position on the vertebrate scale.

A background-color response, pale on a white substrate, dark on a black one, has been described in the hagfish, *Myxine glutinosa*, but the co-ordinating mechanism for this change is unknown.

Lampreys appear to lack a background-response but exhibit a diurnal rhythm in color, dark in the day and pale at night, that in some species is mediated by the pineal gland. Young, in 1935, found that removal of the pituitary abolishes this rhythmical color change in adults and ammocoete larvae of *Lampetra planeri* and the lamprey then becomes permanently pale in color. In the ammocoetes, pinealectomy also abolished this diurnal rhythm but, in contrast to hypophysectomy, the animals were permanently dark. These observations have recently been extended (Eddy and Strahan, 1968) to two species of Australian lampreys. These antipodeal cyclostomes also exhibit a diurnal rhythm in color which stops following hypophysectomy. In larval *Geotria australis*, pinealectomy also abolishes the rhythm but in metamorphosing larval *Mordacia mordax*, the lateral eyes must be removed to see this effect.

As described earlier, the pineal is the site (especially in the dark) of formation of melatonin which in anuran tadpoles is a very potent stimulant of melanin aggregation and so pales the skin. It has been found that the injection of melatonin into larval *Geotria* also results in skin pallor but this effect is absent in *Mordacia*. In addition, if the pineal is transplanted under the skin of *Geotria* a local paling is observed. Melatonin may thus be involved in regulating the rhythmical changes in color seen in the am-mocoetes of *Geotria* both by the production and release of melatonin and by acting as a photoreceptor organ. Following pinealectomy, the am-mocoetes are permanently dark, which may reflect either the lack of an inhibitory effect of melatonin on the release of MSH or, more likely, a direct antagonism to the action of MSH on the melanophores. It also seems likely, in retrospect, that the same mechanism(s) regulates color change in the ammocoetes of *Lampetra planeri*.

The involvement of melatonin in color change in some larval cyclostomes, some amphibians and a chondrichthyean, is most interesting. Melatonin does not seem to have this role in many vertebrates but nevertheless it is present in representatives of all the vertebrate groups. The propensity of

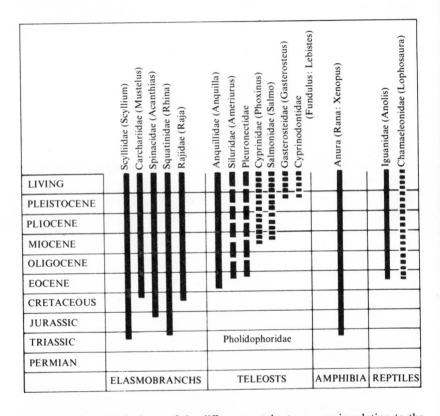

Fig. 7.9. The geological age of the different vertebrate groups in relation to the types of mechanisms (humoral, neural, or both) that they utilize to coordinate their melanophore background color responses.

The solid bars represent the possession of predominant humoral mechanisms; the broken bars, mixed, humoral and neural mechanisms; the small squares, predominantly neural control.

To this diagram could be added the Chelonia (*Chelodina oblonga*) and the Ophidia (*Crotalus*) both of which have a humoral control mechanism and which are identifiable from the Triassic.

It can be seen that humoral mechanisms appear to be the oldest and suggests that neural control is a later acquisition. (From Waring, 1942.)

the pineal to respond to diurnal changes in light makes it a potentially valuable gland for mediating endocrine rhythms which are dictated by changes in the seasons. As we shall see in Chapter 9, the pineal may in this manner contribute to the control of reproductive cycles in some vertebrates.

Evolutional color change mechanisms in vertebrates

From the preceding observations on the mechanisms controlling color change, we may make some guesses about the evolution of this process in vertebrates. Information of a comparative nature about extant species from diverse phyletic groups can be interpreted in a manner that may help us to reconstruct the past. Waring has summarized the available information about vertebrate color change in Fig. 7.9 but states that he has been sternly warned 'about pressing this kind of thing too far'.

The original underlying mechanism coordinating color change in vertebrates would appear to have been a humoral one as this has been observed exclusively in the Chondrichthyes (elasmobranchs) and has persisted in the Anura and Chelonia, all of which can be traced back to the Triassic. Superimposed, and probably subsequent to this, has been the evolution of a neural control of the melanophores which is seen in some teleosts and reptiles. In the teleosts, there is evidence of transitional changes as some species appear to utilize both neural and humoral mechanisms. The eels (Anguillidae) are specially interesting in this respect as they are normally completely dependent on humoral control, but there is also clear evidence for the presence of a neural mechanism that normally does not appear to contribute to color change. We can only speculate as to whether the eel represents a stage in the evolution towards a neural control of color change or is an evolutionary regression from this development. At the bottom of the vertebrate phyletic tree are the cyclostomes that possess an MSH that helps mediate a diurnal rhythm in color change but there is little evidence to suggest that these lowly fishes have an adaptive background color response. We can speculate even less as to whether or not the role of melatonin in controlling color changes in larval cyclostomes, reflects a primeval effect of this hormone, though it might!

Morphological color change

Several hormones influence morphological changes in the color of the integument. They include MSH and sex hormones, and corticotrophin and melatonin may also be involved. Such humoral effects on pigmentation are less well characterized than those of MSH on physiological color changes. Considerable interspecific differences in humoral effects on morphological pigmentation occur that make it impossible to give any generalized definition of their respective roles.

Melanin is formed from tyrosine by a complex chain of reactions which initially involves a copper-containing enzyme called tyrosinase. This enzyme may be in a soluble form in the cytoplasm of the melanocyte but when active it is attached to the melanosomes. A genetic absence of tyrosinase results in albinism though other factors, including ultraviolet light and several drugs, can also contribute quantitatively to changes in pigmentation. Increased levels of integumental melanin are associated with increased tyrosinase activity which can be influenced by hormones.

The morphological pigmentation of the skin, as well as fur and feathers, is in the first instance the result of the formation of melanin in the epidermal melanocytes, where it is attached to the melanosomes. In the skin, each melanocyte is associated with several keratinocytes to which the melanosomes, with their attached pigment, can be transferred. This functional association is called the 'epidermal-melanin unit' (Quevedo, 1972). In birds and mammals, the melanin is passed from the melanocytes associated with the feather tracks or hair follicles to the developing feathers or fur.

Dramatic changes in pigmentation associated with endocrine function have been noted in mammals that show seasonal differences in coat color, in birds that display sexual dimorphism in plumage color and in mammals suffering from endocrine imbalances. Many monkeys display prominent changes in color of parts of their skin associated with the sexual cycle and the areas are called the sex skin. At ovulation the buttocks enlarge and become red in color because of an engorgement of blood in the large venous sinuses. There is also an accumulation of mucopolysaccharides in the skin. This development of the sex skin is under the control of estrogens. These are only a few examples of the pigmentary changes that may occur in vertebrates and which are influenced by the action of hormones.

Hormones and seasonal changes in fur color

As described earlier, short-tailed weasels (*Mustela erminea*) change their coat color from brown to white with the onset of winter (Rust and Meyer, 1968). This change is a photoperiodic response due to changes of the length of the daylight hours and can be prevented by hypophysectomy. The latter operation results in a permanent white coat but the growth of new brown fur can be promoted when the weasels are injected with MSH or corticotrophin. In addition, if the pituitary is transplanted to the kidney, where MSH release is increased due to a lack of hypothalamic inhibition, brown fur also grows on previously white animals.

If melatonin, in a 'slow-release vehicle' of beeswax, is implanted under the skin, weasels that are undergoing a normal spring molt from a white to brown pelage, regrow white, instead of brown, fur. Rust and Meyer (1969) suggest that this is the result of a stimulation by melatonin of the release of MSH-R-IH from the hypothalamus so that the MSH levels drop. This experimental treatment of weasels also prevents the development of the gonads and the onset of normal spring reproductive cycles, so that it is possible that an interference with the normal levels of other hormones, apart from MSH, is also contributing to the effect. Precise experiments of this kind, to determine the role of hormones in seasonal changes of coat color in mammals, have been rare but, as described below, there are several other examples of pigmentary changes in coat color that can be induced by a deficiency or excess of MSH.

Hormones and morphological color change in cold-blooded vertebrates

Cold-blooded vertebrates that undergo physiological color changes have also been shown to increase the levels of melanin in their skin in response to a continual environmental 'black background' stimulation of MSH secretion. Such a change has been observed in amphibians, teleosts and chondrichthyeans. In the goldfish, *Carassius auratus*, corticotrophin increases the cutaneous levels of melanin and the activity of tyrosinase (Chavin, Kim and Tchen, 1963); an effect that cannot be mimicked by mammalian MSH. In contrast, MSH, but not corticotrophin, increased cutaneous melanin synthesis in the killifish, *Fundulus heteroclitus* (Pickford and Kosto, 1957) and one must, therefore, be careful not to draw any general systematic conclusions about the role of such hormones in melanin synthesis in the Teleostei. As the hormone preparations used in the fishes were of mammalian origin the differences in the responses could reflect the degree of similarity of these exogenous hormones to the particular endogenous MSH present in each species of fish.

The effects of changes in the level of endogenous MSH on melanin levels in the melanophores has been observed in the dogfish *Scyliorhinus canicula* (Wilson and Dodd, 1973*b*). Removal of the neurointermediate lobe of the pituitary in this chondrichthyean resulted in an almost complete loss of melanin from the skin. In the converse experiment, when increased circulating levels of MSH were promoted by severing the inhibitory hypothalamic connections to the intermediate lobe, there was an increased concentration of melanin in the skin.

Cold-blooded vertebrates may also exhibit morphological changes in skin color and pattern that are of a rather colorful nature. Some female lizards develop colored, orange and orange-red, spots on various parts of their bodies during the period of the development of their eggs in the body; they are thus called 'pregnancy spots' (Cooper and Ferguson, 1972; Ferguson and Chen, 1973; Medica, Turner and Smith, 1973). Such changes in color have a hormonal basis which has been examined in the collared lizard, *Crotaphytus collaris* and the leopard lizard *C. wislizenii*. The injection of progesterone induces such pigmentation in ovariectomized lizards and estrogen increases the response though it is ineffective alone. The natural levels of these hormones change during the growth of the eggs and their circulating concentrations have been measured in such lizards and can be correlated with the development of the pregnancy spots. The injection of FSH also induces the formation of such pigmented areas in leopard lizards. It would thus appear under natural conditions that a release of this pituitary hormone stimulates the development of the ovarian follicle together with a release of gonadal steroids (see Chapter 9), and these directly mediate the response. The role of pregnancy spots in these lizards is uncertain but it has been suggested that they may deter the males from inappropriate amorous advances.

Hormones and sexual dimorphism in avian plumage color

The mechanisms of the effects of hormones in mediating the seasonal sexual dimorphism in the color of plumage in weaver birds, *Steganura paradisaea*, and the non-seasonal differences in domestic brown leghorn fowl have been studied by Hall (1969).

The male weaver bird grows prominent black feathers just before (pre-nuptial) the breeding season. When areas of white feathers are plucked from the these birds the injection of pituitary ICSH (or LH) results in the appearance of melanin granules in the feather tracts of these areas and a related growth of black feathers. The formation of melanin is associated with an increased activity of tyrosinase in the feather tracts. The effect of ICSH is direct and is not mediated through any action on the gonads as it is still seen in castrated birds.

The male house finch, *Carpodacus mexicanus*, has red or orange feathers on its crown, throat and belly, which is in contrast to the female in which this plumage is brown. When the colored feathers are plucked from castrated males the new, regrown feathers are of the female, brown type which contrasts with the renewal of the colored plumes in intact birds (Tewary and Farner, 1973). It thus appears that, like in the male weaver bird, the more gaily colored plumage of the male is determined by the presence of male sex hormones. It seems, however, that these hormones are from the gonads of the finches, though an indirect action that could involve the pituitary is also possible.

The male domestic brown leghorn fowl has black feathers on its neck and breast while those in the female are a pinkish-brown. In this instance, the coloration of the male plumage is not hormone dependent but that of the female is due to the action of estrogens. The injection of estradiol increases melanin formation and the activity of tyrosinase. The estrogens act directly on the feather tracts as their action is a local one at the site of the injection (Greenwood and Blyth, 1935).

Hormones and pigmentation during endocrine imbalance

Changes in endocrine function associated with the adrenal cortex and pituitary have often been observed in man and experimental animals. Addison's Disease, which results in a deficiency of corticosteroids, is associated with the deposition of excessive pigment in the skin of man and a rise in the blood MSH levels (Thody and Plummer, 1973). Cortico-steroids, by a feedback mechanism, normally prevent an excessive secretion of MSH and this latter hormone promotes the formation of melanin in the skin. Direct injections of MSH have been shown to deepen the skin color of African negroids (Lerner and McGuire, 1961). Pregnancy in women is also often associated with an increased pigmentation in certain areas of the skin and this change is also thought to reflect an extra release of MSH. Adrenalectomy in the prairie deer mouse (*Peromyscus m. bairdii*) results in a darkening of the coat color which is associated with an increase in circulating MSH concentration. This change in release of MSH, in response to adrenalectomy, does not, however, occur in laboratory mice so that a broad generalization cannot be made about this effect (Geschwind, Huseby and Nishioka, 1972). Certain genetic strains in mice respond to injections of MSH, or the implantation of pituitary tumors containing MSH, by developing a dark color in newly grown fur (such as that formed following plucking) (Geschwind *et al.*, 1972). This melanotrophic effect of MSH thus appears to be fairly common but its significance in normal regulation of melanin synthesis still eludes us.

Mechanisms of hormone-mediated changes in integumental melanin distribution

Physiological color change is an alteration in the dispersion of melanin in the melanophores, such as that mediated by MSH and adrenaline, and this is related to the level of cellular cyclic AMP. This nucleotide is formed in the presence of MSH as a result of the activation of adenyl cyclase which is presumably associated with the melanophores. Cyclic AMP stimulates a dispersion of the melanosomes. This response requires the presence of calcium and may involve a microtubular system in the cell but the details of this are unknown (Novales, 1972). Adrenaline, on the other hand, may inhibit adenyl cyclase and decrease the formation of cyclic AMP (α-adrenergic effect) or in some instances, as in the spade foot toad, *Scaphiopus couchi*, and the lizard *Anolis carolinensis*, it also activates the enzyme (β-adrenergic effect) and so mimics the effect of MSH (Abe *et al.*, 1969; Goldman and Hadley, 1969). The neural responses of the melanophores also appear to be mediated by changes in the levels of cyclic AMP (Novales, 1973). The aggregating nerve fibers may exert an α-adrenergic effect and dispersing fibers a β-adrenergic one. The latter effect is observed *in vitro* but its significance *in vivo* is unknown.

The increase in melanin synthesis that occurs in morphological color change appears to be due to an increase in the activity of tyrosinase. As we have seen, this enzyme is associated with the action of MSH in mammals and also ICSH and estrogens in birds. The precise mechanisms by which this change occurs may, however, differ for each hormone. Lee, Lee and Lu (1972) have studied the effect of MSH on a mouse skin tumor that contains a high concentration of melanocytes (a melanoma). The increase in tyrosinase activity in response to MSH in this tumor does not appear to be the result of formation of new enzymes but rather the activation of those already present; possibly by the removal of an inhibitor. In the amphibian skin, a different mechanism may operate, and it is thought that a trypsin-like enzyme is released which activates tyrosinase which is present in the cytoplasm. The latter enzyme is then attached to the melanosome where it initiates the conversion of tyrosine to dopa that eventually leads to the formation of melanin. In birds (Hall, 1969), the action of ICSH in increasing cutaneous tyrosinase activity also does not appear to involve the synthesis of a new enzyme, as its effect is not inhibited by puromycin. The effect of estradiol on pigmentation in the brown leghorn fowl, however, *is* inhibited by puromycin so that the action of this hormone may then be due to an induction of tyrosinase. The many effects of hormones on integumental pigmentation thus may be reflected in a diversity in the mechanisms by which they exert their effects. At the present time, however, it would appear that the activity of tyrosinase is central to their morphological actions.

Has there been an evolution of the role of MSH?

While MSH has well-established effects in mediating color change in cold-blooded vertebrates, its normal role in birds and mammals is uncertain. MSH undoubtedly, in some circumstances and in certain species, stimulates the synthesis of melanin in the skin of mammals but whether this is its normal physiological role is not clear. It is possible that MSH may exhibit such a function especially in species like the weasel that seasonally change the color of their coat. Such an effect is, however, limited to a relatively few species and yet mammals, as well as birds, possess MSH and sometimes in several molecular forms (see Chapter 3). Does it then have other physiological effects in these animals? Despite a widespread search no satisfactory answer to this has emerged. Several other effects of MSH have been noted, such as an ability to promote the regeneration of visual purple in the retina (Hanaoka, 1953) and an antagonism to the hyperglycemic effect of adrenaline (Munday, 1957). Is it possible then that the MSH is merely a vestigial hormone in most birds and mammals? Its tenacious phyletic persistence in the pituitary makes this intuitively difficult for some to believe!

There is also an unanswered question as to whether or not the effects of MSH on melanin dispersion in the melanophore and melanin synthesis in the melanocyte reflect the same basic effect. In other words, does the effect on synthesis represent a distinct and separate evolution of the hormone's role and is the fact that melanin is involved in both merely fortuitous? If these effects are quite distinct it is possible that they have both existed, side by side, for a long time. Birds and mammals may have only lost the melanin-dispersing action together with the responding melanophores. The MSH's, corticotrophin and lipotrophin exhibit many structural similarities to each other and it has even been suggested that they could have arisen from a single 'parent' molecule. A loss of MSH may thus not be genetically simple, even in a species where the hormone may have no present use.

Conclusions

The integument is a very complex tissue that may be involved in several physiological phenomena including osmoregulation, color change, temperature regulation and reproduction. There are many characteristic processes involved in such mechanisms that have a definite systematic distribution (for example, sweat glands, mammary glands and branchial chloride-secreting cells) so that when hormones are involved, as they often are, their effects follow phyletic suite. Such responses must have also arisen at distinct times during vertebrate evolution. It is interesting to observe that there is often a definite relationship between the nature of the particular hormone and the general physiological process involved; sex hormones influence sexual processes in the skin, as well as elsewhere, and adrenocorticosteroids regulate (see Chapter 8) electrolyte movements in sweat glands and across the amphibian skin, as well as in the kidney. Some hormones have a special propensity to mediate processes in the integument. MSH thus influences pigmentation in nearly all groups of vertebrates commencing phyletically with the cyclostomes and on occasions it may even promote formation of melanin in mammalian skin. Its action, however, appears to be rather 'conservative' as it has no other established effect on any other types of process in the body nor for that matter in the integument either. In contrast, prolactin is 'versatile' as, apart from effects at non-ectodermal sites, it influences many integumental processes including molting cycles, the secretions from the mucous and mammary glands, proliferation of the crop-sac and development of the brood-patch in birds, and the control of water and salt movements across the gills of fishes. Some of these effects will be described in the next chapters.

8. Hormones and osmoregulation

About 70% of the body weight of animals is water in which are dissolved a variety of solutes, the presence of many of which is vital for life. Within the body, the solutions inside the cells differ from those that bathe the outside and the composition of each of these solutions must be maintained so as to provide an environment with an electrolyte content and osmotic concentration suitable for life. These intra- and extracellular fluids provide the framework in which life exists.

The physicochemical properties of the body fluids in animals usually differ greatly from those of their external environment. Animals continually suffer exposure to the whims of the exoteric conditions and this will tend to change the composition of their body fluids. In addition, although the intra- and extracellular fluids have identical osmotic concentrations, there are qualitative differences in the solutes they contain, and equilibration, due to diffusion, will tend to occur. Such animals, however, maintain the gradients between their body fluids and the environment; an equilibrium that is maintained as a result of a complex pattern of physiological events. These processes involve the cells, and special tissues and organs that are concerned with osmoregulation. The integration of the functions of these homeostatic tissues relies largely on hormones. The nervous system makes little direct contribution to such regulatory processes though at the cellular level itself considerable autoregulation, independent of hormones, exists. Hormones do ultimately influence some cellular processes of course, but they generally appear to do this in effector tissues like the kidney, gills and gut, which are especially concerned with the overall osmoregulation of the animal. For a more complete account of the role of hormones in osmoregulation the book by Bentley (1971) could be consulted.

Animals occupy diverse osmotic environments; the major ones are the sea, fresh water like rivers and lakes, and dry land. Differences exist between the availability of water and salts within these environments and this is particularly apparent to animals that lead a terrestrial life. Water may be relatively freely available to some terrestrial species that live in areas where rainfall is high, and lakes, ponds and rivers exist in close proximity to where they live. Other animals, however, live in dry, desert regions where water may only be available sporadically and in limited quantities. Salts are freely available to marine animals but in fresh water the supplies are more restricted, and some terrestrial animals may occupy regions where a low salt content in the soil may be reflected in a salt-deficient diet. It is thus not surprising to find that the processes for controlling the water and solute content of the body, called osmoregulation, can differ considerably between species that habitually occupy diverse ecological situations. These differences may be manifested as a tolerance to osmotic changes but are principally seen as differences in the functions of the tissues and organs concerned in maintaining the composition of the body fluids. Not unexpectedly, the evolution of the tissues and organs concerned with osmoregulation has been accompanied by changes in the role of the endocrine glands that help integrate their functioning.

Homeostatic events that contribute to osmoregulation may involve changes in either the rates of loss of water and solutes or in the processes of their accumulation.

Osmoregulation in terrestrial environments (see Fig. 8.1)

Terrestrial vertebrates may lose water as a result of evaporation from the skin and respiratory tract. This loss is greatest in hot, dry air in which little water vapor is present. In homeotherms, evaporation from the respiratory tract and the skin may be increased. In the latter this may be due to the activity of the sweat glands, as a result of the need to dissipate heat. While evaporation will be the predominant route for water loss in animals living in hot, dry conditions, additional quantities pass out of the kidneys in the urine, and out of the gut in the feces. Reproductive processes, such as egg-laying and lactation, will also result in increased water losses. Water is gained in the food, from which substantial quantities may be absorbed across the intestines, and as a result of drinking. In addition, amphibians, like frogs and toads, can absorb water across their permeable skin, from damp surfaces and pools of fresh water. Any excessive water that may be gained in such ways is excreted by the kidneys.

Salt losses in land-living vertebrates occur in the urine and feces and, in mammals, in the sweat gland secretions. Some birds and reptiles possess special glands in their heads called nasal salt glands which have an ability to secrete concentrated solutions of salts. Some regulation of the losses from the sweat glands and gut occurs, and the secretion of the kidneys undergoes a rigorous process of conservation so that salts that are deficient in the body may be conserved. Additional conservation of urinary salts can occur from urine during its storage in the urinary bladder of amphibians and some reptiles. Birds and many reptiles lack a urinary bladder but in such animals the urine passes into the cloaca and up into the large intestine

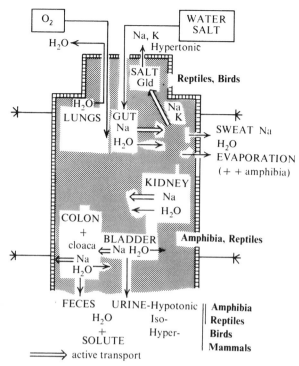

Fig. 8.1. A diagrammatic summary of the pathways of water and salt losses and gains in vertebrates living on dry land. (From Bentley, 1972.)

where some of its contained salts and fluid can be transported back into the blood. Salts are mainly gained in the diet of terrestrial animals and the drinking of brackish water may also result in salt accumulation. The latter process, however, only occurs rarely in nature. Excesses of accumulated salts can be excreted by the kidneys and in many birds and reptiles by the nasal salt glands.

Osmoregulation in fresh water (see Fig. 8.2)

Many species of vertebrates live in fresh water, a solution that is hypo-osmotic to the body fluids and which only contains small amounts of dissolved solutes. Water will thus tend to be gained by osmosis across the integument of such animals. Amphibians have a relatively permeable skin and can take up large amounts of water in this way. Fishes respire with the aid of gills which, apart from allowing the exchange of oxygen and carbon dioxide, are also permeable to water and so are an additional route for the

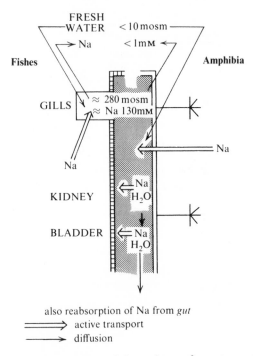

Fig. 8.2. A diagrammatic summary of the pathways for water and salt losses and gains of fishes and amphibians that live in fresh water. (From Bentley, 1972.)

accumulation of water in the body. The skin of reptiles and mammals that frequent fresh water is usually quite impermeable so that little water is accumulated through this channel and, as they breathe with the aid of lungs, they do not suffer the osmotic problems associated with the presence of gills.

Vertebrates living in fresh water may be prone to a greater salt loss than their terrestrial or marine relatives. When the integument is permeable, such as the skin in the Amphibia and the gills of fish, salt loss may be expected to occur as a result of diffusion. Such potential losses, however, are rather limited and are much smaller than may be expected on a simple physicochemical basis. In addition, salt losses may continually occur in the urine because of the necessity for the excretion of water that has accumulated by osmosis.

The solute excretion in the urine is reduced by its reabsorption from the renal tubules. In aquatic amphibians and reptiles, like turtles, and some fishes, sodium is also reabsorbed from the urine that is stored in the urinary bladder.

Salts are principally obtained from the food of aquatic vertebrates. However, additional gains of sodium chloride may be made as a result of their active transport, against electrochemical gradients, across the skin of amphibians and the gills of fish. It has also been suggested that some turtles may actively accumulate sodium across their pharyngeal and cloacal membranes during their irrigation by the external freshwater solution.

Osmoregulation in the sea (see Fig. 8.3)

Most species of fishes as well as a number of reptiles and at least one frog (the crab-eating frog, *Rana cancrivora*) live in the sea. This solution is strongly hyperosmotic to the body fluids of most vertebrates. The exceptions are the hagfishes (Agnatha), the sharks and rays (Chondrichthyes)

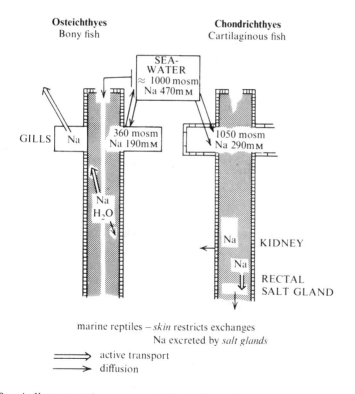

Fig. 8.3. A diagrammatic summary of the pathways for water and salt losses and gains in osteichthyean and chondrichthyean fishes that live in the sea. (From Bentley, 1972.)

and the coelacanth which are isoosmotic, or slightly hyperosmotic, to sea-water. The crab-eating frog is also slightly hyperosmotic to the sea-water in which it lives.

Most of the bony fishes (Osteichthyes) thus tend to lose water by osmosis, especially across their gill membranes. The Chondrichthyes and myxinoid Agnatha, on the other hand, may gain small amounts of fluid in this way. The sodium chloride concentration in the sea-water is much higher than that in the blood of osteichthyean, or even chondrichthyean fishes so that an accumulation of salt will tend to occur.

The mechanisms utilized to maintain osmotic balance in sea-water are varied. Marine teleost fish drink sea-water and much of the salt that is present is absorbed across the gut wall and water follows this solute by osmosis. The salt is excreted by special cells called 'chloride-secreting cells', or 'ionophores', present in the gills. Divalent ions are either excreted in the feces or urine. The gain of salts by marine chondrichthyeans is small compared to that of teleosts, and this is excreted in the urine and as a concentrated solution from a tissue unique among the vertebrates, the rectal salt gland. This salt-secreting organ is situated in the nether regions of the gut. Marine reptiles appear to have a relatively impermeable integument that restricts the gain of salt but excesses, such as may be gained in the food, can be excreted by the cephalic salt glands that are often modified tear, or orbital, glands.

It can be seen that the osmotic problems of vertebrates differ considerably and depend on the environment where they live as well as the anatomical and physiological wherewithal that is conferred by their phylogeny. The maintenance of osmotic homeostatis is dependent on a variety of tissues and glands, some of which, like the kidney, are present in all the major phyletic groups of vertebrates, and others, like cephalic salt glands and the rectal gland, which have a more restricted distribution. The activity of many of the organs and tissues involved in osmoregulation are controlled by hormones. A summary of these is given in Table 8.1.

Osmoregulation in vertebrates is dependent on the active participation of such tissues as the kidneys, gills, skin, urinary bladder, gut and certain salt-secreting glands. These tissues contribute to the excretion and conservation of water and salts, and their roles and physiological significance are not the same in all species.

The hormones that influence osmoregulation most directly are the neurohypophysial hormones, principally vasotocin and vasopressin (ADH), the adrenocorticosteroids, the catecholamines and prolactin. Corticotrophin and angiotensin are indirectly involved because of their roles in controlling the release of adrenocorticosteroids. Angiotensin may also have a more direct effect on some membranes while the urophysis and corpuscles of Stannius, which have putative endocrine functions, may also be involved.

Active transport and secretion of ions, especially sodium, potassium, bicarbonate and chloride, across, or from the epithelial membranes that make up the tissues that effect osmoregulation, are basic to their physiological function. Such membranes are osmotically permeable to water, which can pass across them with an ease that may vary, depending on the membrane and the physiological conditions. The adequate functioning of these osmoregulatory tissues is ultimately dependent on their blood supply. Hormones may thus influence the activity of osmoregulatory tissues by actions at several sites:

(1) Hormones may alter the processes of active sodium and chloride transport and the secretion of hydrogen ions, bicarbonate and potassium. Cortisol and aldosterone can alter sodium and potassium movements across many epithelial membranes. Vasotocin and ADH may also promote such transmural sodium transport. Catecholamines can increase transport and secretion of chloride ions in several tissues and also can inhibit the effects of ADH. All these hormones act directly on the cells involved.

(2) Osmotic and diffusional movements of water and sodium across epithelial membranes can be changed by hormones. Vasotocin and ADH may increase the permeability to water of the renal tubule (in some species) as well as amphibian skin and urinary bladder. Prolactin can reduce the permeability of the gills of certain teleost fishes to sodium, thus limiting diffusional losses of this ion in fresh water.

(3) The catecholamines and neurohypophysial peptides can alter the diameter of blood vessels and so may influence the functioning of osmoregulatory tissues by virtue of their vasoactive actions. The urine flow, especially in non-mammals, may be influenced by changes in the rate of filtration of plasma across the glomerulus and this process can be altered by these hormones. Alterations in secretion and absorption of ions across the gills of fishes may also be changed by such hormonally mediated variations in the regional blood flow.

While this chapter is principally concerned with hormones, the role of nerves in osmoregulation should also be mentioned. Neural integration is not common, though it does occur. The nasal salt-secreting glands in birds and reptiles are stimulated to secrete as a result of the stimulation of autonomic cholinergic nerves. Corticosteroids may contribute to the well-being of the underlying processes but do not directly stimulate secretion. The sweat glands of mammals secrete in response to the need to dissipate heat and this usually occurs following stimulation of autonomic cholinergic or adrenergic nerves (as described earlier, Chapter 5). However, sweat gland

TABLE 8.1. *Target organs for osmoregulatory-type responses to hormones in vertebrates*
The responses in *italics* appear to be physiological ones while others are either only pharmacological or the evidence for their normal, in-vivo role is as yet equivocal. The responses are not necessarily present in all members of the orders of vertebrates that are indicated.

Target organ	Phyletic distribution of target organ	Stimulatory hormone	Nature of response	Phyletic distribution of responsiveness
Kidney	All vertebrates	Vasotocin	*Decreased glomerular filtration rate (GFR) and increased renal tubular water reabsorption = decreased urine flow (antidiuresis)*	Amphibians, reptiles and birds
		Vasopressin	Decreased GFR = antidiuresis	Teleost (eels)
			Increased GFR = diuresis	Some teleosts and lungfishes
			Increased renal tubular water reabsorption = antidiuresis	Mammals
		Prolactin	Decreased renal Na excretion	Mammals (rat, man)
			Increased urine flow (increased GFR, decreased renal tubular water reabsorption)	Some teleosts
Urinary bladder	Most vertebrates (except birds, some reptiles and many fishes)	Vasotocin	*Increased water and Na reabsorption*	Amphibians (mostly anurans)
		Aldosterone	Increased Na reabsorption	Amphibians and reptiles
		Prolactin	Reduced water permeability, increased Na reabsorption	Teleost (Starry flounder)

Site	Hormone	Effect	Animal group
Gills (Fishes and larval amphibians)	Cortisol	*Increased outward Na secretion*	Marine teleosts
		Increased inward Na absorption	Freshwater teleosts
	Vasotocin	Increased inward or outward movements of Na	Freshwater (inward) or marine (outward) teleosts
	Aldosterone	Uptake across larval amphibian gills?	
	Adrenaline	Decreased secretion by Cl-secreting cells	Marine teleosts
	Prolactin	*Decreased Na diffusion outwards and water accumulation (inwards)*	Freshwater teleosts
		Decreased Na extrusion	Marine teleosts
Skin (All vertebrates)	Vasotocin	*Increased water and Na absorption*	Some amphibians
	Aldosterone	*Increased Na absorption*	
	Angiotensin	As above	
	Prolactin	Decreased permeability to water and Na	Urodele amphibians (?)
Sweat glands (Mammals)	Aldosterone	*Reduced Na loss in sweat*	
	Adrenaline	*Increased secretion during exercise*	
Salt glands (Some birds and reptiles (nasal and orbital glands) and chondrichthyeans (rectal glands))	Cortisol	Facilitates secretion	Birds and reptiles
	Cortisol	Reduces secretion	Chondrichthyeans
	Aldosterone	Reduces Na secretion	Lizard
	Prolactin	Increases secretion	Bird
	Vasotocin	Increases secretion	Bird

TABLE 8.1 – *continued*

Target organ	Phyletic distribution of target organ	Stimulatory hormone	Nature of response	Phyletic distribution of responsiveness
Salivary glands	Mammals	Aldosterone	*Decreases Na and increases K loss*	
		Adrenaline	*Dries up secretion*	
Intestine	All vertebrates	Cortisol	*Increased NaCl absorption*	Teleosts
		Aldosterone	Increased NaCl absorption in colon	Mammals, amphibians
			As above	Mammal (rat)
		Angiotensin	Decreased salt and water absorption	Teleosts
		Prolactin	Increased fluid absorption	Mammal (rat)

secretion during exercise may depend on circulating adrenaline. The vasoconstrictor tone of blood vessels is primarily dependent on the activity of adrenergic nerves which can thus, indirectly, alter the functioning of tissues. This effect is sometimes observed in the kidney of animals but is probably not a usual physiological mechanism.

The role of hormones in osmoregulation

Mammals

These vertebrates have a complement of hormones, similar to other vertebrates, that can influence osmoregulation, though the roles of such secretions may differ somewhat from those in non-mammals.

Vasopressin is unique to the mammals. It reduces urinary water losses (antidiuresis) as a result of an increased osmotic reabsorption of water from the kidney tubules. Vasopressin's phyletic forebear, vasotocin, as we have seen, has a slightly different chemical structure that confers on it a pronounced ability to contract smooth muscle such as in the oviduct and uterus and also, when injected, to promote contractions of the myo-epithelial cells in the mammary glands. Injected vasotocin also has an antidiuretic action in mammals and it could thus conceivably function in such a physiological role. Vasopressin, however, lacks the prominent effects that vasotocin has on non-vascular smooth muscle contraction and so exerts a more specific action in the body. Its evolutionary perpetuation in mammals is therefore not surprising. Vasopressin does not appear to have any other physiological role, on other organs, in mammals. It can, however, exert other effects, such as increasing the blood pressure, contracting the uterus and raising blood sugar levels, when it is injected in large amounts.

Adrenocorticosteroids play an important role in controlling sodium and potassium metabolism in mammals. The absence of the adrenal cortex in mammals quickly results in death, resulting mainly from losses of sodium and an accumulation of potassium. Aldosterone is the most effective of the adrenocortical hormones that exhibit actions on sodium and potassium metabolism in mammals, though the others, especially corticosterone, can also exert such effects. Sodium excretion from the kidney, sweat glands and salivary glands is reduced while potassium loss is increased; there is a drop in the ratio of sodium/potassium in the secreted fluids. It also seems likely that aldosterone can promote sodium reabsorption from the large intestine (this has been demonstrated *in vitro*) and the mammary gland ducts (Yagil, Etzion and Berlyne, 1973). The osmoregulatory effects of the corticosteroids in mammals are thus all directed to the same general purpose:

sodium conservation and potassium excretion, and seem to involve at least five different target tissues.

Adrenaline has a less prominent role in osmoregulation. Its action in stimulating sweat gland secretion has already been mentioned. In addition, adrenaline can antagonize the release, and the effects, of ADH on the kidney. Such inhibition is an α-adrenergic action that can be demonstrated in experimental animals, though its possible physiological importance is not yet clear.

Angiotensin, apart from initiating the release of aldosterone, has been shown to promote sodium reabsorption from the kidney tubule and the rat colon. Another interesting effect of angiotensin is its ability, when injected, to promote drinking in rats, as well as some other mammals (Fitzsimons, 1972). Drinking is elicited by the sensation of thirst which arises in the brain in a number of circumstances, including a reduction in the volume of the extracellular fluids. This latter response is reduced if the kidneys are removed, suggesting that the renin–angiotensin system may be involved. Indeed, the injection of small amounts of angiotensin II into the region of the 'thirst center' in the anterior diencephalon of the brain promotes drinking. The physiological importance of these effects of angiotensin is not yet clear.

Birds

Birds possess osmoregulatory hormones that are similar to those of other non-mammalian tetrapods.

Vasotocin acts as an antidiuretic hormone comparable in its effect to vasopressin in mammals. Vasotocin increases water reabsorption from the renal tubule of birds and in slightly higher, but still physiological, concentrations also decreases the GFR (Skadhauge, 1969). The hormone thus has two effects on the avian kidney, both of which decrease urinary water loss. Two types of nephrons have been identified in the kidney of the desert quail, *Lophortx gambelii* (Braun and Dantzler, 1972, 1974), a mammalian-type, with a loop of Henle and a reptilian-type that lacks this tubular segment. Glomerular filtration across the reptilian-type nephron is more variable than in the mammalian-type and the former cease functioning when excess sodium chloride is administered. The reptilian-type nephron is the site where vasotocin acts when it decreases the GFR in the desert quail.

It is interesting that vasotocin probably has another physiological role in birds, as it contracts the oviduct and so can assist oviposition. Vasotocin may thus have two physiological, but unrelated, roles and this situation may also occur in reptiles and amphibians. As described earlier (Chapter 5), vasotocin, when injected, also has a hyperglycemic and hyperlipidemic effect in birds, and it can also stimulate secretion from the nasal salt glands. The physiological spectrum of vasotocin's action could thus be even larger than just an involvement with osmoregulation, but the importance of such non-osmoregulatory effects in the body are in doubt.

The *adrenocorticosteroids*, of which birds possess aldosterone and corticosterone, reduce renal sodium loss and facilitate potassium excretion, just as in mammals. Adrenalectomy, in ducks, also reduces the ability of the nasal salt glands to secrete hypertonic salt solutions. Corticosterone, but not aldosterone, can restore this deficiency in adrenalectomized ducks. Corticotrophin and corticosteroids also enhance salt gland secretion when they are injected into intact ducks. These effects, however, are probably indirect and the result of an elevated glucose concentration in the blood (see Peaker, 1971; Phillips and Ensor, 1972).

The immediate stimulus for secretion of the avian salt glands is a neural one that is initiated by a hypertonicity of the plasma. Hormones may, however, impinge their effects on this process. Adrenaline inhibits the salt gland's response to hypertonic saline which probably reflects its vaso-constrictor effect. Hypophysectomy abolishes the secretory response of the salt gland and this deficiency can be partly restored by the injection of corticotrophin or prolactin. Injections of prolactin into normal ducks also stimulates salt gland secretion. This action of prolactin is an interesting one for, as we shall see later, this hormone has an osmoregulatory function in some teleost fishes. It is, however, likely that its action in birds is indirect, due to its role in maintaining an optimal intake of food and water, especially following hypophysectomy (see Ensor, Simons and Phillips, 1973).

The adrenal cortex of birds is influenced by the amount of salt in their diet and this effect can be seen in nature. It has been observed (Holmes, Butler and Phillips, 1961) that birds living in environments near the sea or supplies of brackish water, where their salt intake may be high, have larger adrenal glands than species that habitually have fresh water to drink.

The *prolactin* stores in the pituitary are also influenced by the bird's salt intake. Herring gulls, *Larus argentatus*, given salt solutions to drink suffer a depletion in the stores of prolactin in their pituitaries, which is associated with an elevated osmotic concentration of the plasma (Fig. 8.4). The functional significance, if any, of these observations is unknown; for, as we have seen, we do not know of any clear role for prolactin in avian osmoregulation. It is interesting that prolactin levels in the pituitaries of rats also decline in response to dehydration (Ensor, Edmondson and Phillips, 1972). In some teleost fishes, prolactin is released in response to hypoosmotic conditions (see later) and this hormone serves an important role in their osmoregulation. The depletion of pituitary prolactin in birds could represent a non-specific response (stress) or the activation of a

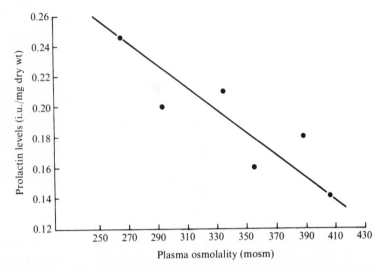

Fig. 8.4. The relationship of plasma osmolality and the storage of prolactin in the pituitary gland of juvenile herring gulls or black-backed gulls (*Larus argentatus* and *L. fuscus*). These birds were given saline-solutions of different concentrations to drink. As the concentration of this drinking water was increased the plasma osmolality rose while the prolactin storage declined. (From Phillips and Ensor, 1972.)

pathway for the hormone's release, that is a phyletic survivor from ancestral forms. However, prolactin could serve a physiological role in osmoregulation that is, as yet, to be defined.

The possible interrelations of nerves and hormones in influencing the electrolyte metabolism of birds that possess nasal salt glands is summarized in Fig. 8.5. The avian adrenal, as in mammals, can be controlled by both the adenohypophysis and a renin–angiotensin system.

The cloaca of birds appears to play a role in their osmoregulation. Urine passes into this segment of the gut, from which it travels up into the large intestine which is the site of fluid reabsorption. It seems likely that this reabsorption may be influenced by hormones. Vasotocin does not seem to be involved but some indirect evidence suggests that corticosteroids may increase salt absorption, as in the mammalian colon. However, the evidence for this is still only fragmentary (Crocker and Holmes, 1971).

Reptiles

Reptiles are poikilothermic, a process that profoundly influences their osmoregulation, and represent a substantial metabolic departure from the

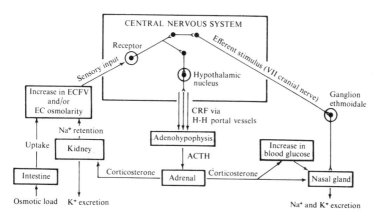

Fig. 8.5. A diagrammatic summary of the physiological and endocrine processes that appear to be involved in controlling sodium and potassium excretion in marine birds. It should be noted that not all species of birds possess a functioning nasal salt gland and that the evidence for the scheme depicted above is largely based on observations on the domestic duck and gulls. (From Holmes, 1972.)

birds and mammals. These animals live in diverse osmotic habitats including the sea, fresh water, and terrestrial situations ranging from deserts to tropical rain forests.

Reptiles possess *vasotocin* and this, when injected, can decrease urine flow by a dual action (as in birds), namely an increase of the renal tubular reabsorption of water and a decrease of the GFR (see LeBrie, 1972).

The role of the *adrenocorticosteroids* in reptilian osmoregulation is not clear. Adrenalectomy has been variously reported to cause no change in plasma electrolytes, result in a potassium accumulation or sodium loss or, as in birds and mammals, to have both of these effects. These diverse observations may be the result of the longer periods of time that poikilotherms often take to respond to altered physiological and environmental circumstances. Injected aldosterone can increase renal tubular reabsorption of sodium in the snake, *Natrix cyclopion*, but only after it has received additional sodium chloride which, presumably, decreases endogenous levels of the steroid hormone (LeBrie, 1972). On the other hand, the injection of aldosterone into a saline-loaded lizard, *Dipsosaurus dorsalis*, has no effect on renal electrolyte excretion (Bradshaw, Shoemaker and Nagy, 1972). *Dipsosaurus dorsalis*, like many birds and other reptiles, has a functional nasal salt gland, but injected corticosteroids do not initiate secretion from this gland (Shoemaker, Nagy and Bradshaw, 1972). Indeed, the sodium secretion is abolished by aldosterone injections. It has been suggested that in the lizard salt gland, aldosterone may promote sodium reabsorption from the tubules in the glands as it does in the mammalian kidney.

When excesses of sodium chloride are administered to the lizards, *Dipsosaurus dorsalis* and *Amphibolurus ornatus*, corticosteroid levels in the plasma increase (see Bradshaw, 1972). This effect contrasts with what is seen in mammals and amphibians where a sodium deficiency is associated with increased release of corticosteroids. In these species of lizards, the increased level of corticosteroids facilitates urinary sodium losses by decreasing renal sodium reabsorption, a response that is opposite to that seen in mammals where renal sodium reabsorption is increased.

The role of the adrenocorticosteroids in controlling sodium and potassium metabolism in reptiles is not completely understood. Some of the responses are reminiscent of those in other phyletic groups of vertebrates but some are quite different. One cannot, at present, draw any unified picture about the role of adrenocorticosteroids in the Reptilia.

Aldosterone may act at extrarenal sites in some reptiles. The urinary bladders of reptiles are the site of active sodium reabsorption from the urine that is stored there. This salt transport can be increased, *in vitro*, by aldosterone in the tortoise, *Testudo graeca*, and the freshwater turtle, *Pseudemys scripta* (Bentley, 1962; LeFevre, 1973). It is unknown whether this effect of aldosterone exists normally in these chelonians but it could reflect an evolution of this hormone's osmotic role. As we shall see, aldosterone also stimulates sodium transport across the amphibian urinary bladder.

The amphibians

Osmotically, the amphibians are a very interesting group as they bridge the gap between the fishes and the amniotes. Phyletically, they represent the first terrestrial vertebrates yet they are still largely dependent on the ready availability of fresh water and have aquatic larvae. The Amphibia, therefore, are a group of considerable interest both with respect to osmotic regulation and the endocrine mechanisms they utilize for this process.

Vasotocin has an antidiuretic effect in most, but not all, amphibians. Mesotocin, the other amphibian neurohypophysial hormone, has little effect on the kidney. As in the birds and reptiles, vasotocin initiates both the reabsorption of water from the renal tubules and a decrease in the GFR (Sawyer, 1972a). The tubular response may, however, be lacking in the urodele amphibians.

The magnitude of the antidiuretic effect of the neurohypophysial hormones also varies in different species. Vasotocin, for instance, does not reduce urine flow in the South African clawed toad, *Xenopus laevis*. This toad is aquatic, a situation where such a response to released vasotocin would be physiologically inappropriate and could even lead to death resulting from hyperhydration. Other aquatic amphibians, like the mud-puppy, *Necturus maculosus*, the mud eel, *Siren lacertina*, and the congo eel, *Amphiuma means*, exhibit antidiuretic effects after injections of vasotocin but the responses are small and the amounts of hormone required are large so that the physiological significance of this effect is doubtful (Bentley, 1973). Young tadpoles do not exhibit water retention (reflecting an anti-diuresis) in response to injected vasotocin, but as bullfrog (*Rana catesbeiana*) tadpoles approach metamorphosis such a response becomes increasingly apparent, though it does not reach its full expression until after meta-morphosis has occurred (Alvarado and Johnson, 1966).

Vasotocin exerts several other interesting effects with respect to the osmoregulation of amphibians. The skin of amphibians is permeable to water, which moves across it by osmosis. In anurans (frogs and toads), water crosses the skin much more readily when the tissue is stimulated by vasotocin. This hormone appears to make the integument less waterproof; an action that is is also seen on the renal tubule. This cutaneous effect of vasotocin is not seen in urodeles (newts and salamanders) nor in some anurans, including tadpoles and *Xenopus laevis*. The ability of vasotocin to increase the skin's osmotic permeability appears to be greater in species that normally occupy dry, rather than relatively wet or damp, habitats. Amphibians do not drink, and an increased rate of water accumulation across the skin may aid rehydration in some species, such as those from desert areas, where water is only available sporadically.

Amphibians usually possess a large urinary bladder in which they can store water equivalent to as much as 50% of their body weight. This water can be reabsorbed in times of need and so constitutes a store that may be very useful to some species. Vasotocin, or dehydration which releases this hormone into the blood, increases the rate of water reabsorption across the urinary bladder of many amphibians. This effect is, however, mainly seen in anurans; the urodeles examined (except for the fire-salamander, *Salamandra maculosus*) lack this response.

The crab-eating frog, *Rana cancrivora*, lives in the sea-water in coastal mangrove swamps in Southeast Asia. As described earlier, this interesting amphibian maintains its blood plasma at a hypertonic concentration with respect to the external solution by retaining additional salts and urea. The skin of these amphibians is not responsive to vasotocin, but this hormone increases the permeability of the urinary bladder to urea (as well as water) and so, by permitting its reabsorption from the urine, apparently con-tributes to the conservation of this solute in the body (see Dicker and Elliott, 1973). A cutaneous response would be a disadvantage to such animals as it would increase the rate of water loss if they entered hypertosmotic solutions.

The tetrapod urinary bladder has no true embryological analogue in the fishes and appears to have first evolved in the Amphibia. It is interesting that some species of this group utilize it for water storage, and possibly urea conservation, with which use it has also acquired a responsiveness to vasotocin. A comparable effect on the bladder has not been demonstrated in any other vertebrate group so that this endocrine adaptation is unique.

When amphibians, in water, are injected with neurohypophysial hormones, they gain weight due to a water retention. This action is known as the 'Brunn effect', or 'water balance effect', and is due to a stimulation by vasotocin at the three distinct sites described above: the kidney, the skin and the urinary bladder. A single hormone thus has multiple effects, all of which are directed to the same physiological purpose, namely the conservation of water.

As in other vertebrates, the neurohypophysial hormones can be shown, when injected *in vivo* or *in vitro*, to have other actions including a hyperglycemic effect and an ability to contract the oviduct. However, the physiological significance of these effects is not clear.

The *catecholamines*, including adrenaline, have ubiquitous effects on tissues. One of these, which has been shown *in vitro*, is an ability to antagonize the osmotic effects of neurohypophysial hormones (Handler, Bensinger and Orloff, 1968). Injected adrenaline has also been shown to increase cutaneous water uptake by the toad, *Bufo melanostictus* (Elliott, 1968). The action of vasotocin on the osmotic permeability of membranes is mediated by the adenyl cyclase–cyclic AMP system. Catecholamines can inhibit the formation of cyclic AMP which is a manifestation of their α-adrenergic effects. In addition, these hormones can increase the levels of cyclic AMP, which is a β-adrenergic effect. These actions of adrenaline and noradrenaline on membrane permeability are therefore not unexpected though their physiological significance is unknown in the Amphibia. Catecholamines could, however, modulate the actions of vasotocin.

The sodium metabolism of amphibians can also be regulated by hormones that can act at several different sites in the body. Vasotocin, *in vitro*, stimulates sodium transport across the skin, from the external media to the blood, and its reabsorption from the urinary bladder. These effects, though prominent, do not seem to persist for a prolonged time and are difficult to reconcile with normal physiological regulation of sodium in the intact animal. *Aldosterone* has more persistent effects in increasing such sodium transport across the skin, the urinary bladder, and the colon, in frogs and toads (Crabbé and De Weer, 1964; Cofré and Crabbé, 1965). It seems likely that aldosterone also acts on the urinary bladder of urodeles but the evidence for this is equivocal. As this corticosteroid is released in response to sodium-depletion in amphibians and is effective at low concentrations, it seems likely that it normally adjusts sodium transport at these sites. As we have seen, aldosterone stimulates sodium reabsorption from the kidney tubules in mammals, birds and possibly reptiles, but despite frequent attempts to demonstrate it, this action appears to be lacking in amphibians. This is an interesting endocrine situation, as aldosterone makes its initial phyletic appearance in the lungfishes and the amphibians, yet the important renal action of this hormone does not appear to have evolved until much later.

The *renin–angiotensin* system is present in the Amphibia and, as in other tetrapods, contributes to the release of aldosterone. The renin concentration in the kidney of salt-depleted frogs increases (Capelli, Wesson and Aponte, 1970). When renin from frog kidney is injected back into frogs, the aldosterone (but not corticosterone) concentration in the plasma rises (Johnston *et al.*, 1967); however, the plasma renin activity has been measured in bullfrogs and shown to *increase* following intravenous infusions of sodium chloride solutions (dehydration decreases it) which is opposite to the response observed in mammals (Sokabe *et al.*, 1972).

The involvement of the renin–angiotensin system in the specific control of aldosterone release may be present in the Amphibia and this effect could be one of its earliest evolutionary manifestations. There appear, however, to be differences from the physiological situation in mammals. It has even been suggested that angiotensin may constrict the efferent glomerular arterioles and so control the GFR in lower vertebrates, including amphibians. The possible role of the renin–angiotensin–aldosterone system in lungfishes, where aldosterone first appears on the phyletic scale, has not been investigated and this would be particularly interesting.

The 'water-drive' effect of injected *prolactin* in newts, *Notophthalmus viridescens* (see Chapters 3 and 7), is associated with changes in the skin, including a 'thickening' and an increased secretion of mucus. It would not be surprising if this second metamorphosis, from life in a terrestrial to an aquatic environment, were associated with osmoregulatory adjustments. The injection of prolactin plus thyroxine into these newts, when they are in their terrestrial phase, has been found to result in a decrease in the the permeability of the skin to water and sodium (Brown and Brown, 1973). This change could facilitate their osmotic adaptation to an aqueous environment and, as we shall see later (p. 302), the decrease in the permeability of the skin to sodium has some similarities to the effects of prolactin on the gills of some teleost fish. A phyletically closer analogy has been observed in another amphibian, the mudpuppy, *Necturus maculosus*, which possesses external gills. When this neotenous urodele is hypophysectomized it loses sodium at increased rates and its serum sodium concentration declines (Pang and Sawyer, 1974). The precise route for such sodium loss has not been described but, as in teleosts, it can be prevented by the injection of prolactin.

The fishes

The fishes are phyletically very diverse and contain several distinct groups that osmoregulate differently. These include the Osteichthyes (bony fishes), the Chondrichthyes (elasmobranchs, cartilaginous fishes) and the Agnatha (cyclostomes or jawless fishes, lampreys and hagfishes). Most of the available information about the role of hormones in osmoregulation applies to a single order of the Osteichthyes, the Teleostei. The role of the gills in osmoregulation in fishes has been thoroughly summarized by Maetz (1971).

Neurohypophysial hormones

The well-known antidiuretic effect of vasotocin, that is seen in tetrapods, does not seem to occur in most fishes. Indeed some fishes, but not all, exhibit a diametrically opposite response; the urine flow is increased. This diuresis is seen in some teleosts, like the goldfish, *Carassius auratus*, and the eel, *Anguilla anguilla*. It is also interesting that such a diuretic response is very prominent in two of the extant lungfishes, the African lungfish, *Protopterus aethiopicus*, and the South American lungfish, *Lepidosiren paradoxa* (Sawyer, 1972*b*). The diuretic effect of vasotocin in fishes reflects its action in increasing the glomerular filtration rate. This effect, as suggested by Sawyer, is not very different, basically, from the mechanism with which vasotocin produces the reduction in GFR that is seen in tetrapods. In the latter, vasotocin constricts the *afferent* glomerular arteriole while, to have a diuretic action, a similar effect, but on the *efferent* arteriole, would result in the observed increase in the GFR. Thus, a vascular effector site may merely have shifted from one branch of the glomerular arteriole to the other. Not all the glomeruli may be responsive to vasotocin as there is plenty of evidence in the fishes that a change in the GFR can reflect an alteration in the number of functioning glomeruli (glomerular recruitment) rather than a change in the hemodynamics in individual nephrons.

To complicate the situation further it has recently been shown that, while large doses of injected vasotocin have a diuretic action in freshwater European eels, relatively small amounts (less than 10^{-10} g/kg body weight) have an antidiuretic action as in tetrapods (Babiker and Rankin, 1973; Henderson and Wales, 1974). This response is the result of a decreased

GFR and presumably reflects changes in the opposite direction from that of glomerular recruitment; tubular water reabsorption is unchanged. It thus appears that the eel afferent glomerular arteriole may also respond to vasotocin and there may be a balance between the hormonal effects at this site and the efferent arteriole; the over-all response depending on the concentration of the peptide hormone. An action at the latter site may be able to override an effect on the former.

It should be emphasized that the diuretic effects of vasotocin do not occur in all fishes or even all teleosts and the antidiuretic effect has only been observed in freshwater European eels. This makes it difficult to envisage whether or not the neurohypophysial peptides really have a physiological action on the kidney or whether, on injection, the exogenous hormones are merely exerting their well-known pharmacological effects on blood vessels. Indeed vasotocin has not even been identified in the blood of fishes. A diuretic effect of vasotocin could, however, be useful to species living in fresh water, as it may facilitate the excretion of water that is accumulated by osmosis, while an antidiuretic effect may be useful when the fishes are bathed in hyperosmotic solutions like sea-water. In lungfishes that undergo a period of estivation, enclosed in a cocoon in the dried mud at the bottoms of lakes and rivers where they live, the initiation of a diuresis could aid excretion of both accumulated solutes and water taken up by the fish during its 'awakening' in fresh water. Indeed, African lungfishes from which the pituitary has been removed cannot survive once they are replaced in water.

Neurohypophysial peptides, when injected, have also been shown to increase the turnover of sodium chloride in some teleosts, such as eels, when they are transferred from fresh water to sea-water. Such hormones may also increase active ion uptake in some freshwater teleosts. It is possible that these hormones have a direct effect on the permeability of the gill epithelium which would be analogous to some of their actions in tetrapods. Alternatively, it has been shown (Maetz and Rankin, 1969) that regional changes in the branchial blood flow occur which could mediate alterations in ion transfer; thus, adrenaline which increases the blood supply to the respiratory areas of the gills, produces a decline in salt excretion from the chloride cells. This change may be the result of a shunting of blood away from the central part of the gill filaments that contain the chloride cells. Neurohypophysial peptides (and acetylcholine) have the opposite action on the blood supply in the gills (see Fig. 8.6) and so could facilitate the functioning of the chloride cells. We can thus conjecture that the vasoactive effects of the neurohypophysial peptides may be phyletically older than their direct actions on the permeability of epithelial membranes (Maetz and Rankin, 1969). Such vascular effects can be seen in *some* fishes where

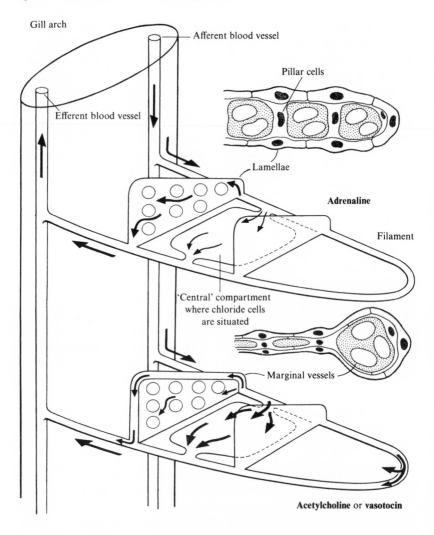

Fig. 8.6. A hypothetical model illustrating how hormones may alter the distribution of the blood flow in the gills of teleost fish and so alter their respiratory and osmoregulatory functions. Adrenaline increases the blood flow to the *lamellae* of the gills, which have a predominant respiratory function, by causing a relaxation of the *pillar cells* (PC). Acetylcholine or neurohypophysial peptides (including vasotocin and isotocin) constrict the lamellae (possibly by contracting the pillar cells) and divert blood to the central compartment. The chloride-secreting cells are situated in the interlamellar region of the central compartment so that their function is facilitated by the presence of the neurohypophysial hormones. (From Rankin and Maetz, 1971.)

they mediate changes in renal and gill functions that influence osmo-regulation.

Vasotocin has no effect on the urine flow in Agnathan fishes (or at least in *Lampetra fluviatilis*). The rate of sodium loss in the urine is, however, increased (Bentley and Follett, 1963) while the branchial losses of sodium are unchanged. The effects of vasotocin on osmoregulation in chondrichthyean fishes do not appear to have been investigated.

Differing physiological roles of vasotocin in eliciting an antidiuresis in tetrapods and a diuresis, or antidiuresis in some teleost fishes, would necessitate different mechanisms for regulating the release of this hormone;

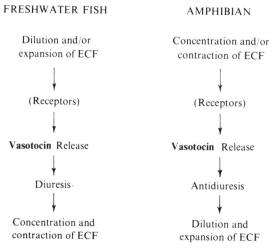

FRESHWATER FISH   AMPHIBIAN

Dilution and/or expansion of ECF	Concentration and/or contraction of ECF
↓	↓
(Receptors)	(Receptors)
↓	↓
Vasotocin Release	**Vasotocin** Release
↓	↓
Diuresis	Antidiuresis
↓	↓
Concentration and contraction of ECF	Dilution and expansion of ECF

Fig. 8.7. A hypothetical scheme showing the possible stimuli that may effect the release of vasotocin in fishes as compared to the typical tetrapod release reflex (such as exemplified in amphibians). The releasing-stimuli may be diametrically opposite to each other, for while *concentration* (or contraction) of the extracellular fluid (ECF) brings about vasotocin release in amphibians, in the freshwater fishes this would need to occur in response to a *dilution* if vasotocin were to mediate a diuresis physiologically. (From Sawyer, 1972*a*.)

thus, in freshwater fish, vasotocin may possibly be released in response to a dilution, or expansion, of the body fluids, while in tetrapods (and possibly the sea-water eel), it is secreted following a concentration of the extracellular fluids. The possible contrast in these two mechanisms in fish and amphibians is summarized in Fig. 8.7, though it should be noted that the release mechanism in the fish is still hypothetical.

The contrasting roles of vasotocin in altering urine flow via different mechanisms, involving the renal tubule or glomerulus, are shown in Fig. 8.8.

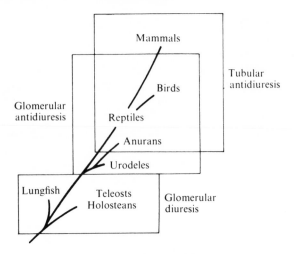

Fig. 8.8. A schematic diagram showing the phyletic distribution of the responses of the kidney to neurohypophysial hormones in vertebrates.

An exception may occur in the European eel (Teleostei) where small amounts of vasotocin may also have an antidiuretic effect (see text). (From Sawyer, 1972*b*.)

Vasotocin can elicit a glomerular antidiuresis in non-mammalian tetrapods and a glomerular diuresis in fishes. To this should be added a glomerular antidiuresis in the eel. In addition, vasotocin can promote water reabsorption from the renal tubules (a tubular antidiuresis) from all tetrapods (possibly with the exception of the urodeles and crocodilians).

Catecholamines

It is uncertain if the catecholamine hormones have a physiological role in the osmoregulation of fishes but adrenaline, when injected, can alter the movements of water, sodium and chloride across the gills (see Pic, Mayer-Gostan and Maetz, 1973). In teleosts in sea-water, injected adrenaline reduces the active extrusion of sodium and chloride from the gills. It increases branchial permeability to water in either fresh water or sea-water. The effects on ion movements can be prevented by α-adrenergic blocking drugs and that on water by β-adrenergic inhibitors. β-Adrenergic receptors mediate the increases of blood flow to the central lamellar regions of the gills (see the last section) and this effect could be contributing to the osmotic change. A more direct effect on the permeability of the gill epithelium to water is, however, considered to be a more likely mechanism. The ionic responses to adrenaline also appear to be due to a direct effect on the tissue resulting from an inhibition of the activity of the chloride-secreting cells.

The over-all effects of injected adrenaline on fluid balance in teleosts are a hypernatremia and hyperosmolarity in sea-water and an accumulation of water in fresh water; these changes are consistent with the observed responses of the gills. Stress, such as associated with laboratory handling or forced swimming, results in elevated levels of catecholamines in the blood of teleosts and also produces disturbances in fluid balance. It thus seems likely that catecholamines may influence osmoregulation in unusual circumstances but it is not clear whether or not they have such a role in more equitable conditions.

Adrenocorticosteroids

The corticosteroids, mainly cortisol, have a far better established effect on osmoregulation in fishes than do the neurohypophysial peptides. Again, this is predominantly seen in teleost fishes and in this group most of the experiments have been carried out on eels from Europe, *Anguilla anguilla*, North America, *Anguilla rostrata*, and Asia, *Anguilla japonica*. Apart from their ready availability and the feasibility of performing surgical procedures on them, eels can osmoregulate in either fresh water or the sea and during their normal lives migrate between these two environments. In sea-water, radioactive ion flux measurements indicate that teleosts, including eels, have a very large turnover of sodium amounting to as much as 50 to 60% of the total present in their body every hour, but in fresh water this is less than 1% per hour. In sea-water, this ion exchange is the result of a rapid transfer of salts across the gills and the drinking of sea-water which is absorbed from the fish's gut. The excess sodium chloride is excreted across the gills by the chloride cells. The kidney plays little part in the excretion of the excess solutes in teleosts as it lacks an ability to form a hypertonic urine. The influx of sodium across the gills in sea-water appears to be a passive process possibly involving exchange diffusion; the sodium that enters being exchanged for sodium leaving the body. The absorption of sea-water across the gut depends largely on active transport of sodium and this is related to the presence of the enzyme Na–K activated ATPase. This enzyme has also been localized in the chloride cells (Kamiya, 1972). The maintenance of adequate levels of Na–K ATPase in the gut, gills and kidneys depends on the action of cortisol which, in turn, is regulated by corticotrophin from the pituitary (Epstein, Cynamon and McKay, 1971; Pickford *et al.*, 1970). When eels enter sea-water there is an increase in the concentration of cortisol in their blood which persists for several days (Hirano, 1969; Ball *et al.*, 1971; Forrest *et al.*, 1973*b*). After this time, however, the steroid level declines so that it is similar to that in eels adapted to fresh water. In teleost fishes, the corticosteroids contribute to osmoregulation, as in other vertebrates. Removal of the adrenals in freshwater eels results in a decline in plasma sodium but in sea-water, there is an accumulation of this ion. This effect can be overcome by injecting cortisol (Chan *et al.*, 1967; Butler *et al.*, 1969; Mayer *et al.*, 1967; Henderson and Chester Jones, 1967). It is interesting, however, that the principal steroid involved is cortisol which, in the tetrapods, mainly influences intermediary metabolism. In teleosts, cortisol may fulfill both physiological roles.

Although the corticosteroids influence osmoregulation in the fishes, as well as the tetrapods, the mechanisms involved differ. The use of the gills is a piscine (and possibly larval amphibian) prerogative and they are thus a site for the steroid's action that is confined to the fishes. In addition, the relative importance of the effects of corticosteroids on the gut of teleosts living in sea-water may be greater than their action at this site in tetrapods or even in freshwater teleosts. The mechanisms by which the hormones

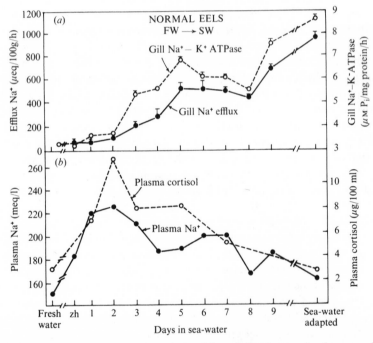

Fig. 8.9. A diagram showing the relationships of the changes in plasma sodium and cortisol concentrations (*b*), the efflux of sodium and the branchial Na–K ATPase in North American eels (*Anguilla rostrata*) following transfer from fresh water to sea-water (*a*). It can be seen that the cortisol concentrations in the plasma initially rise and this is followed by increased Na–K ATPase levels in the gills, which parallels added sodium efflux and the resulting decline in the plasma sodium concentration. (From Forrest *et al.*, 1973*a, b.*)

facilitate sodium transport may also differ, for in the tetrapods there is little evidence to suggest that corticosteroids exert their acute effects on sodium transport by increasing the levels (or inducing formation) of Na–K ATPase. Instead it seems more likely that aldosterone increases the activity (or rate of turnover) of the sodium pump (directly or indirectly) rather than by increasing its total capacity, such as would be expected if an increase of Na–K ATPase occurred. In fishes, cortisol may still have an additional, acute, type of effect like that of aldosterone in tetrapods, as there is evidence to suggest that when fish are transferred from fresh water to sea-water corticosteroid-mediated changes occur independently of any alteration in the level of Na–K ATPase (see Forrest *et al.*, 1973*a*). The temporal relationship between plasma cortisol and sodium levels, Na–K ATPase concentration in the gills and the branchial efflux of sodium in eels during adaptation to sea-water is shown in Fig. 8.9.

There is little information about the action of corticosteroids in non-teleost fish. Adrenocorticosteroids appear to have no effect on renal electrolyte losses in freshwater lampreys though they can decrease the rate of sodium loss across the gills (Bentley and Follett, 1963). The possible effects of corticosteroids in electrolyte balance in chondrichthyean fishes has not been very extensively studied, though in the shark, *Hemiscyllium plagiosum*, secretion from the rectal salt gland may be decreased by injected cortisol (Chan, Phillips and Chester Jones, 1967). The control of the rectal salt gland secretion is, however, not understood and it is uncertain how this effect of cortisol (which is not an homologous hormone in elasmobranchs) is related to the over-all regulatory mechanism.

Pituitary hormones; corticotrophin and prolactin

The pituitary has an essential role to play in influencing osmoregulation in fishes. Its effects are principally due to two of its hormones, corticotrophin and prolactin (or paralactin). The importance of the effects of these two hormones, however, differs considerably between species of fish and depends on whether they are living in fresh water or the sea. Hypophysectomy thus may have little osmotic effect on some species though in others dramatic changes may occur, especially if they are living in fresh water.

When freshwater eels are hypophysectomized they can survive for several weeks though there is a slow depletion of their electrolytes (Fontaine, Callamand and Olivereau, 1949; Butler, 1966). When placed in sea-water, these eels cannot osmoregulate properly and they accumulate excess sodium. Hypophysectomy results in a considerable lowering of plasma cortisol levels which can then be elevated by injecting corticotrophin (Hirano, 1969); thus, a pituitary interrenal (or adrenal) control-axis exists in the fishes, the interruption of which can upset osmoregulation (see Maetz, 1969). This is manifested by low levels of Na–K ATPase and reduced fluid and ion movements across the gills and gut which can be substantially corrected by injected cortisol (Epstein, Katz and Pickford, 1967; Butler and Carmichael, 1972). This hormone, however, does not completely restore osmotic balance, suggesting that other factors may be involved.

The osmotic deficiencies resulting from surgical removal of the pituitary are also due, apart from a lack of corticotrophin, to the absence of prolactin. Pickford and Phillips, in 1959, found that when killifish, *Fundulus heteroclitus*, were hypophysectomized they were able to survive in salt water but they soon died when placed in fresh water. If, however, they were injected with mammalian prolactin, their survival in fresh water was considerably prolonged. No other hormones were found to have this effect. Many species of fishes die in fresh water following hypophysectomy though others, like the goldfish, eel and trout, can survive for considerable periods of time. The importance of the pituitary for survival in fresh water varies considerably among the teleost fishes; thus, 18 species of the order Antheriniformes were found to be unable to survive hypophysectomy if kept in fresh water though many of the order Ostariophysi survive (Schreibman and Kallman, 1969).

Death in fresh water following hypophysectomy was found in *Poecilia latipinna* to be accompanied by a considerable loss of sodium which was prevented by the injection of prolactin (Ball and Ensor, 1965, 1967). This observation has been confirmed in other species of fish and is due principally to an excessive loss of sodium across the gills (see Lam, 1972; Ensor and Ball, 1972). The lack of such osmotic sensitivity in some fish, like eels and goldfish, to hypophysectomy seems to be the result of the more restricted permeability of their bodies to sodium, but even in these species prolactin can be shown to decrease branchial sodium loss (efflux). The mechanism of the effect of prolactin on the ionic permeability of the gills is uncertain.

Circulating levels of prolactin are probably low in fish adapted to sea-water. The importance of this is emphasized by the observation that when sea-water-adapted *Tilapia mossambica* are injected with this hormone their plasma sodium concentration increases (Dharmamba *et al.*, 1973). This treatment, if continued, would probably kill these fishes. The accumulation of sodium is the result of a reduced rate of sodium chloride secretion from the branchial chloride-secreting cells, possibly as a result of an inhibition of Na–K ATPase. The reduced activity of the chloride cells helps effect the adaptation of these fish to fresh water but in sea-water such an action would be disastrous.

Prolactin also influences the permeability of teleosts to water. The results *in vivo* have been a little contradictory as they usually indicate that the hormone increases branchial osmotic permeability but, depending on the species studied, a decrease may also be observed. In-vitro observations on the gills of goldfish also suggest that prolactin reduces permeability to water (see also Ogawa, Yagasaki and Yamazaki, 1973). In intact fish, it is often difficult to decide which effect and site of action is the primary one; thus, prolactin can increase the urine flow and this could be either a direct action on the kidney or the result of an increased branchial permeability to water. Prolactin is thought to have a direct effect on the kidney in fishes, mediated by an increased GFR or a reduced tubular reabsorption of water, or both. Apart from the kidneys, prolactin can also reduce the transfer of fluid across the fish intestine and urinary bladder (Utida *et al.*, 1972). Prolactin may restrict the osmotic permeability of membranes at three distinct sites: the kidney tubules, the intestine and the urinary bladder, and a similar effect on the gills may also occur but this is controversial. Prolactin may have a physiological action in preventing over-hydration of fish in fresh water by facilitating the excretion of water.

Mammalian prolactin, when injected, is effective in promoting changes in water and salt metabolism in teleost fishes but this does not constitute proof that the endogenous hormone also acts in this way in the fishes. Teleost prolactin (or paralactin) can elicit many of the same osmotic actions as the exogenous mammalian hormone. The amount of prolactin stored in the pituitary of the teleost *Poecilia latipinna* is six times greater in fish living in fresh water than in those in salt water, while the activity of the pituitary *eta* cells indicates that the hormone is being released in the freshwater fish (Ball and Ingleton, 1973). The pituitary glands of *Poecilia*, adapted to one-third sea-water, contain about three times the quantity of prolactin as sea-water-adapted fish (see Fig, 8.10). When the one-third sea-water-adapted fish are transferred to fresh water, the level of stored prolactin initially declines because of its release but the rate of synthesis increases so that after about eight days the amount stored in the pituitary rises to about six times the concentration seen in sea-water-adapted fish. Transfer of *Poecilia* from fresh water to sea-water has little immediate effect on the glandular stores, but these do show a gradual decline, presumably due to a decreased rate of synthesis.

Prolactin appears to have an important role in the migration of stickle-backs, *Gasterosteus aculeatus*, between the sea and rivers (see Lam, 1972). These fish normally spend the autumn and winter in the sea but in the spring migrate into rivers to breed. Sticklebacks caught in winter, in the sea, soon die if they are placed in fresh water but if they are first injected with prolactin they survive for much longer. This increased survival ability is also seen in fish in the autumn if they are kept under conditions of long

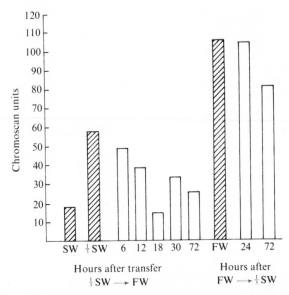

Fig. 8.10. The prolactin (in 'chromoscan units') content of the pituitary glands of the teleost *Poecilia latipinna* following transfer from one-third sea-water (SW) to fresh water (FW), and then to one-third sea-water. After transfer from one-third sea-water to fresh water the prolactin storage initially declines but subsequently increases and after 8 days reaches a level that is about six times greater than in sea-water-adapted fish. Cross-hatched bars show initial control values. (From Ball and Ingleton, 1973.)

day-length. It seems that, due to photoperiodic stimulation (the lengthening of the day) there is an increase in the activity of the prolactin cells which prepares the fish for its future migration into fresh water, during which the hormone is released into the blood.

These interesting observations on the role of prolactin in teleostean osmoregulation have drawn attention to comparable effects in other vertebrates. Among fishes little is known about the role of prolactin in osmoregulation of non-teleosts. In three species of chondrichthyeans, the water uptake across the gills is decreased following hypophysectomy and is restored by injected prolactin (Payan and Maetz, 1971). This effect is similar to that observed in goldfish but is not consistent with the observations on teleostean gills *in vitro*. This effect in chondrichthyeans, as described earlier, could be the result of a primary diuretic action on the kidney rather than the gills. The possible role of prolactin in maintaining optimal hydration in birds and its effects in newts and mudpuppies have already been described and there are even some reports about its osmotic effects in mammals. It has, for instance, been found that prolactin may help maintain hydration in rats during lactation, possibly via an antidiuretic action and by facilitating fluid absorption from the gut (Ensor, Edmondson and Phillips, 1972; Ramsey and Bern, 1972). Urinary sodium retention has also been observed in rats following the injection of crude prolactin preparations (Lockett and Nail, 1965) and the same effect has been shown to occur in man (Horrobin *et al.*, 1971).

The urophysis, corpuscles of Stannius and juxtaglomerular cells

These tissues have a glandular appearance and have a putative endocrine function in some fishes. Their structure and distribution among the fishes has been described earlier (Chapter 2). There are several types of evidence suggesting that these glands may have an osmoregulatory function including:

(i) Changes in their histological appearance in different osmotic circumstances.

(ii) Deficiencies in the fish's capacity to adapt to fresh water or sea-water following surgical removal of the tissues.

(iii) The ability of injected extracts of these glands to alter the fish's balance of water and electrolytes.

At this time no unequivocal conclusions as to the suggested osmotic importance of these tissues is available. This lack of consensus is largely the result of variations in the responses of different species of teleosts to various experimental procedures; nevertheless, the urophysis, corpuscles of Stannius and teleostean juxtaglomerular cells contain biologically active substances and it remains possible that these materials may act as 'hormones' at sites unconnected with osmoregulation.

The cichlid euryhaline fish, *Tilapia mossambica*, and the stickleback, *Gasterosteus aculeatus*, suffer an increased mortality, following transfer from fresh water to saline solutions, if their urophyses are removed (Takasugi and Bern, 1962; Ireland, 1969). This effect has not been demonstrated in all species though it is possible that the rapid regeneration of this tissue may contribute to the observed differences. The histological appearance of the urophyses of *Tilapia mossambica* suggest a depletion of the contained neurosecretory material when these fish are kept in sea-water, as compared with fresh water. The Hawaiian o'io, *Albula vulpes*, also displays such histological variations in urophysial activity (see Fridberg and Bern, 1968). The rate of electrical firing of the urophysial nerve cells has also been shown to be altered in response to osmotic changes. Both a decrease and an increase in spontaneous electrical discharge have been observed in response to hypotonicity and they depend on the particular species examined.

Considerable advances have been made in characterizing the biologically active materials present in the teleostean urophysis (see Zelnik and Lederis, 1973). These substances have been separated into several components with characteristic chemical and pharmacological actions. They are proteins and polypeptides and have been classified in the following way:

(i) Urotensin I, which when injected decreases the blood pressure of rats.

(ii) Urotensin II, which contracts various smooth muscle preparations, including the trout urinary bladder, can increase the blood pressure as well as the urine flow in eels.

(iii) Urotensin III. This promotes sodium uptake across the gills of goldfish, an effect that has not been observed in other species.

(iv) Urotensin IV, which increases water transfer across the toad urinary bladder (*in vitro*) and has other similarities to vasotocin with which it may be identical.

None of these substances has yet been identified in the circulation of fish but urotensin II is discharged from the urophysis under *in-vitro* conditions, which supports the possibility of its hormonal nature (Berlind, 1972*a*). The venous effluent of the urophysis passes into the renal portal blood vessels via the caudal vein, and so any urophysial secretions are in a potentially excellent situation to influence kidney function.

When the euryhaline marine teleost, *Gillichthys mirabilis*, is placed in fresh water the urophysial content of urotensin IV declines but urotensin II is unchanged (Berlind, Lacanilao and Bern, 1972). Two other proteins, that are characteristically present in the urophysis of *Gillichthys*, also show a considerable decline if these fish are kept in fresh water for six days.

Somewhat sporadic evidence thus suggests the possibility that the urophysis may influence osmoregulation in fish. It may, however, have other possible endocrine functions. Urotensin II contracts the smooth muscle of the urinogenital tract, including the sperm duct in *Gillichthys mirabilis* (Berlind, 1972*b*). It has thus also been suggested that the urophysis may have a role in reproduction of fishes, such as in promoting spawning.

The probable role of the corpuscles of Stannius in calcium metabolism in teleost fishes has been described earlier (see Chapter 6). The hypercalcemia that is observed in eels following Stanniectomy is accompanied by a decline in plasma sodium concentration and a rise in potassium (M. Fontaine, 1964). These changes in sodium and potassium in the plasma can be corrected by injecting aldosterone, though this corticosteroid does not appear to be present in fishes. It is, however, likely that other corticosteroids, such as cortisol, which can act as a mineralocorticoid in teleosts, may also have this effect.

Several possible mechanisms have been considered that could be responsible for the effects of Stanniectomy on sodium and potassium metabolism in teleosts:

(i) Because of initial confusion as to distinction of the morphology of the corpuscles of Stannius from the interrenal tissues, it was once proposed that the corpuscular tissues may secrete adrenocorticosteroids. Despite much contradictory evidence, it is now considered that this is not so, though these tissues may metabolize some steroids (but so do many other tissues) (Colombo, Bern and Pieprzyk, 1971). Nevertheless, as we have seen, the profound changes in sodium and potassium balance that accompany Stanniectomy can be compensated for by the injection of corticosteroids, and in the eel this operation is accompanied by a decrease in the plasma cortisol concentration (Fenwick and Forster, 1972). This decline is always accompanied by a hypercalcemia which is consistent with the evidence of Leloup-Hatey (1970) that indicates that the elevated calcium concentrations inhibit the action of corticotrophin and corticosteroidogenesis. The actions of the corpuscles of Stannius on steroid metabolism and osmoregulation thus appear to be indirect effects resulting from an absence of their hypocalcemic-hormone secretion.

(ii) Extracts of the corpuscles of Stannius, when injected, elevate the blood pressure of rats and eels in a manner suggesting that they contain a renin-like substance that can promote the formation of angiotensin (Chester Jones *et al.*, 1966; Sokabe *et al.*, 1970). Removal of the corpuscles of Stannius results in a decrease of the blood pressure of freshwater eels to levels that are normally seen when these fish are adapted to sea-water. As the decline in the GFR observed in sea-water is probably the result of the drop in blood pressure it is possible that the two events may be related. This interesting suggestion, however, does not appear likely as Stanniectomy (in North American eels, *Anguilla rostrata*) is not accompanied by a significant change in the GFR (Butler, 1969). In addition, plasma renin activity has been shown to *increase* in eels placed in sea-water (Sokabe *et al.*, 1973) which, if anything, would be expected to increase, not decrease, the blood pressure.

(iii) It is possible that a renin–angiotensin system exists in the teleosts which controls the formation and release of adrenocortical steroids in a manner comparable to its effects on aldosterone in mammals (Chester Jones *et al.*, 1966). The kidney of teleost fishes contains renin in relatively large amounts compared to the corpuscles of Stannius so that it is considered unlikely (Butler, 1969) that the levels in the latter tissues are of critical importance. Indeed renin-like substances have been identified in a number of other tissues so that their presence in the corpuscles of Stannius may be a 'red herring'. There is no conclusive evidence that renin is involved in the osmoregulation of fishes.

The nature of the processes controlling osmoregulation in fishes is incomplete and much of the information about the role of hormones in this process is still speculative. It is thus still an exciting field of study for the comparative endocrinologist.

Conclusions

The regulation of the water and salt content of vertebrates is primarily mediated by the kidney but several other glands and tissues are also involved. Many of these 'accessory' osmoregulatory organs have a distinct systematic distribution; for instance, gills in fishes, nasal salt glands in birds, and sweat glands in mammals. The osmoregulatory functions of most such organs are controlled by hormones which each tend to contribute to this process in a specific manner. Thus the neurohypophysial hormones increase osmotic permeability whether it be in the tetrapod kidney or the amphibian skin and urinary bladder. Similarly adrenocorticosteroids help regulate sodium metabolism by increasing transmural sodium transport in the renal tubule of mammals, the skin and urinary bladder of amphibians, the sweat, salivary and mammary glands of mammals, and the gills of many fishes. Thus the same type of hormone is often concerned (or is 'utilized') with coordinating the same general physiological process, though in somewhat different ways in various groups of vertebrates. There are, however, exceptions and systematically unique features, like the role of the gills in osmoregulation of fishes, is accompanied by what may be a unique type of action of a particular hormone; in this instance the ability of prolactin to control the permeability of the branchial (and other) epithelial membranes in teleosts. In mammals, this hormone appears to be principally concerned with the regulation of lactation though it is now suspected that it may also be capable of influencing osmoregulation in some tetrapods. The physiological significance of the effects in the latter is at the time of writing still in doubt and could represent a 'vestigial' endocrine response.

9. Hormones and reproduction

The reproductive process is not essential for the life of the individual, though it may make it more interesting, but it is necessary for the perpetuation of the species. In many so-called lower forms of life, reproduction may be an asexual process. A notable disadvantage of this type of reproduction is a diminution in the chances of genetic variability, and the transmission of such inherited changes to other individuals, so that evolutionary adaptation is hampered. The effects of the absence of sexual propagation is more likely to be apparent in vertebrates, that take a relatively long period of time to reproduce. The lapse of time between the generations may thus be large compared, for instance, to that in unicellular organisms. Reproduction is a complex process and this is especially true in species that occupy environments where the conditions are variable and large physicochemical changes occur. The young, developing animal is not usually as adaptable as the adult to such changes in the environment, and so must either be protected from these deviations by the parents or be produced on occasions that are most suitable to its more limited physiological capabilities. In vertebrates both conditions usually prevail; the embryo may develop to a quite advanced stage before becoming independent of the parent and it is usually produced during a season when such conditions as the temperature and food and water supply are favorable.

Reproduction in vertebrates thus involves considerable physiological coordination. The sexual process that requires the union of the sperm and ova necessitates complex physiological, social and morphological arrangements to ensure that these gametes each ripen at a similar time, and that the two sexes then meet and effect their union. The growth and differentiation of the fertilized egg often involves complex parental care, which may occur *in utero*, within the parent itself, or in an egg that is specially produced to meet the potential needs of the embryo. Care of the young often continues for a period of time following such initial development in the egg or *in utero*. The foregoing events may not be possible, or successfully accomplished, except during certain seasons of the year when the conditions are favorable.

In vertebrates, the coordination of all the processes outlined above involves hormones and the degree of complexity of their actions directly reflects the intricacies of the reproductive processes in a particular species.

The endocrine control of reproduction in man is thus more involved than in a jawless fish, like the lamprey. The basic pattern, however, is remarkably uniform and involves the endocrine secretions of the pituitary and the gonads. The influences of the hypothalamus and the median eminence on the pituitary gonadotrophin release is vital in most groups of vertebrates, but this control seems to be lacking in the cyclostomes and chondrichthyeans. The pituitary gonadotrophic hormones in vertebrates are chemically analogous but have undergone evolutionary changes in their structures. The gonadal steroids on the other hand are identical throughout the vertebrate series. One of the most notable endocrine differences among the vertebrates is the ability of the placenta of eutherian mammals to produce hormones that are similar to those secreted by the gonads and the pituitary. Otherwise the evolutionary changes are largely a matter of detail. These adaptations include modifications of the gonadal ducts, such as may assist in the processes of fertilization, the production of different types of eggs and the internal incubation of the embryo. A plethora of secondary sex characters, involving such morphological factors as size, color and scent glands have appeared in the different vertebrate groups and these help to ensure that the sexes meet and mate in the breeding season. Also involved in the mating procedure are a multitude of different patterns of pre-nuptial behavior. The precise manner by which the time of breeding is controlled also varies in different vertebrates and may involve differences in the length of the gonadal cycle and adaptations to the receipt of different environmental signals, such as light, which is predominant in birds, and temperature, that seems to be more important in reptiles. Variation in the functioning of the hypothalamus, pituitary and gonads may reflect such differences in the manner of timing of breeding.

There are several major differences in the patterns of reproduction in vertebrates that have considerable effects on the endocrine control of reproduction. These concern the manner by which fertilization is accomplished and the site where the embryo differentiates and grows.

Life in an aqueous medium, such as the sea, lakes and rivers, provides a relatively stable physicochemical environment for ova, sperm and the fertilized eggs. Many fishes ensure their reproduction by producing vast numbers of ova, often millions at a time that are extruded, before fertilization, into the external solution. The reproductive activities of the male are coordinated to this oviposition so that vast numbers of sperm are released amongst the eggs and *external fertilization* occurs. This process, which also takes place in most amphibians, has the advantage of simplicity but is only possible in aqueous situations. A general physiological corollary of external fertilization is that usually the eggs are small and contain relatively few nutrients for the support of the young. The animal thus, nutritionally, can afford to produce the vast numbers of ova necessary to assure fertilization on a scale that is adequate for the survival of a sufficient number of the young. Primarily terrestrial species, like the reptiles, birds and mammals, must of necessity resort to *internal fertilization*. This process requires the production of fewer eggs but more intimate contact and collaboration between the sexes. Internal fertilization also occurs among fishes including some bony fishes, all of the chondrichthyeans and some amphibians.

There are also considerable differences, again with important endocrine repercussions, between the eggs of different vertebrates and the processes that assist in their successful transposition into viable young. These differences are partly related to the nature of the fertilization process (due to the number of eggs that must be produced) and to life in a terrestrial environment. Eggs that are produced by most fishes and amphibians are highly prone to evaporation and so cannot readily survive on dry land. Some frogs deposit such eggs in damp burrows but this is unusual. Birds, most reptiles and prototherian mammals, such as the platypus, produce eggs that are covered with a protective shell, that limits evaporation, and they contain large amounts of nutrients that are sufficient to sustain the young until it reaches a stage of development when it can fend for itself. Similar eggs with a horny shell and large amounts of nutritive yolk are also produced by the chondrichthyean fishes and hagfishes. The production of eggs from which the young develop in the external environment is a process called *oviparity* and the eggs are termed alecithal and megalecithal eggs depending on the amounts of yolk nutrients that they contain. In many species, including some chondrichthyean and teleost fishes (where the ovum is often fertilized within the follicle in which it develops), amphibians and reptiles, the eggs may be retained for prolonged periods in the oviduct during which time the young develops in a relatively protected and secluded situation. This is called *ovoviviparity*. A more intimate contact between the eggs and the wall of the oviduct or uterus may occur whereby the developing young can exchange respiratory gases and even gain fluids and nutrients. This condition is called *viviparity* and as nutrients may be gained from the parent, the eggs usually contain far less yolk. Viviparity has evolved many times in nature and is present among chondrichthyeans, teleosts (where the young are usually contained within a hollow ovary), amphibians, reptiles and mammals. There are many endocrine variations that result from these different ways of providing for the development of the young including hormonal influences on the maturation and formation of the different types of eggs, the morphological development and physiological behavior of the oviducts and the triggering, at the appropriate time, of the expulsion of the egg (oviposition) or young (parturition).

The eggs and young of many species receive little or no parental care once they are separated from the mother. In some species, however, they are cared for and this may be necessary for their survival. Some teleost fishes are known to deposit their eggs in specially prepared nests which they protect and over which they may circulate water. Others such as the teleosts *Tilapia mossambica* keep a brood of hatched young in the fastnesses of a large mouth from which the young can emerge or retreat to. Several species of frogs (*Pipa pipa, Gastrotheca marsupiata*) and the sea-horse (*Hippocampus*) keep young in a pouch, or marsupium, on their backs while frogs incubate them in modified vocal sacs. Some snakes and most birds personally incubate their eggs and the care with which birds feed and protect the newly hatched young is well known. Birds usually collect food, which they present to their young and often this is predigested. Pigeons and doves produce a special pasty secretion, that contains a high concentration of fat and protein, from their crop-sac, the so-called pigeon's milk. The formation of such a special milk secretion with which to feed the young is a characteristic systematic feature of mammals. Such processes, whether it is 'tender loving parental care', 'broodiness' in birds, or lactation in mammals, are all largely controlled by hormones.

The reproductive apparatus of vertebrates

The gonads of vertebrates have a dual function, as they not only produce the germ cells but also some of the hormones that control the reproductive process; thus the testis in addition to being the site of formation and maturation of the sperm also produces androgens, principally testosterone and androstenedione in the interstitial tissue (or Leydig cells) and, probably, also in the Sertoli cells. The ovary contains vast numbers of germ cells (primordial follicles) some of which, following a period of growth and maturation, ripen into ova. The follicles in which this latter process occurs are also the site of formation of estrogens. Following extrusion of the mature ovum (ovulation) from its follicle, the tissue 'heals' and this may involve an invasion and proliferation of lutein cells so that a corpus luteum is formed. In many species, this structure is the site of formation of progesterone which can also sometimes be produced by the interstitial tissue of the ovary. A more detailed description of the structure of the gonads and the hormones that they produce is given in Chapters 2 and 3.

Associated with the gonads are the duct systems through which the germ cells are delivered to the outside of the animal. Discrete gonadal ducts are absent in the cyclostomes where the eggs and sperm are shed directly into the body cavity from which they escape to the exterior through pores that are formed in the region of the cloaca.

In the teleost fishes, the ovarian ducts represent extensions of the gonadal tissue but in other vertebrates, they are modified Mullerian ducts that differentiate in the embryo under the influence of estrogens to form the oviducts or uterus. The oviducts and uterus are surrounded by a sheath of smooth muscle fibers that, by rhythmical contractions, can propel objects, like eggs, towards the exterior. This musculature may be relatively weak, as in the amphibians, or, as in mammals, be cʳ able of very strong contractions and make up a major portion of the uterine wall. In mammals, this muscle layer is called the *myometrium*. The contractility of such muscle can be influenced by hormones, especially those from the neurohypophysis, that stimulate their contraction. Underlying the muscles of the gonaducts is an inner lining of cells that in mammals is referred to as the *endometrium*. This inner layer of tissue contains numerous glandular cells and may be modified in various ways so that it contributes to the well-being of the egg and, if internal fertilization occurs, the survival of the sperm and its union with the ovum. In oviparous species, segmental differences in function may occur along the length of the oviduct, such as are associated with the formation of albumin and the secretion of a hard outer shell. In amphibians, the jelly-like secretion so characteristic of clumps of frogs' spawn is secreted by glandular cells in the oviduct. In ovoviviparous and viviparous vertebrates, the lining of the oviduct is modified so as to furnish an appropriate environment for maintaining the retained egg or to allow for the implantation and development of the blastocyst and the formation of a placenta. The activity of the surrounding musculature is reduced on such occasions. The female gonaducts (oviduct or uterus and vagina) thus undergo considerable structural change during the reproductive cycle which is mediated by the actions of estrogens and progesterone.

The sperm are conveyed from the testis along the vas deferens, which is also a tube surrounded by an outer layer of smooth muscle, during which time they may be mixed with secretions from certain accessory sex glands that include the prostate or prostate-like glands. The maintenance of the structure and function, and cyclical changes of the male gonaducts, their associated accessory sex glands as well as the external genitalia, such as the penis, are due mainly to the action of testosterone. In the absence of this hormone, structural and physiological degeneration of such tissues take place.

Secondary sex characters in vertebrates

The secondary sex characters are so named because they are not primarily involved in the formation and delivery of the sperm or ova. They nevertheless may play an important part in the prenuptial and nuptial events and contribute to the behavioral and functional synchronization necessary for the fertilization of the ripened ova, the mechanical success of copulation and the survival of the young. The secondary sex characters differ in each sex and contribute to the dimorphism of the male and female. The differences in appearance between the sexes are basically controlled genetically and the expression of them may be influenced by the actions of sex hormones. Broad differences, such as those of size, are usually independent of the continuous action of sex hormones while in other instances only the initial differentiation of a sexual character during early life may depend on hormones. Hormones are, however, not necessarily continuously needed to maintain such organs after their differentiation; thus, the changes in the larynx of boys at puberty, that results in a deeper voice, require the presence of testosterone though subsequent castration does not result in a return to the prepuberal soprano condition. In other instances, however, a continuous supply of hormones may be required to maintain a secondary sex character, as is seen in the instance of the penis in man and the breasts in women. Some secondary sex characters may undergo periods of development and involution that correspond to the changes in the sexual cycle and the differences in the rates of hormone production that occur during these periods. The seasonal development and subsequent shedding of the antlers of deer are well-known examples of this but there are numerous others. Dodd (1960) and Parkes and Marshall (1960) give an excellent account of these structures in cold-blooded vertebrates and birds.

In *cyclostome fish*, the endocrine control of secondary sex characters has been studied in lampreys (see Larsen, 1965, 1969, 1973). One cannot distinguish, from their external appearance, between the male and female lamprey during the early part of their autumnal breeding migration into the rivers. With the approach of spring and the onset of breeding, an anal fin appears in the female while the dorsal fins of the male heighten. Such morphological changes do not occur if the animals are hypophysectomized or gonadectomized. There is also a swelling of the cloacal region in both sexes just before breeding and the urinogenital papilla grows larger in the male. These latter characters do not, however, appear to depend on the presence of steroid sex hormones.

The *chondrichthyean fishes* also display few dimorphic secondary sex characters. In some species, the most notable difference is the presence of a pair of copulatory organs called claspers in the external cloacal region of the male. These rod-like organs develop at puberty and their size can be increased by the administration of testosterone but this effect is not large.

The *osteichthyean fishes* show a considerable range in dimorphic sexual differences. In some species it is difficult to detect such variations, but in others there may be prominent differences in size. The male is sometimes much smaller than the female; variations in color may occur and there may be differences in the size of the fins. Such diversity may only arise, or be accentuated, during the breeding season. The dorsal fins of the bowfin, *Amia calva*, thus become a brilliant green color prior to breeding while the belly turns a pale green. In the female the latter is white. The anal fins sometimes become enlarged in fishes such as *Gambusia affinis* and *Xiphophorus*, where it is called the 'sword' and functions as an organ for copulatory intromission called a gonapodium. The injection of testosterone into the female fish may result in the development of secondary sex characters just like those in the male.

During the breeding season, the male South American *lungfish*, *Lepidosiren paradoxa*, develops long finger-like out-growths from the fore- and hind-limbs. These organs are bright red in color due to a rich blood supply and it has been suggested that they may function as respiratory organs. The eggs of these lungfish are laid in burrows where the oxygen tension of the water is low. It has been surmised that the male, who guards this nest and fans the eggs may secrete oxygen from these organs into the oxygen-poor water of the burrow. The growth of these, so-called, limb-gills can be induced by the injection of testosterone (Urist, 1973) which confirms their nature as that of a secondary sex organ.

Amphibians often display prominent sexual differences during the breeding season. The bright orange coloration and the dorsal crest of the male crested newt, *Triturus cristatus*, can be induced in non-breeding animals following the injection of testosterone. The development of the nuptial thumb-pads in frogs is under hormonal control. Testosterone not only stimulates the development of these tissues but, in *Bufo fowleri*, can also promote the development of the vocal sacs in immature males and prompt them to give their characteristic mating calls or croaks.

It is usually rather difficult to distinguish between the sexes in *reptiles*. The males of many lizards, however, possess appendages about their heads and throats that have a fan-like appearance and can be erected for display. Dorsal crests are also present in some lizards. The tails of male turtles often grow longer than those of the females and aid in copulation. Some reptiles, especially snakes, possess erectile peni that differentiate from their cloacal tissues. The males in many species of snakes and lizards also possess a special secretory segment in the distal part of the renal tubule, called the 'sexual segment' and the development of this is under androgenic control.

The colorful dimorphic differences in the plumage of *birds* is well known, and generally (though not always) appears to be under the control of estrogens in the female; thus, the bright coloration of the male in domestic fowl, quail and pheasants is not dependent on androgenic hormones but the duller, more conservative plumage of the female is under estrogenic control. The injections of estrogen into male birds can stimulate the formation of female plumage. The color of the beaks of birds also is influenced by sex hormones during the breeding season. The beak of the male house sparrow is normally black in the breeding season and this coloration can be induced at other times of the year, or in castrates, by administering androgens and FSH or LH and FSH (Lofts, Murton and Thearle, 1973). The domestic fowl possesses a red fleshy structure, called the comb, on its head and this is much more developed in roosters than in hens but it regresses following castration (caponization). The comb of the domestic fowl is extremely sensitive to the presence of androgens which induce its hypertrophy. This response in capons has been widely used to identify androgenic materials, and indeed provided the first unequivocal evidence for the presence of androgenic hormones in the mammalian testis.

It is unnecessary to recall the secondary sex characters in man. In other *mammals* differences in size and coloration commonly occur. The red 'sex

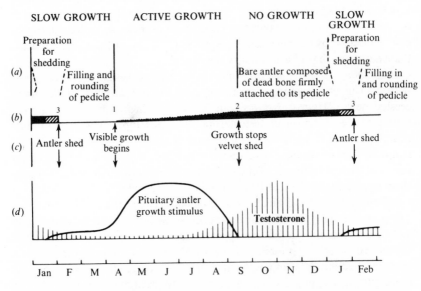

Fig. 9.1. The seasonal cycle in the growth of the antlers of the male Virginia deer (*Odocoileus v. borealis*) in relation to testosterone. This is controlled by light mediating the (hypothetical) release of pituitary hormone(s).
  (*a*) The three major phases; growth is slow, fast, and then ceases.
  (*b*) The changes in the physical size of the antlers during these periods in (*a*). .
  (*c*) The principal events in the cycle.
  (*d*) The increases in the pituitary growth stimulus which rises in the spring and subsides in the autumn when it is inhibited by the rising levels of testosterone. The precise nature of the pituitary stimulus is not clear but is probably a gonadotrophin(s). (Modified slightly from Amoroso and Marshall, 1960, taken from Waldo and Wislocki, 1951.)

skin' of the buttocks of some female monkeys during the sex cycle has been referred to in Chapter 7. Scent glands, that are under androgenic control, are also common in many mammals. The wild boar and male members of the cat family secrete odoriferous materials (pheromones) into the urine from special cells in the kidney and these glands are controlled by testosterone. The antlers of deer start to grow in the spring, apparently under the influence of pituitary hormones. In the autumn, when these animals breed, the antlers lose their covering of 'velvet', stop growing and come under the control of the rising testosterone levels in the blood. When the concentrations of this hormone subsequently decline the antlers are shed, usually at about the end of January (Fig. 9.1).

Finally it should be stressed that the behavior of vertebrates during the breeding season is also a secondary sex character (a most important one) that is influenced by androgenic and estrogenic hormones (see Goy and Goldfoot, 1973). The often bizarre (as it may appear to us) behavior of animals during courtship and mating is usually the result of the actions of the sex hormones and can often be initiated by the injection of these.

Periodicity of the breeding season; rhythms in sexual activity

Most vertebrates only breed periodically but, nevertheless, at fairly precise times of the year. In temperate zones, this more usually occurs in the spring but in some species, like deer, sheep, goats, badgers and grey seals, it occurs in the autumn. In equatorial regions where the climate and food supply are relatively similar throughout the year, breeding may often take place at any time. Similarly, some domesticated species, like the laboratory rat, the domestic fowl and man may breed throughout the year; a situation that appears to reflect continuously favorable circumstances.

An ability to reproduce during predictable seasons of the year clearly may be of considerable advantage as the young can then be produced at a time when such factors as the environmental temperature and the food and water supply are adequate. The chances for the survival of the young will thus be enhanced.

How is such precise timing possible? In temperate zones, the environmental conditions that prevail in a certain season are usually fairly predictable; thus, the animals can be expected to take their 'cues' and make their reproductive preparations on a basis of the solar calendar. Changes in the length of the day are a direct reflection of these events so that the length of the periods of light and darkness may furnish an excellent calendar to work by. Indeed, such photoperiodic stimulation is basic for the control of the reproductive cycle in most vertebrates. The first clear indication that light influences vertebrate gonadal function was made in

1925 by a Canadian zoologist called William Rowan. He found that the gonads of the junco finch, which normally enlarge when the days grow longer in the spring, could be stimulated to grow, even in winter when the birds were subjected to artificially prolonged periods of light. Other factors, however, can also impinge on the onset of the reproductive cycle and even override it. These include temperature, the nutritional condition of the animal and the related availability of supplies of food and water. There is also evidence for the presence of an internal inherent rhythm in the sexual activity of some species. It is often difficult to disentangle these various factors and to decide which is predominant.

The effects of light on reproduction have been studied in many species of birds but fewer mammals and cold-blooded vertebrates. Preparations for

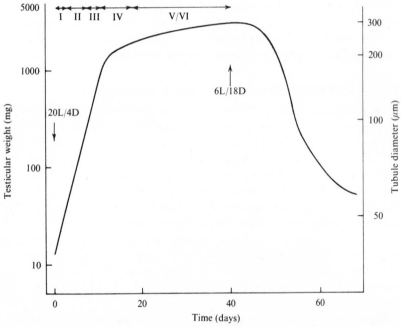

Fig. 9.2. Photoperiodically stimulated growth of the testes of the Japanese quail. For the first 40 days the birds were subjected to long, daily photoperiods of 20 h light and 4 h dark (20L/4D) and after this to short day-length of 6 h light and 18 h darkness. The diameter of the seminiferous tubules (for a given testis weight) is given on the scale on the right.

At the top of the diagram (Roman numerals) the changes in the development of the sperm are given; I = spermatogonia only, II = spermatogonia dividing, a few spermatocytes, III = numerous spermatocytes, IV = spermatocytes and spermatids and V, VI = spermatids and mature sperm. (Modified from Lofts, Follett and Murton, 1970.)

spring breeding often commence about the end of December when the length of the daylight hours starts to increase. As shown by Rowan, these conditions can be copied in the laboratory and dramatic increases in the activity of the gonads can then be shown to occur; thus, in the Japanese quail subjected to long-day photoperiods of 20 hours light and 4 hours darkness, the weight of the testes increases from 8 mg to 3000 mg in about three weeks (Fig. 9.2). The subsequent substitution of a short-day photo-period, of six hours light and 18 hours darkness results in a decline in the weight of the testes. The gain in testicular weight is due mainly to an increase in the length and diameter of the seminiferous tubules though increases in the activity of the Leydig and Sertoli cells also take place. Comparable increases in development also occur in the ovaries of birds.

Studies in mammals have been made on the laboratory rat which, if kept in continuous light, suffers deficiencies in the development of its repro-ductive system and eventually becomes infertile. Alternate periods *per se* of light and dark also appear to contribute to gonadal stimulation in animals. The breeding cycle of the ferret has also been shown to be dependent on the length of the daily period of light. Shielding the eyes from light, or cutting the optic nerves, usually abolishes the effect of such photostimulation, receptors for which appear to be present in the eye. Such ocular receptors, however, are not always vital as it has been shown, for instance, that the domestic duck still exhibits its periodic breeding behavior even after it is blinded. It has been suggested that breeding activity is due to direct photo-stimulation of parts of the brain, a process that may occur through their translucent skull, or it may reflect an endogenous rhythmical cycle that is inherent.

The reproductive cycles of all birds or mammals do not necessarily respond to light. Such photostimulatory effects are absent in rabbits, guinea-pigs, ground-squirrels and guinea fowl. These differences in response to external stimuli may reflect the effects of domestication or, possibly, in the case of guinea-pigs and guinea fowl, their origin from equatorial regions where animals do not experience large changes in day-length. Tropical deer that normally breed all the year round also persist in this habit after many years in Europe even though they experience cold winter conditions. Deer from equatorial regions that normally have a seasonal cycle also persist in their pattern of reproduction when moved to Europe.

Amoroso and Marshall (1960) have classified animals into those that have 'a long day' and 'short day' breeding season. Long-day animals, which breed in spring, include most birds, as well as horses, donkeys, ferrets, cats and racoons. Goats, deer and sheep are short-day species that breed in the autumn when the day-length is declining.

The effects of light in stimulating development of the gonads and the timing of reproduction are not seen in the absence of the adenohypophysis or when the hypothalamic connections to the median eminence are cut. Differences in the length of the daily photoperiods of light and darkness control the release, via the optic nerve, of LH/FSH-RH from the median eminence which, in turn, initiates the release of gonadotrophins from the pituitary. The gonadotrophins, FSH and LH (and sometimes also pro-lactin), exert their various effects on the development of the germ cells and the formation and release of the gonadal steroid hormones. While a distinct FSH and LH exist in mammals and birds it appears that in most other vertebrates there is a single molecule (see Chapter 3) that has sufficient of both activities to control gonadal development.

As mentioned above, the reproductive rhythms of all animals are not responsive to light. The environmental temperature, for instance, may also play an important role. While birds and mammals often will not breed in extremely hot or cold conditions, thermal changes are usually not of great importance in determining breeding in such homeotherms. In poikilotherms, however, such effects may be more significant. Spallanzani (1784) considered that reproduction in reptiles and amphibians may be related to the environmental temperature and this still seems to be correct though light may also contribute. Licht (1972) has carefully analyzed the role of temperature in controlling reproduction in reptiles and considers that it supplies the most important stimulus. Such stimuli could be acting at several sites.

(*a*) A direct action on the brain could influence the release of hormones from the median eminence.

(*b*) Temperature could be exerting a direct action on gonadotrophin formation and release in the pituitary itself.

(*c*) When the temperature is increased it has been shown that the responsiveness of the target tissues to gonadotrophins increases. This is shown in Fig. 9.3 where the responses of the ovaries and oviducts of the lizard, *Xantusia vigilis*, can be seen to increase considerably at higher temperatures.

(*d*) It is possible that changes in body temperature may indirectly alter the levels of hormones by changing the rate of their inactivation.

Temperature has also been shown to influence the reproductive cycle in fishes.

In poikilotherms the effect of temperature may be of a 'permissive' nature for in the presence of a low body temperature metabolism is depressed and could be at such a low level that an action of light, or other stimulating factors, may be ineffective.

The availability of food and water can have dramatic effects on the breeding cycle. Many birds that live in the dry desert areas of Africa and

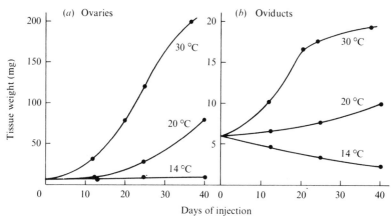

Fig. 9.3. The effects of the environmental temperature on the responses of the ovaries and oviducts of lizards (*Xantusia vigilis*) to the injections (on alternate days) of ovine FSH. It can be seen that at 14 °C there was little change in the weight of the tissues, there was a rather small effect at 20 °C while at 30 °C the growth of the ovaries and oviduct were marked. (From Licht, 1972.)

Australia (so-called xerophilous species) rapidly come into breeding condition following unpredictable seasonal rains. Breeding in most amphibians, even those from temperate regions, is also finally determined by rain and the availability of water in their breeding ponds. The South African toad *Xenopus laevis*, is thought to be unresponsive to light; reproduction is determined by an optimal nutritional condition and the availability of water. Domestic animals such as sheep are often fed a special protein-rich diet to bring them into breeding condition. It should be remembered, however, that not all species breed when they are in their best physiological condition as this may occur shortly after prolonged periods of hibernation or estivation, as seen commonly in amphibians and lungfishes, or at the end of a prolonged fast that follows a migration, such as in salmon and lampreys. At present it is not known how such a nutritional state and the availability of food and water influence breeding.

The breeding cycle of vertebrates is also influenced by a variety of ill-defined factors that for want of better knowledge are sometimes called psychological effects. These can be seen quite dramatically in many animals kept in captivity where they do not breed despite the fact that they otherwise appear to be in excellent physiological condition. This deficiency may be the result of the absence of certain environmental 'cues' such as sufficient social contact with other members of the species and an inability to perform a ritual courtship display. Social influences can be very important and it has been seen that reproduction is promoted in colonies of sea-birds when the numbers grow past a critical level. The mechanism for such effects is unknown but would appear to be mediated by the central nervous system and the hypothalamus.

While the hypothalamus and median eminence usually exert a major influence in controlling reproduction, as referred to earlier, this does not appear to occur in all vertebrates. In lampreys, the pituitary is essential for reproduction but it can be transplanted to other parts of the body, away from the region of the hypothalamus and breeding can still occur (Larsen, 1973). There is similarly no evidence that the hypothalamus controls reproduction in the chondrichthyes though gonadotrophins from the ventral lobe of the pars distalis are essential (Dodd, 1972a).

The activity of the pineal gland may also influence reproductive cycles though in this instance the evidence is principally confined to the laboratory rat. The rat pineal produces melatonin during the hours of darkness and this hormone can exert an inhibitory effect on reproduction (see Chapters 2 and 3). Pinealectomy (see Reiter and Sorrentino, 1970; Quay, 1970; Wurtman, Axelrod and Kelly, 1968) in young rats thus hastens their sexual maturation and in adults may increase the weight of the gonads. When hamsters that are normally exposed to long-day photoperiods are placed in darkness, their testes normally decrease in weight, from 3000 mg to 500 mg and this regression can be prevented by pinealectomy. The effects of pinealectomy on gonadal function in birds have been inconsistent, possibly reflecting the surgical trauma associated with this operation (Ralph, 1970). Recent experiments (Oishi and Lauber, 1974) on Japanese quail have failed to demonstrate any effect of removing the pineal on the growth of the gonads. Goldfish exposed to long-day photoperiods show an increase in the weight of the gonads in the months of January to May (but not at other times) and this effect is increased more than two-fold when the fish are pinealectomized (Fenwick, 1970). On the other hand, pinealectomy delays the maturation of the gonads in lampreys where pineal secretion may have a progonadotrophic action! (Joss, 1973).

Melatonin, that is secreted by the pineal, when administered to rats, mice and weasels decreases the responsiveness of the gonads to light stimulation. This gonadal effect has also been observed in the Japanese killifish (Urasaki, 1972), the domestic fowl and quail. The pineal may thus, through the action of secreted melatonin, exert an antigonadotrophic effect. Its site of action is uncertain but is probably the brain and this may lead to a decreased release of LH/FSH-RH from the median eminence. More direct effects, however, have not been excluded. Considerable differences exist in the responses of vertebrates to pinealectomy or the administration of melatonin (the results have been called 'inconsistent') so that at this time one cannot make a general statement as to the pineal's role in vertebrates. The daily rhythmical changes in the synthesis of melatonin, however, suggest that it could function as a 'biological clock' mediating daily or seasonal rhythms including reproduction. The pineal could thus add to or modify the role of the median eminence. It must be emphasized that at present the evidence for such an effect is equivocal. A particularly discordant note are the observations of Brown-Grant and Östberg (1974). They found that, in laboratory rats, denervation of the pineal, which abolishes its cyclical activity, did not interfere with the rats' normal ovarian cycles. They suggest that previous results, which they were unable to confirm, could have been the result of surgical trauma to the animals.

The nature of the stimuli that control reproduction are complex and we do not yet fully understand how they exert their effects. The endocrines, in close association with the brain, principally mediate the response of the reproductive system to such stimuli. The eminent British physiologist, F. H. A. Marshall, was the first to emphasize the importance of such an interrelationship in controlling breeding. Some years ago, he summarized the situation (Marshall, 1956) as follows: 'that (the) generative activity in animals occurs only as a result of definite stimuli, which are partly external and partly internal, while the precise nature of the necessary stimuli varies considerably in different kinds of animals according to the species, and still more according to the group to which the species belong'.

Maturation of the gametes – the gonadal cycles

As we have seen, animals come into breeding condition at different times of the year depending on the stimuli they receive, and react to, both from the external and their internal environments. If these 'cues' are sufficiently appropriate and are processed correctly, then breeding will be attempted. This process involves a complex series of changes in the body that are, to a considerable extent, mediated by altering the concentrations of hormones in the blood. The sperm and the ova then mature, or ripen, in preparation for their eventual union. As these preparations are proceeding, the changing levels of hormones contribute to the other physiological changes that are necessary to assure the fertilization of the ovum and, if this process is successful, the continued development of the egg and the embryo.

Such cycles in gonadal activity are relatively simple in the male when compared to those in the female. Sperm that can fertilize the ovum may be continually available for a period of many weeks, or even, as in man and feral pigeons, at all times of the year. The female, however, only produces ova available for fertilization periodically and, if not fertilized, they usually survive for less than a day. Such a periodic production of ova is an important event as it may not then occur again for many months. To mark this somewhat unique occurrence and make it clear to the male that he is at last acceptable, the female may send out various external signals and even actively seek male company. These 'signs' include 'calling', as in the cat, the production of a scent, as in the urine of the bitch, and the adoption of certain inviting sexual postures.

In mammals this period of sexual receptivity by the female is commonly called 'heat', or by physiologists, *estrus*. The preparatory period which precedes this is proestrus but if the animal is in a quiescent state, when no ova are being produced that are available for fertilization, it is called anestrus. The period during which the ova are being specially prepared for fertilization is called the *estrous cycle* which varies from four days in the laboratory rat to 27 days in kangaroos and (in its equivalent form, the human menstrual cycle) 28 days in women.

Many animals only experience a single estrous cycle in a year (called monoestrous) while others may have several such waves of ova production (polyestrous) spread out over several months of the breeding season or even for the entire year. Whether or not a single estrous cycle will be succeeded by another often depends on whether or not fertilization has occurred. If not, then there may be (though not always) another chance for successful reproduction within the over-all range of the general breeding season.

Testicular cycles in vertebrates

While certain male domestic animals exhibit continual spermatogenesis and sexual readiness throughout the year, this is not usual except in vertebrates from tropical regions and man. Seasonal breeding in a species is accompanied by a periodic maturation of the sperm (as well as the ova) along with such accessory and secondary sexual characters that facilitate its delivery on an appropriate occasion. Sperm may be available at all times during the reproductive season or mature in a single or several succeeding waves. The cystic type of spermatogenesis, where large numbers of sperm develop in unison inside envelopes that eject their contents into the seminiferous tubules, is most usual in amphibians and fishes (anamniotes) and is especially suited to those species where massive numbers of sperm are suddenly required for external fertilization. In reptiles, birds and mammals (amniotes), sperm mature from cells in the lining of the seminiferous tubules and this may be a more or less continuous process though it may also occur in waves. This acystic spermatogenesis is thought to be more suited to internal fertilization which may be attempted several times during a breed season.

The maturation of the sperm may proceed in several different ways which are dictated by whether or not the species is a seasonal breeder and whether it is poikilothermic or homeothermic.

Post-nuptial spermatogenesis is the more usual situation in seasonal poikilothermic breeders. This is illustrated by the frog *Rana esculenta*. Spermiation normally occurs in March, in Northern Europe and this is associated with a decline in testicular weight. Soon after this, however, the weight of the testis again increases and spermatogenesis proceeds throughout the summer but is halted during hibernation in winter (see Fig. 9.4) though it gradually increases again in the spring. The major spermatogenetic events thus occur in the summer preceding the breeding season and following the nuptial pairing in the spring of that year. Such a pattern of testicular activity is seen in many fishes and reptiles though considerable variations can occur. In some instances, sperm may mature fully prior to the winter hibernation, in other species, spermatogenesis may be halted at some

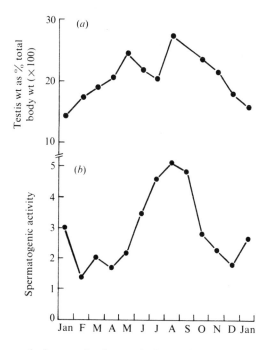

Fig. 9.4. Seasonal changes in the testicular weight (*a*) and spermatogenetic activity (*b*) in the European frog *Rana esculenta*. The decline in testicular weight that commences in May reflects spermiation during the breeding season. This sperm is that formed during the previous summer. Subsequently to this, spermatogenesis proceeds during the succeeding summer months but declines with the onset of winter hibernation. (From Lofts, 1964.)

intermediate stage of development and go on later, in the spring, or again sometimes it merely slows down in winter and proceeds more slowly.

In homeotherms, *prenuptial spermatogenesis* is usual. Testicular activity following the breeding season, during the winter months, may be slight but there is a rapid increase in activity when the spring nuptials become imminent. This pattern is usual in mammals and birds that breed periodically, though some species (such as bats) may store mature sperm in the epididymis for several months; during a period of hibernation for instance. Most birds exhibit characteristic 'refractory' periods following the breeding season, when the testes fail to respond to photoperiodic stimuli or administered gonadotrophins. The reptiles show a considerable diversity in testicular cycles. Chelonians usually exhibit amphibian-like post-nuptial spermatogenesis but the Lacertilia have several different testicular cycles (Fig. 9.5) and a pre-nuptial spermatogenesis is common.

The cyclical patterns of spermatogenesis described above are also termed *discontinuous spermatogenesis* in contrast to *continuous spermatogenesis*. The latter, apart from being present in some domestic temperate species, is common in animals that live in tropical areas where climatic conditions are relatively favorable at all times of the year. The frog, *Rana esculenta*, indeed

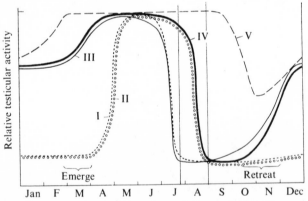

Fig. 9.5. A diagrammatic representation of the various patterns in seasonal development of the testes of lizards.

I through IV represents variations in the relative testicular activity of lacertilians that live in temperate regions: V, some tropical *Anolis* lizards.

'Emerge' and 'retreat' indicate the times that temperate species leave and enter winter hibernation.

The vertical lines indicate the period when most temperate lacertilians show no spermatogenetic activity.

It can be seen that pre-nuptial spermatogenesis occurs in types I and II (in spring) but in types III and IV spermatogenetic activity commences in the autumn. (From Licht, 1972.)

has a continuous spermatogenetic cycle in warm Mediterranean regions while, as described above, it has a discontinuous cycle in the more northern parts of Europe. This frog has thus been classified as a *potentially continuous breeder* or a continuo–discontinuous type. The environmental temperature determines which pattern will persist in such species. It should be noted, however, that it is not possible to alter a discontinuous spermatogenetic cycle to a continuous one simply by raising the temperature; it will not, for instance, change in *Rana temporaria* where the tissues appear to undergo an inherent rhythm in their ability to respond to hormonal stimuli. Variations that occur in the patterns of spermatogenetic activity are illustrated in Table 9.1 which summarizes differences among amphibians from different geographic areas.

The testicular cycle (like the ovarian cycle) is controlled by the adeno-hypophysis. Removal of the pituitary abolishes such cyclical activity and results in a regression of the testis that involves both the germ cells, in the seminiferous tubules, as well as the endocrine interstitial tissue (Leydig cells). In mammals, this is thought to involve the action of FSH on the seminiferous tubules and LH (in this instance more appropriately called ICSH) on the interstitial tissue. As we have seen (Chapter 3) in reptiles, amphibians and fishes a single gonadotrophin, incorporating both of these activities, may be present. While LH stimulates the production and release of testosterone the mode of action of FSH is less certain. It is usually necessary for the full maturation of the sperm but this may not be a direct effect. Hypophysectomized mammals, birds and fishes, in which the testes atrophy, can produce sperm following the administration of testosterone. It is surprising, however, that testosterone apparently cannot restore spermatogenetic activity in amphibians or at least in *Rana pipiens* (see Basu, 1969; Lofts, 1968).

It should be pointed out that the administration of testosterone to intact animals has often been observed to result in testicular regression and an inhibition of spermatogenesis. This paradox is apparently due to an inhibition by the androgen of the release of endogenous gonadotrophins. A parallel direct effect on the seminiferous tubules has not, however, been excluded. The inhibitory effects of testosterone on spermatogenesis thus appear to reflect an overabundance of this steroid.

Testosterone is, nevertheless, usually necessary for the maturation of the sperm but it is not yet clear whether FSH acts solely by stimulating the production of an androgen. Unfortunately the spermatogenetic effects of the administration of testosterone are often variable. It seems likely that FSH may act on the Sertoli cells to produce androgens that in turn mediate the maturation of the sperm (see Lofts, 1968).

Spermatogenesis is a prolonged and complex process that requires pre-

TABLE 9.1. *The types of spermatogenetic cycles exhibited by anuran amphibians that live in different geographical regions.* (From Basu, 1969)

Species	Habitat	Remarks
Continuous type of spermatogenesis		
Bufo arenarum	S. America	
Bufo paracnemis	S. America	
Bufo granulosus d'orbignyi	S. America	
Bufo melanostictus	India, Java	Lunar periodicity from Java reported
Telmatobius schreiteri	Andes mountains (high altitude)	Low winter temperature cannot affect the cycle
Hyla raddiana andina	Andes mountains (high altitude)	
Rana erythraea	India	
Rana grahami	India	
Rana hexadactyla	India (Pondicherry)	
Leptodactylus ocellatus reticulatus	S. America	
Leptodactylus prognathus	S. America	
Leptodactylus laticeps	S. America	
Physalaemus fascomaculatus	S. America	
Pseudis paradoxa	S. America	
Pseudis mantidactyla	S. America	
Rana cancrivora	Java	
Discontinuous type of spermatogenesis		
Rana temporaria	Europe	
Leptodactylus asper	S. America	
Leptodactylus bufonis	S. America	
Phyllomedusa sauvagii	S. America	
Rana pipiens	USA	High temperature causes spermatogenetic continuity
Hyla crucifer	USA	
Pleurodema bufonina	S. America (Patagonia), Australia	
Continuo–discontinuous type of spermatogenesis		
Rana esculenta	Europe	In winter spermatogenesis goes only up to spermatid formation
Rana gracea	S. America	
Rana ocellatus typica	S. America	
Rana tigrina	India (Calcutta)	
Rana nigromaculata	Japan (Niigata), China	

Species	Habitat	Remarks
Leptodactylus ocellatus typica	S. America	Two interruptions in spermatogenesis during cold winter and high summer
Variable spermatogenetic cycle as per geographic distribution		
Rana esculenta	Europe	Discontinuous cycle
ridibunda	Mediterranean region	Continuous cycle
Discoglossus pictus	Europe	Discontinuous cycle
	Mediterranean region	Continuous cycle

and postnatal maturation of the gonacytes, mitotic divisions of the spermatogonia and meiotic reduction divisions to form the spermatocytes, spermatids and the final (spermiogenesis) differentiation of spermatozoa. Androgens, and possibly FSH, are required for certain of these steps to proceed in a normal manner but there is considerable interspecies variation as to the stages of sperm maturation at which these hormones act. The endocrinology of gametogenesis is an important subject about which we know little. Dodd (1960, 1972*b*) has summarized what is known about this process in vertebrates. In the rat, testosterone may be necessary for early pre- and postnatal development of the gonacytes and it also promotes the meiotic division of the spermatocytes later on. FSH is required for the maturation of the spermatids. This pattern is, however, not the same even among the mammals and the information that is available is rather meager. In lampreys (Cyclostomata), hypophysectomy has little effect on the final stages of the maturation of the sperm. When this operation is performed in late winter or spring spermiation still occurs, but if hypophysectomy is carried out earlier, in October for instance, there is a considerable delay and spermiation may not take place at all (Larsen, 1973). In chondrichthyean fishes, hypophysectomy also inhibits the earlier stages (meiotic divisions of the spermatogonia) of sperm maturation and the same situation seems to apply among the teleosts. It seems likely that such effects of hypophysectomy are mediated by insufficient androgens though the lack of a direct effect of FSH may also be important. One cannot, with the limited and varied information available, make any generalization as to which stages of spermiogenesis are hormone-dependent in vertebrates. There are instances in fishes, as well as reptiles, birds and mammals, where the later stages of maturation are apparently dependent on endocrine stimulation. The situation in amphibians is especially confusing as in some instances spermatogenesis appears to be occurring at a time when the endogenous production of testosterone is low and exogenous testosterone may exert a direct inhibitory effect on the early stages of spermatogenesis. The comparative endocrinology of gametogenesis needs, and certainly merits, further exploration.

Little direct information is available about the circulating levels of testosterone and gonadotrophins in seasonally breeding animals. Changes in the concentrations of such hormones are inferred from histological examination of secondary sexual characters. The Leydig cells, and their analogues the boundary cells show a seasonal pattern in their histological appearance (see Lofts, 1968). In the periods that precede breeding, these cells enlarge and accumulate lipids and cholesterol. These cells become depleted of these materials at the height of the breeding season and this is thought to reflect the secretion of androgens which utilize lipids as their substrates. These histological changes are associated with breeding behavior, the development of secondary sex characters and can be imitated by the injection of gonadotrophins. Such changes have been observed in fishes, amphibians, reptiles, birds and mammals. An example of this can be seen in Fig. 9.6 where the height of the thumb-pad epithelia in the frog *Rana esculenta* can be seen to decline in June when the lipid and cholesterol content of the interstitial tissue is greatest. A seasonal pattern in the ability

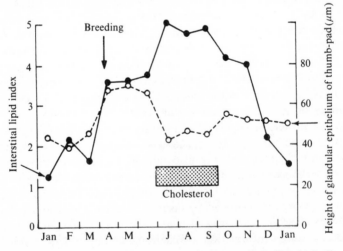

Fig. 9.6. Seasonal changes in the amount of lipid observed, histologically, in the interstitial cells of the testis of the frog *Rana esculenta*. This reaches a maximum in mid-summer, after breeding has occurred, when the cellular cholesterol levels are also greatest. These changes are thought to reflect a decline in the synthesis of androgens (which utilize the lipids as substrates for their formation). This change is consistent with the decline in the development of the thumb-pad which is under androgenic control. (From Lofts, 1964.)

of the testis of the cobra to convert progesterone to androgens, *in vitro*, is shown in Fig. 9.7. This androgen synthesis reaches an initial maximum during breeding in May but drops subsequently as the testes atrophy.

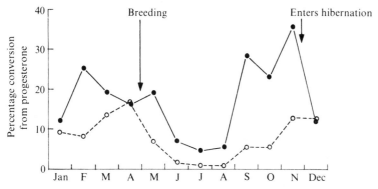

Fig. 9.7. Seasonal changes in the ability of the testis of a snake, the cobra, to convert (*in vitro*) progesterone (which acts as a substrate) to androgens. Testosterone production declines following breeding in May but rises again in late summer and autumn, during post-nuptial spermatogenesis, only to decrease once more as the snakes go into winter hibernation: ●, testosterone; ○, androstenedione. (From Lofts, 1969.)

When spermatogenesis is again initiated in the autumn the rate of progesterone to androgen conversion again increases but declines again with the onset of winter hibernation. Direct measurements of the plasma androgen concentrations have been made in the Australian lizard *Trachysaurus (Tiliqua) rugosa* (Bourne and Seamark, 1973). These reptiles breed in spring when the testicular weight is about 1300 mg, compared to 180 mg in summer. During the breeding season, the androgen concentration in the plasma is 33 ng/ml but it is only 10 ng/ml at other times of the year. A 10-fold increase in the plasma testosterone levels has also been observed in starlings (*Sturnus vulgaris*) during the breeding season and this was associated with an increased activity of the testicular interstitial cells (Temple, 1974). A periodic decrease in the hormone-secretory interstitial, as well as the spermatogenetic, tissue occurs in all the non-mammals that breed periodically. In mammals, however, there is usually a permanent hormone-secretory tissue in the testis.

The Sertoli tissue, which has been identified in all groups of vertebrates, has for a long time excited speculation as to its function. The histological appearance of this tissue shows changes in parallel to those of the Leydig cells (see Lofts, 1968) and spermatogenesis. An accumulation of lipids, in the Sertoli cells, follows spermiation but these materials are depleted when spermatogenesis is occurring. In animals that normally breed continually, like laboratory rats, hypophysectomy results in an accumulation of lipids in the Sertoli cells and this is thought to reflect a lack of their stimulation by FSH. It is now widely accepted that the Sertoli cells produce androgenic steroids that may influence spermatogenesis, and FSH probably acts to stimulate the secretion of such hormones. The cyclical changes that occur in the appearance of the Sertoli cells thus presumably reflect changes in the endogenous gonadotrophin levels in the blood.

Ovarian cycles in vertebrates

The maturation of the ovum in the ovary and its extrusion (ovulation) and passage into the oviduct or uterus involves the coordinated activity of FSH, LH, sometimes prolactin, and the secretion of estrogens, progesterone and, possibly, even small amounts of androgens from the ovary. The ovarian cycle results from increases and declines in the circulating concentrations of these hormones and this is largely the result of their interactions in stimulating or inhibiting each other's release, through a negative- and positive-feedback to the median eminence and hypothalamus.

These hormonal rhythms have only been closely analyzed in mammals and even these results are usually confined to more domesticated species.

Placental mammals

Three general patterns have been identified in the ovarian cycle of placental, or eutherian, mammals and these have been described (*a*) in the sheep, pigs and cattle, (*b*) in the laboratory rat and (*c*) in man.

Sheep

The ewe usually comes into estrus in the autumn, as a result of stimulation by shortening periods of daylight, and if pregnancy does not occur will continually produce ova at intervals of about 16 or 17 days until the following spring. Some breeds of sheep, like the merino, may breed for longer periods of the year. The estrous cycle of the ewe lasts for about 17 days and the hormonal changes that occur are summarized in Fig. 9.8. The onset of estrus is taken as time zero in the ovarian cycle and this lasts for about 24 hours, ovulation occurring towards the end of this time. Following ovulation, the blood supply to the ruptured Graafian follicle increases and the granulosa cells luteinize to form the corpus luteum. This structure reaches a maximum size on about day 8. Luteinization of the follicle is initiated by the action of LH and the secretion of progesterone is also stimulated by this hormone. LH is luteotrophic, an effect that is seen in most mammals. In some species, including the rat, mouse, rabbit and possibly the sheep, prolactin may also have a luteotrophic effect. Sometimes the two hormones act in conjunction with each other. Progesterone secretion from the corpus luteum rises until about day 11 of the cycle and then on day 13 undergoes a precipitous decline.

Accompanying these events is the development of the Graafian follicles and the maturation of the ova. This process proceeds under the influence of FSH and the estrogens that are secreted by the follicular cells which are

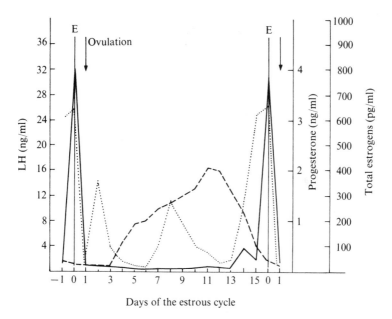

Days of the estrous cycle

Fig. 9.8. The estrous cycle of the ewe. Estrus (E) occurs at time *zero* and is followed by ovulation. The concentrations of plasma progesterone (from the jugular), estrogens (ovarian vein) and LH (from the jugular or ovarian vein) are shown in relation to these events. It can be seen that on day 12 there is a decline in progesterone which is accompanied by climb in estrogen concentration that initiates a 'surge' in release of LH that results in ovulation: dashed line, progesterone; dotted line, total estrogens; solid line, luteinizing hormone (jugular or ovarian vein). (From Hansel and Echternkamp, 1972.)

also stimulated by LH. LH thus appears to have a general steroidogenic effect on the ovarian tissues. During the preovulation phase of the cycle, estrogen levels are moderate but as can be seen in Fig. 9.8 they may display some periodic changes. The LH level is low but sufficient to maintain the secretion of steroid hormones. The estrogens and progesterone that are produced act on the accessory sex organs, especially the uterus and vagina, to get them into 'tip-top' condition for the prospective fertilization, the implantation of the egg, and pregnancy. The release of LH is kept low in the preovulatory period as a result of a negative-feedback inhibition of its release that is exerted by progesterone on the hypothalamus.

Between days 13 and 16 of the cycle dramatic changes take place in the hormonal concentrations that result in ovulation. There is a rapid decline in progesterone that reflects a breakdown of the corpus luteum. LH levels are still adequate to maintain this tissue but it loses its sensitivity to the hormone. Estrogen levels then climb upwards and by a positive-feedback action on the hypothalamus bring about a massive release of LH that causes the follicle to rupture and extrude its ovum. The effect of the estrogen via the median eminence is reflected in a decline of stored luteinizing hormone-releasing activity in the ewe's hypothalamus (Crighton, Hartley and Lamming, 1973).

The hormonal basis for estrous behavior has recently been questioned. It had commonly been assumed that it was solely a reflection of the actions of estrogens. The estrogen level, however, usually drops considerably just prior to estrus; the preovulatory surge is in fact rather temporary. There is now evidence to indicate that not only estrogen but also progesterone, as well as a small but significant release of an androgen, androstenedione, from the ovary, may contribute to estrous behavior.

After ovulation the LH stimulates the follicle granulosa cells to luteinize and if fertilization does not occur the cycle will then recommence. In the event of fertilization and an ensuing pregnancy, the corpus luteum, as we shall see later, will persist for a much longer time and contribute to the events of the gestational period.

The corpus luteum can thus be seen to play a commanding role in the estrous cycle and it has been called 'the clock'. The reason for the decline in the activity of the corpus luteum during the latter part of the estrous cycle has only recently been elucidated. It has been known for many years that, when the uterus of guinea-pigs is removed, the corpus luteum persists for a much longer period of time. This effect can also be seen in the ewe, as well as the cow and sow, but not in women, the rhesus monkey, dog, badger nor marsupials (Anderson, 1973). The non-pregnant uterus, in some species, appears to produce a substance that has been called a *'luteolysin'* that causes the corpus luteum to atrophy. There is some evidence to suggest that this may be a prostaglandin. Whatever its nature, it is interesting that this effect is not seen if the ovary is transplanted to a region that is distant from the uterus, such as the neck. The ovarian arterial blood and the uterine venous blood pass by each other in closely apposed vessels and it seems that this special vascular arrangement functions as a unique pathway by which the luteolysin can get to the ovary before it becomes diluted, or destroyed, in the general circulation. It is important, however, to remember that this effect is not seen in all species nor has such a role for a luteolysin been accepted by all (see Nalbandov, 1973). Instead, it has been proposed that the death of the corpus luteum results from a lack of LH due to competition for this trophic hormone by newly developing ovarian follicles.

Finally, before leaving the ewe, some comment should be made about the modern techniques that have made such investigations possible. For many years information about the estrous cycles in animals depended on histological examination of the ovary and its accessory tissues, the uterus and vagina. It was also possible to measure urinary steroid excretion but identification and measurement of blood steroid concentrations were not generally feasible or were unreliable. Applications of the science of immunology to endocrinology, however, changed all this. The basic technique for the radioimmunoassay of hormones, including the sex hormones, was developed initially by Solomon Berson and Rosalyn Yalow in New York. This procedure, which uses radioactively labelled antibodies to various hormones, made it possible accurately to identify and measure the small amounts of a variety of hormones present in the blood. In addition, antisera to many hormones have been developed, including anti-LH and anti-progesterone, which, by their ability to block, or not, certain processes in the reproductive cycle, facilitate understanding the gonadal mechanisms that are coordinated by the pituitary and steroid hormones.

The laboratory rat

The female rat has an ovarian cycle that is broadly similar to that of the ewe but there are some notable differences (see Fig. 9.9).

The rat estrous cycle lasts for four days and the events are normally regulated on the basis of a diurnal rhythm. This facilitates experiments on these animals as it is known that ovulation always occurs just after midnight and other changes can also be timed with remarkable accuracy (Fig. 9.10).

The adoption of such a rhythm is vital to the rat as the corpus luteum, which, as we have seen, acts as a 'clock' in many other mammals, is small and does not persist for long during its estrous cycle. Progesterone is present throughout the estrous cycle but it does not appear that changes in its levels contribute to the timing of ovulation. Instead there appears to be a 'timed' estrogen release on the morning of proestrus sometime before 10.00 a.m. This steroid promotes, as in other mammals, LH release but in contrast to other species this is immediately followed by a marked elevation in the plasma progesterone concentration. As this precedes the formation of the corpus luteum it presumably arises as a result of a luteotrophic effect

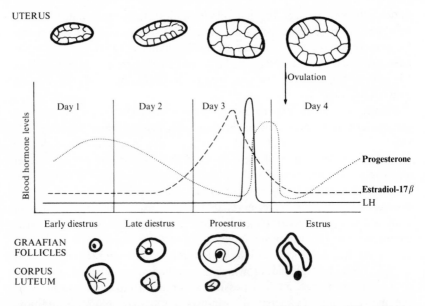

Fig. 9.9. The estrous cycle of the laboratory rat. The blood hormone levels; progesterone, estradiol-17β and LH are shown in relation to estrus and ovulation.

Also included are a representation of changes in the stages of development of the uterus, Graafian follicles and corpus luteum. It can be seen that the corpus luteum does not persist throughout the entire estrous cycle of the rat and that the preovulatory increase in progesterone is due to the secretion of this hormone by the ovarian interstitial tissue.

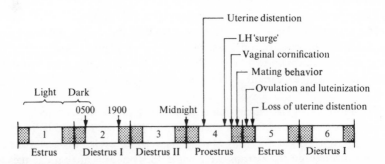

Fig. 9.10 The principal events in the rat estrous cycle in relation to the time of the day.

This cycle is precisely timed on the basis of a diurnal rhythm. Ovulation can be seen to occur shortly after midnight. Other events including; mating behavior, the LH 'surge' and the development of the uterus. (From Armstrong and Kennedy, 1972.)

of the LH on the ovarian interstitial tissue. This progesterone is vital to the onset of estrus and subsequent changes that occur in the uterus, which becomes less distended while the endometrium becomes more glandular. Progesterone does not appear to be important for ovulation which cannot be prevented by anti-progesterone serum.

The corpus luteum persists only if pregnancy or pseudopregnancy (as a result of copulation without fertilization) occurs, otherwise it does not have a significant role in controlling the rat ovarian cycle.

Man

The human ovarian, or menstrual, cycle lasts for 28 days and is quite distinct from that in non-primate mammals. A similar ovarian cycle also occurs in monkeys and apes. The corpus luteum persists for a more prolonged period of the cycle than in the rat but not for the whole of it, as in the ewe, cow or sow. The timing of the events of the human ovarian cycle is arbitrarily taken from the initiation of menstruation. This process is due to the discharge of superfluous structural remnants and secretions of the endometrium and contains some blood. In effect, menstruation represents the termination of the previous ovarian cycle and the life of the corpus luteum and lasts for about 4 to 5 days.

The preovulatory period of the human ovarian cycle thus takes place in the absence of a corpus luteum so that progesterone secretion is relatively small. 17α-OH-progesterone levels rise but this metabolite has little progestin activity. The follicle ripens under the influence of FSH and secretes estradiol which controls the release of this gonadotrophin by a negative-feedback inhibition in the hypothalamus. As in other mammals, a sudden surge in LH, initiated by the secreted estrogen, results in ovulation, usually on about day 16. There is also a sudden rise in FSH release at this time which may contribute to ovulation. No precise period of estrous behavior exists in the human female. She is receptive to the male at any time of the cycle. The ruptured follicle starts to luteinize after ovulation and progesterone secretion, due to the action of LH, rises but subsequently declines when the corpus luteum later degenerates and menstruation then occurs. In the event of fertilization, however, the corpus luteum persists. The reason for the premenstrual decline in the activity of the human corpus luteum is unknown and this degenerative process can proceed normally following hysterectomy; apparently it does not involve the action of a uterine luteolysin.

At least three different types of ovarian cycles thus exist among placental mammals. The principal differences involving the role of the corpus luteum and progesterone.

Ovulation

The mechanisms of initiation of ovulation, however, may also differ among the mammals. In the examples described above, ovulation takes place in response to an internal programming that controls hormone release so that ovulation is then said to be *spontaneous*. In other mammals, however, ovulation can be *induced* as a result of copulation and sexual excitement. This latter type of ovulation is known to occur in such species as the rabbit, cat, ferret, mink and racoon and it is suspected that it may also sometimes occur even in women. In such species, estrogen is released from the developing ovarian follicles which are under the influence of FSH (see Schwartz, 1973). This estrogen indicates when the follicles are ripe and results in mating behavior. The latter is in contrast to spontaneous ovulators in which progesterone is also necessary. If copulation takes place, this initiates a surge of LH release, as a result of neural stimulation of the hypothalamus and pituitary, and ovulation occurs. This event is accompanied by a rise in progesterone levels and takes place several hours after coitus when the sperm are ensconced in the oviduct. In non-mammals, the situation is less clear, however, the mere presence of the male or even some substitute may be all that is necessary to initiate ovulation. Apart from gallinacious birds like the domestic fowl, as well as domestic geese and ducks, most birds do not produce eggs in the absence of the male. It has, however, been reported that some birds, such as pet parrots, will lay eggs if suitably stroked and tickled. Copulation may thus not always be necessary and courting behavior and sexual display may be effective stimulants of ovulation.

Following parturition, several species of eutherians including the rabbit, ferret, mink and racoon come into a post-partum heat when they copulate and this, as indicated above, results in ovulation. Copulation not always needed to precipitate ovulation in these circumstances for, as we shall see in the next section, post-partum ovulation is common in marsupials where it is a spontaneous event and occurs at a time that merely reflects an extension of the normal estrous cycle.

Delayed implantation

Pregnancy usually persists for a precise and predictable period of time. Some interesting and, at first, mystifying exceptions have, however, been encountered. Animals that conceive in the autumn and deliver their young in spring can, on some occasions, such as when the length of the daylight period is artificially increased, produce their young much earlier. There have been other instances described, especially in kangaroos, where a female has been taken into captivity and without any contact with a male has, many months later, given birth to a young one. Faced with the necessity for an explanation, some people were even forced to consider the possibility of virgin birth! The cause is, nevertheless, quite a reasonable one. In a number of mammals especially the mustelid Carnivora (such as weasels and sable) and macropod marsupials (kangaroos), development of the fertilized egg can sometimes cease when a blastocyst, containing about 100 cells, has been formed. This blastocyst lies dormant for a time that may extend for several months, but can be subsequently stimulated to continue development. The delay is called an *obligatory* one when it is determined by external conditions, such as light, as seen in badgers, pine-marten, weasels and the roe deer. In other species, such as the mouse, rat and macropod marsupials it is *facultative* and controlled by more physiological events. As will be described in more detail on p. 344, this inhibition results from the effects of suckling and lactation.

Pregnancy

An excellent account of the role of hormones in this process is given by Heap (1972) and Heap, Perry and Challis (1973). When the fertilized egg is retained in the oviduct or uterus and the subsequent development of the young occurs at this site, within the female, pregnancy is said to be occurring. This term is usually assumed to include the viviparous condition but may also encompass ovoviviparity. The internal incubation of the young is also called *gestation*. The condition of pregnancy appears to have reached its highest state of organization in placental mammals though little information is available about this process in non-mammals. Pregnancy is not a uniquely mammalian phenomenon as it occurs in some chondrichthyeans, teleosts, reptiles and amphibians, though not in birds. Gestation may occur for quite long periods of time in placental mammals but this is not unique as it may extend for two years in some viviparous sharks and is of four years duration in the ovoviviparous urodele, *Salamandra atra*.

As we have seen, the hormonal preparation of the mammalian uterus for the reception, fertilization and implantation of the egg is initially stimulated by estrogens and progesterone, the latter usually having the subsequent dominant action, though both steroids act in simultaneous collaboration. Subsequently during pregnancy, these favorable uterine conditions need to be maintained and even modified from time to time as the fetus grows and is eventually delivered to the outside world. The necessary supplies of hormones are then not only altered qualitatively but increased quantities may also be required. These added needs have been met in various ways by the placental mammals and principally involve the function of the pituitary, the ovary, the placenta and the uterus.

Progesterone, to use an oft-quoted phrase, is called 'the hormone of pregnancy' but substantial, though usually smaller, amounts of estrogens are also used during gestation. These gonadal steroids maintain the endometrium and contribute to the considerable expansion that occurs in the myometrium during pregnancy. The hypertrophy of these muscles results from the stretching of the walls of the uterus and the induction, by estrogens, of new contractile proteins. Contractions of the uterus are not usually desirable during pregnancy and the responsiveness of the myometrium to stimulation is reduced by progesterone. The contractile effects of oxytocin are, for instance, usually reduced by pretreatment of the uterus with progesterone while estrogens have the opposite effect and enhance the responses to this neurohypophysial hormone. Such effects have not been demonstrated in all species but are very reproducible in some, like the rabbit. A most important role of progesterone in pregnancy in placentals is the inhibition of the estrous cycle and ovulation. This effect results from a negative-feedback inhibition of the release of LH from the pituitary and may be required when the periods of gestation exceed the length of the normal estrous cycle. Corpora lutea also persist in many viviparous and ovoviviparous non-mammals and although their precise role is uncertain it is suspected that they may also have a comparable role in such animals.

The problem of how to supply the added hormonal requirements of pregnancy has been met in various ways by different species of placental mammals. Estrogens and progesterone are typically secreted by the vertebrate ovary. The corpus luteum is usually the principal ovarian source of progesterone but normally this structure does not persist for longer than the estrous cycle. As we shall see, this situation even occurs in pregnant Australian marsupials. The period of gestation in these animals is similar to that of their estrous cycles so that a prolongation of the life of the corpus luteum is unnecessary. In the placental mammals, which have relatively longer periods of gestation, the corpus luteum persists for a much longer time and often remains functioning throughout the entire period of pregnancy. This extended survival is the result, in some species, of an inhibition of the effects of uterine luteolysins, due to the presence of extra material in the uterus and the stimulating actions of mixtures of luteotrophic hormones. These hormonal combinations may consist of FSH, LH, prolactin, and gonadotrophins that may be produced by the placenta. The precise hormonal content of this so-called '*luteotrophic complex*' differs considerably from species to species. Its function is to extend the normal lifetime of the corpus luteum and to promote the secretion of progesterone.

The production of progesterone by the ovary may be facilitated in various ways. In some species (the horse) additional corpora lutea may form, but in others (human and cattle) only a single corpus luteum is usually present. Animals that produce several young at a time have a correspondingly greater number of corpora lutea available for the production of progesterone. During pregnancy, the secretion of progesterone by individual corpora lutea may be increased by the action of the luteotrophic complex. In addition, the amount of available progesterone is the net result of its rate of production and destruction. As described in Chapter 4, proteins that bind steroid hormones are present in the plasma and when they are in such a bound condition the rate of the destruction of these hormones is reduced. During pregnancy, the formation of such steroid hormone-binding proteins in the liver may be increased, probably as a result of stimulation by estrogens. There are considerable interspecific differences in the physiological patterns that ensure adequate progesterone in pregnancy.

In some species, such as the rabbit, ovariectomy during pregnancy always results in prompt abortion. In other species, such as the sheep, guinea-pig, rat and human this operation does not necessarily result in a loss of the fetus. The placenta in these species produces sufficient gonadal steroids to support the uterus, though the supply may be inadequate during early pregnancy. There are considerable interspecific differences in the ability of the placenta to produce hormones. The placenta of the rabbit and goat, for instance, does not produce any steroid hormones, or luteotrophins, while the human placenta produces large quantities of all these materials. The pig placenta produces estrogen but not progesterone, while the sheep produces large amounts of progesterone.

The appearance of the placenta as a temporary endocrine organ that helps to supply the hormonal requirements of pregnancy is a fascinating physiological adaptation. Such a role has not been described in non-mammals or even in marsupial mammals. It is not possible to draw any orderly phyletic line as to the distribution of this hormone-secreting tissue in placentals and it could have evolved separately on several occasions to suit the needs of each particular species. In recent years, it has become apparent that tumorous tissues in mammals may produce a variety of hormones that normally arise from discrete endocrine glands. Perhaps there is some analogy between such tumors and the evolution of an endocrine placenta!

Parturition

The delivery of the young is a very precisely timed event that, although it undoubtedly involves hormonal changes is not very well understood. During pregnancy, the progesterone-dominated uterus is in a relatively quiescent state and it is reasonable to suspect that removal of such a progesterone-block may precipitate contractures that result in parturition. Plasma progesterone levels are quite low at birth in some animals, like sheep and ferrets, but in others, like women and guinea-pigs, this is not so. Estrogen levels often rise dramatically on the approach of parturition and this may sensitize the uterus and oppose the progesterone-block. Oxytocin is released from the neurohypophysis and stimulates contraction of the uterus. This hormone facilitates the delivery of the young but in some species parturition can still occur in the absence of oxytocin. Its role is therefore not always essential in mammals. Another polypeptide hormone, of uncertain chemical composition, *relaxin*, is released from the ovary and is also present in the placenta of rabbits and guinea-pigs. Relaxin promotes the relaxation of the ligaments of the pubic symphysis to allow passage of the young through what has been described as 'this triumphal arch!' This effect (on the pelvic ligaments) is detectable in most mammals but is most prominent in rabbits and guinea-pigs.

The desire to care for the young is also a hormonally mediated effect. The process of lactation from the mammary glands is normally (except it seems in man) an important part of this process in all mammals. The role of the hormones in lactation has been described in Chapter 5.

Marsupial mammals

The marsupials, that include the American opossum and large numbers of Australian mammals, like kangaroos, have a unique form of reproduction that is accompanied by some remarkable endocrine innovations. The initiation of the study of the reproductive endocrinology of the Australian marsupials was principally due to the efforts of H. Waring on the occasion of his academic migration to Australia in 1948.

Systematically, the marsupials are distinguished from the placentals by the absence of a true placenta; only a yolk-sac placenta is present so that the physiological connection with the parent is more tenuous than in placentals. Young marsupials are born in a relatively immature state comparable to that of embryos in the later part of the gestational period of placentals. The newborn young of marsupials are suckled on the teat where they undergo a considerable part of the development that would normally occur within the uterus of placentals. In many, but not all, marsupials this takes place in an external pouch or marsupium.

According to the opinion of G. B. Sharman, who is well informed in these matters, marsupials and placentals probably evolved from a common oviparous ancestor. Sharman's (1970) account of marsupial reproduction should be consulted by those who wish to extend their knowledge of this process. An even more comprehensive account has been provided by Tyndale-Biscoe (1973).

Australian marsupials usually breed from mid-summer to early winter: January to June in the Southern Hemisphere. This can, however, be modified so that, in periods of drought, reproduction will be inhibited while in favorable conditions it may occur at almost any time of the year. While there are many similarities between the endocrine control of reproduction in placental and marsupial mammals there are also some remarkable differences. The gonadal steroids, estrogens and progesterone, mediate the pre- and post-ovulatory changes in the reproductive tract during the estrous cycle of marsupials. Little is known about the nature of the pituitary gonadotrophins in marsupials except that some effects seen in

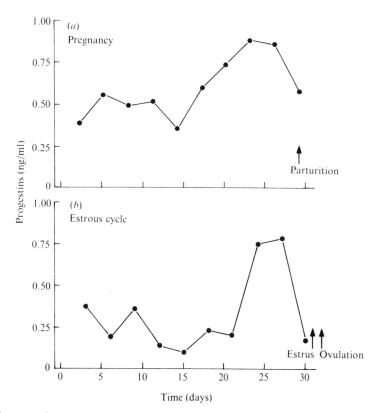

Fig. 9.11. Changes in the plasma progestin levels during pregnancy (*a*) and the estrous cycle (*b*) of a marsupial, the tammer wallaby *Macropus eugenii.*

The commencement of pregnancy or the estrous cycle was initiated by removing the suckling young from the pouch, thereby initiating the development of the fertilized blastocyst (see text) or the next reproductive cycle. Progesterone levels commenced to rise on about day 15 and declined just prior to estrus and ovulation, or parturition. (From Lemon, 1972.)

placentals can be mimicked by the injection of the placental hormone preparations. In marked contrast to placentals, pregnancy in marsupials does not interfere with the concurrent estrous cycle and the maturation and ovulation of the egg. This egg is usually produced at the normal time. Certain marsupials, the macropods or kangaroos, also display an interesting form of delayed implantation or more correctly called in this instance an *embryonic diapause* that differs from the process that is occasionally seen in placentals. The process of lactation in marsupials also has some rather unique features.

The estrous cycle of marsupials lasts for about 28 days and, as in placentals, consists of an initial period of follicular growth that is accompanied by the development of the uterus and vagina, at first under the influence of estrogens and then, in its secretory phase, by progesterone. Estrus follows a sudden decline in the level of progesterone (Fig. 9.11), reminiscent of that seen in some placentals, and lasts for several days during which ovulation occurs. The precise endocrine stimuli for ovulation have not been elucidated but it is spontaneous and not the result of any external sexual or copulatory stimuli. Fertilization is followed by the development of the egg into a blastocyst and if the animal is not already lactating, pregnancy will occur. A corpus luteum is formed in the ruptured follicle and persists for the period of time that is usual in the estrous cycle; its life is *not* prolonged by the pregnancy. An extended life for the corpus luteum is not necessary in marsupials as the period of gestation is usually nearly identical to the time of the normal estrous cycle (which continues to occur concurrently with the pregnancy!); nevertheless, the progesterone levels in pregnancy are greater than those in the normal estrous cycle (Fig. 9.11) which seems to reflect a hypersecretion from the ovary.

Ovulation of the egg, that ripens during pregnancy, occurs at various times in relation to parturition. The relationships of the estrous cycle and pregnancy are summarized in Fig. 9.12. Ovulation may occur *prior* to parturition, as in the swamp wallaby, *Wallabia bicolor*, where the period of gestation is 35 days compared to only 32 days for the estrous cycle. In this marsupial, pre-parturition ovulation is followed by copulation. If fertilization takes place, a blastocyst develops which if, lactation then occurs, lies dormant (see later). In other species like *Megaleia rufa*, parturition is closely succeeded by ovulation, post-partum copulation, and the formation of a blastocyst. In the grey kangaroo, *Macropus giganteus*, the period of gestation is much shorter than the estrous cycle, just as seen in the Australian brush possum, *Trichosurus*, and other non-kangaroos, and pre-scheduled future ovulation is then inhibited by the suckling stimulus provided by the young. If, however, the young is removed, ovulation follows nine days later. In the latter part of lactation of the grey

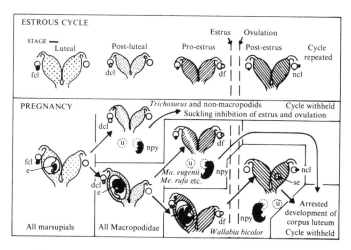

Fig. 9.12. The estrous cycle and pregnancy in marsupials. A diagrammatic summary of the size and functional relationships of the ovary and the uterus.

Estrous cycle: fcl, functional corpus luteum; dcl, degenerating corpus luteum; df, developing Graafian follicle; ncl, new corpus luteum.

Pregnancy (additional abbreviations): e, intrauterine embryo; npy, newborn pouch young attached to teat; se, segmenting egg.

Three different patterns in the reproductive cycle of the marsupials are shown; non-macropods like *Trichosurus* (Australian possum) and macropods (kangaroos) which have two types of cycle shown by; *Macropus eugenii* (the tammar wallaby and *Megaleia rufa* (the red kangaroo), and *Wallabia bicolor* (the swamp wallaby).

For detailed discussion see the text. (From Sharman, 1970. Copyright © 1970, by the American Association for the Advancement of Science.)

kangaroo this inhibition may decline so that ovulation and fertilization may occur, though while the young is in the pouch, the fertilized egg does not develop further than the blastocyst stage.

When the young kangaroo leaves the pouch the development of a dormant blastocyst can then proceed and pregnancy thus continues. The young kangaroos, however, remain with the mother and continue to suckle from outside the pouch; thus, the female kangaroo may have one young in the pouch, and another, much older young, 'at heel'. The two young then feed from different teats and the composition of the milk that each feeds on is quite different notwithstanding the fact that the endocrine secretions that are available to both glands are identical.

The delayed implantation in macropod marsupials follows the division of the fertilized egg to a stage when 80 cells are present. This blastocyst, in contrast to the placental one, is surrounded by a shell membrane and a layer of albumin; in which state it can survive for several months. It lies in the uterus, in the branch opposite to that where the preceding pregnancy occurred. The temporary inhibition of the development of the blastocyst depends on the suckling stimulus from the young kangaroo in the pouch. Once suckling declines the blastocyst then starts to develop further. The nature of the inhibitory stimulus is thought to result from neural stimulation of the pituitary, as a result of the suckling. Ovariectomy does not have any effect on the dormant blastocyst but the injection of estrogen and, especially, progesterone can initiate its development. It appears that the corpus luteum of lactation, that is formed from the follicle that gave rise to the dormant blastocyst, is relatively quiescent during lactation and its subsequent development and rapid secretion initiates the succeeding pregnancy. Secreted progesterone appears to accomplish this change by initiating, and so synchronizing, both the further development of the blastocyst and the luteal phase of the uterus; the latter providing an environment that is necessary for the growth of the embryo. The nature of the inhibition of the corpus luteum is uncertain but it is interesting that the injection of oxytocin, which is normally released in response to suckling, can prevent the development of the corpus luteum and the blastocyst in kangaroos that have been deprived of their suckling young. Oxytocin may thus have evolved a physiological role that has not been described in other vertebrates. It is, however, interesting that injected oxytocin shortens the life of the corpus luteum in cows and this effect depends on the presence of an intact uterus (Anderson, 1973).

The suppression of the normal estrous cycle that characteristically occurs in lactating marsupials also is related to the suckling stimulus (Fig. 9.13). In non-macropods, like *Trichosurus* (and *Macropus giganteus* in the early stages of lactation) this inhibition is due to a direct inhibitory action on the ovary that is probably mediated through the pituitary. In most macropod marsupials, there is an interesting deviation from this pattern as the corpus luteum of lactation appears to be essential for the response. It is thought that, following suckling, a release of a pituitary hormone stimulates the corpus luteum which releases a secretion that inhibits the ovary. It is uncertain what this secretion may be; if it is progesterone it presumably must be a minimal release so as to not disturb the dormant blastocyst. The nature of the pituitary stimulus is uncertain; it cannot be mimicked by the injection of eutherian LH or prolactin.

Parturition in marsupials does not appear to be such a dramatic event as it is in mammals which bear much larger young. It, nevertheless, appears to be hormone-dependent as it is prevented by ovariectomy. This effect may reflect the absence of estrogens, which in a sudden dramatic pre-parturition surge seem to contribute to the birth of the young in placentals. The progesterone concentration in the blood also increases, and then

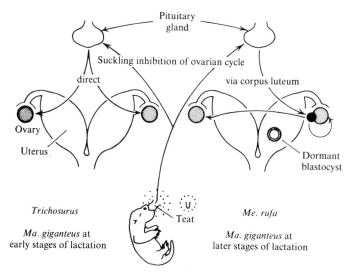

Fig. 9.13. The mechanism by which suckling may inhibit the ovarian cycle in marsupials.

Trichosurus vulpecula and *Macropus giganteus* in early stages of lactation, return to estrus soon after the young is removed from the pouch. In these animals inhibition seems to result from a direct inhibitory effect on the ovary, mediated. by the suckling stimulus.

Megaleia rufa and *Macropus giganteus* at later stages of lactation do not return to estrus until about a month after the young is removed from the pouch. In these marsupials a functional corpus luteum is necessary for the inhibition to occur which, as a result of suckling–pituitary stimulation, releases an 'inhibitory factor'. (From Sharman, 1970. Copyright © 1970 by the American Association for the Advancement of Science.)

decreases, just before parturition (Lemon, 1972) and it is possible that this also has a role to play in delivery of the marsupial young. The circulating progesterone levels at parturition, however, may differ considerably, depending on the species, for, as we have seen, the length of gestation varies in relation to the estrous cycle so that the birth of the young occurs at various stages in the development of the corpus luteum.

Whether or not oxytocin is also involved in parturition, as in placentals, is unknown but it is present in the marsupial pituitary and it contracts the uterus of the wallaby, *Setonix brachyurus, in vitro* (Heller, 1973). This tissue is most sensitive in the late stages of pregnancy. Oxytocin, also promotes milk let-down when injected into kangaroos, an effect it also shares with the placentals.

The reproductive pattern in marsupials shows distinct differences from that of placental mammals and appears to be well adapted to their manner of life. Contrary to some popular opinion about the 'lowly' state of development of these animals, their reproduction is an extremely efficient process. The embryonic diapause of the kangaroos constitutes an excellent 'insurance' to continued reproduction so that if a young is lost, or when it is weaned, another pregnancy follows with little delay.

Monotremes

These mammals are confined to Australia and New Guinea and are remarkable as they produce large, shelled eggs which they care for. In the spiny anteater, *Tachyglossus*, the egg is lodged in a pouch for hatching while platypuses lay their eggs and tend them in burrows. The young are fed by lactation in a typical mammalian way and injected oxytocin, which is an homologous hormone in monotremes, initiates milk let-down (Griffiths, 1965). Little is known about the endocrine processes that control reproduction in these very interesting mammals. They are monoestrous. Like other mammals, but unlike birds, they possess a prominent corpus luteum. The egg undergoes some development while in the oviduct, equivalent to 38 to 40 hours' incubation in the chicken.

Non-mammals

Precise information about ovarian cycles in non-mammals is meager compared to that in mammals. Much of the available knowledge is based on morphological and histological observations on the ovaries and the accessory and secondary sexual characters, especially the oviduct. Such information is related to endocrine changes on the basis of the abilities of injected, exogenous, hormones to mimic or prevent such changes. These experimental approaches, while suffering from obvious limitations have, however, demonstrated that differences indeed exist between the ovarian cycles of different non-mammalian vertebrates. With newly available radioimmunoassay procedures for measuring hormone levels in the blood, the precise role of hormones in the reproductive life of non-mammals is now being investigated more rigorously. At the present time the birds have received the most attention but these new techniques will clearly also be extended to the reptiles, amphibians and fishes.

There are several salient areas about which endocrine information in non-mammals promises to be especially interesting. These include:

(*a*) The evolution and role of the corpus luteum, especially in viviparous and ovoviviparous species.

(*b*) The mechanism by which a single gonadotrophin (thought to be present in reptiles, amphibians and fishes) controls the ovarian cycle.

(*c*) The possible physiological role of the neurohypophysial peptides in influencing the contractility of the oviducts.

It is usually somewhat difficult to make a strict comparison between the ovarian cycles of mammals and non-mammals. This partly reflects a lack of information about the latter but the timing of the events also often differs rather radically. Thus, birds usually take many weeks of preparation to come into breeding condition when ovulation becomes possible. This latter process then occurs at regular intervals of about 24 hours which can proceed for several weeks, or in the domestic fowl, with some minor breaks, for up to 300 days of the year. This almost daily ovulation cannot be strictly compared to the entire estrous cycle of mammals but may be more synonymous with very short estrous cycles or rapidly successive periods of estrus. In other species of fish and amphibians, a single massive (sometimes referred to as 'explosive') ovulation may occur but in the meantime the eggs may be held in readiness for some time; ovulation thus does not always appear to be an irrevocable event in a strictly pre-timed program.

Birds

The ovarian cycle of most birds is primarily under photoperiodic control. The daily changes in light are the primary stimulus which directs the over-all endocrine preparations for the breeding season (see Chapter 4). As described earlier (see Fig. 4.15), in quail subjected to a long-day photoperiod, a release of gonadotrophin-releasing hormone, from the median eminence, results in a discharge of pituitary gonadotrophins. Plasma LH levels have been shown to increase in Japanese quail that are photoperiodically stimulated by long day-lengths (Nicholls, Scanes and Follett, 1973). It also seems likely that FSH is released during photoperiodic stimulation but this has not yet been directly measured. The changes in the levels of pituitary gonadotrophins indicate that a release occurs at a precise time each day, which, in the quail, takes place in the evening after dusk. This daily rhythmical release of gonadotrophins promotes the growth of the ovary and the maturation of the follicles. Estrogens are known to stimulate the growth of the avian oviduct (progesterone may also contribute to this increase) and secondary sexual characters and are released during such preparations for reproduction.

When the bird is finally ready to breed, ovulation may begin. In domestic fowl, ducks and geese, this is spontaneous but in other birds the presence of the male is usually necessary so that ovulation may then be said to be induced. There are several other factors that determine whether or not ovulation will occur in birds. If the newly laid eggs are continually removed from the nest some birds will continue to lay more eggs (*non-determinate layers*). A house sparrow has thus been stimulated to produce 51 eggs in a season; ovulation apparently continued until the ovary was exhausted of suitable follicles. The domestic fowl, that can produce 300 eggs in a year, is an even more dramatic example of this phenomenon. How such birds recognize the number of eggs in the 'clutch' is unknown but it has been suggested that this may be the result of a tactile stimulus or sight. Other types of birds (*determinate layers*) produce a set number of eggs and changing the number in the nest does not modify ovulation.

In the domestic fowl, LH, as in mammals, initiates ovulation but the mechanism controlling the release of this hormone is not clear. The injection of progesterone promotes ovulation while estrogens delay it; effects that are in direct contrast to those seen in mammals (Fraps, 1955). The normal rhythmical release of LH, which commences about eight hours prior to ovulation is, however, preceded by an increase in plasma estradiol (Fig. 9.14*b*) just as in mammals. To complicate the interpretation of this, the rise in LH is also accompanied by increased levels of progesterone and the release of the gonadotrophin never *precedes* that of the steroid (Fig. 9.14*a*). It is thus likely that both estrogens and progesterone are involved in the process of ovulation in birds; possibly LH release is initiated by estrogens but other processes that also determine the extrusion of the egg from the follicle may be influenced by both types of steroids in a manner that is reflected by the injection of these hormones.

It has been shown in the domestic fowl, that, as long as there is an egg in the oviduct, further ovulation is inhibited. This effect can be mimicked by placing an irritant, such as a piece of thread, in the oviduct. Such a condition can be prolonged for about three weeks and as no regression in the ovary or oviduct occurs secretion of FSH and estrogens is thought to be unimpaired. The inhibition of ovulation by the presence of an egg in the oviduct is thought to be the result of a neural stimulus which may inhibit LH release. The injection of progesterone or LH into such birds overcomes this inhibition and promotes ovulation.

The precise site of origin of the circulating progesterone in birds is uncertain. Birds do not possess a corpus luteum but the ovary nevertheless secretes progesterone. This steroid may be formed by the follicles themselves, the interstitial tissue or the corpora atretica. The role of progesterone in the ovarian cycle is also not clear. In mammals and other vertebrates, this steroid inhibits the release of LH and LH-like gonadotrophins by its negative-feedback inhibition of the hypothalamus. Such an effect would be unexpected in birds if, as suspected, progesterone stimulates the release of LH. As birds lack a corpus luteum, LH also cannot exert its usual luteinizing and luteotrophic actions though it is possible that it may have comparable effects at other sites in the ovary. There is need, at this stage,

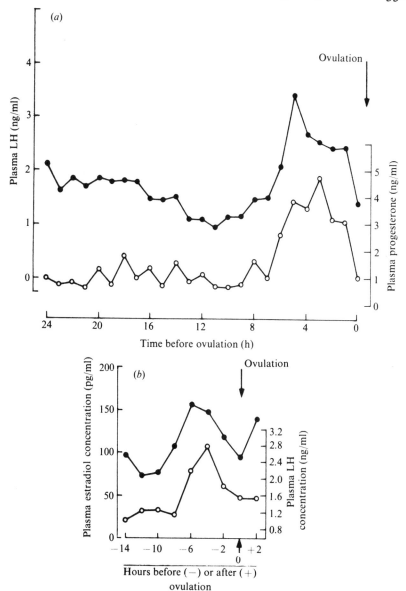

Fig. 9.14. The ovarian cycle in the domestic fowl.

(a) Changes in the plasma levels of LH (●) and progesterone (○) during the ovulatory cycle. The relationship between the rises in the levels of progesterone and LH are not clear; the release of LH apparently does not initiate a release of progesterone as it never precedes it. (From Furr *et al.*, 1973.)

(b) Changes in the levels of LH (○) and estradiol (●). The rise in the level of estradiol occurs about 2 h before that of LH. (From Senior and Cunningham, 1974.)

for a note of caution as it should be recalled that the endocrine observations on the avian ovarian cycle are nearly all confined to domestic species, especially *Gallus domesticus*.

The egg-laying cycle of the domestic fowl is thought to occur in the following manner (van Tienhoven and Planck, 1973). The eggs are laid in a clutch, or 'sequence', three to five in number, which are each produced at intervals of about 26 h. There is a period of 40 to 48 h between the laying of each clutch. Ovulation and oviposition are controlled by an endogenous rhythm lasting for 26 to 28 h the commencement of which is normally timed according to the photoperiod. Depending on such stimulation the length of the cycle can be retarded or advanced by about 2 h and so may vary from 24 to about 30 h. In continual light, other periodic events, such as the times of feeding and fluctuations in temperature, can be used to initiate the egg-laying cycle. Such external stimuli appear to sensitize regions of the hypothalamus and median eminence which, in response to stimulation by circulating progesterone, and possibly estrogens, secretes LH-releasing hormone. LH is then released from the pituitary which results in ovulation 8 h later. Subsequent formation of the egg takes place in the oviduct and the timing of the events there seems to depend on photoperiodic stimulation working in conjunction with the ruptured follicle. Removal of the latter tissue from the ovary results in a retention of the egg in the oviduct. Normally oviposition occurs 13 to 14 h after ovulation. It is possible that contractions of the oviduct that occur during oviposition are assisted by vasotocin which is released from the neurohypophysis.

Avian reproduction has certain unique features that do not occur in mammals and that have resulted in some special endocrine arrangements. As described earlier, oviparous vertebrates respond to estrogens by the formation, in the liver, of a special calcium-binding phospholipoprotein that is incorporated into the egg. This effect is especially important in species, like the birds, that produce eggs with large yolks. It is also usual for birds to make some special arrangements for the incubation of the eggs and, after hatching, for the care of the young. This maternal instinct, which is also often displayed by the male, is influenced by the endocrines. The development of 'brood patches' on the ventral surface of the skin can be promoted by the injection of prolactin and estrogens. These featherless, richly vascularized areas facilitate the transfer of heat from the parent to the eggs. 'Broodiness', or the desire to incubate the eggs and care for the young can also be promoted by injections of prolactin. This hormone suppresses gonadal function in birds and it is thought that broodiness may in fact be the result of a withdrawal of gonadal steroids (see Parkes and Marshall, 1960). At present, it is not clear whether or not prolactin normally exerts such an effect in hens. The action of prolactin in stimulating a secretion from the crop-sac of doves and pigeons (with which they feed their young) has been described earlier (Chapter 5).

Birds thus show some interesting deviations and novelties in the use of hormones for integrating their reproductive processes. An evolution of the role of certain hormones has clearly occurred in this interesting offshoot from a reptilian stock.

Reptiles

The Reptilia contain oviparous, ovoviviparous and viviparous species. Unlike in birds, a distinct corpus luteum is formed following ovulation and although the available evidence strongly suggests that it secretes progesterone it seems unlikely that this hormone contributes to the maintenance of pregnancy. A further distinction from birds and mammals is seen in the mechanism of ovarian control which apparently is mediated by a single gonadotrophic hormone. It has, however, recently been shown that two chemical principles, one with LH-type and the other FSH-type activity, can be separated in extracts from a chelonian (the snapping turtle) pituitary (Licht and Papkoff, 1974a). It remains possible that a single gonadotrophin exists in other groups of the reptiles and if so it apparently can perform the functions that are carried out by both of the mammalian gonadotrophins.

The changes that occur during the ovarian cycle of the ovoviviparous lizard, *Sceloporus cyanogenys*, are summarized in Fig. 9.15. The ovary starts to grow in October or November and this is accompanied by the development of the oviduct. These changes can be prevented by hypophysectomy. The gonadotrophin stimulates gonadal growth and the secretion of estrogen. Implantation of small pellets of estrogens into the region of the median eminence reduces the growth of the oviduct. This effect is probably the result of a lower rate of ovarian estrogen secretion, due to the inhibition of gonadotrophin release by a negative-feedback inhibition. Ovulation is also prevented by such an estrogen implant. Mammalian FSH has been shown to promote ovulation in several species of lizards (Licht, 1970) and the endogenous gonadotrophin no doubt also has this effect.

Following ovulation in *Sceloporus cyanogenys* the corpus luteum develops and this is correlated with a three-fold increase in the circulating progesterone concentration. This elevated hormone level persists in the plasma until parturition when it declines. A similar pattern in circulating progesterone levels has also been observed in the viviparous snake, *Natrix sipedon* (Chan, Ziegel and Callard, 1973). Direct evidence that reptilian corpora lutea can produce progesterone has been obtained following *in-vitro* incubation of such tissue obtained from the snapping turtle, *Chelydra*

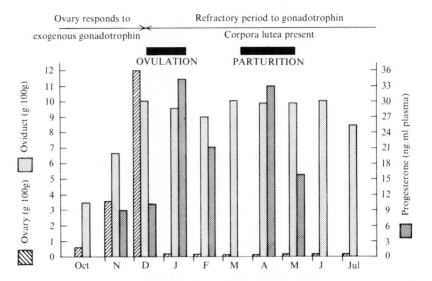

Fig. 9.15. The annual ovarian cycle of the ovoviviparous lizard *Sceloporus cyanogenys*.

The ovaries and oviducts start to develop (under the influence of gonado-trophin) in October and ovulation may occur in December to January. Gestation lasts for about 12 weeks and the young are delivered in late March to mid-May. Corpora lutea persist during pregnancy and the plasma progesterone levels rise, but then decline in late summer following parturition. (From Callard *et al.*, 1972.)

serpentina (Klicka and Mahmoud, 1972). In *Sceloporus cyanogenys* the circulating progesterone levels are reduced following hypophysectomy, suggesting that there is some pituitary control over this hormone, but in pregnant lizards relatively high concentrations continue to persist so that if a luteotrophic effect is present it is apparently not vital (Callard *et al.*, 1972). In addition, the implantation of pellets of progesterone into the region of the median eminence results in a depression of the circulating progesterone concentration (Callard and Doolittle, 1973). This suggests the presence of a negative-feedback inhibition of the release of a trophic hormone and is accompanied by a decrease in the growth of the ovary and oviduct. The latter effect suggests that ovarian estrogen secretion is also inhibited.

Inhibition of gonadotrophin release in reptiles can thus apparently be inhibited by both progesterone and estrogens but how this can be reconciled with the orderly release of a single gonadotrophin that controls several gonadal functions is unknown and it remains possible that a second hormone will be identified, as in the snapping turtle.

Corpora atretica, formed by the dissolution of unovulated follicles, may also contribute to the control of reptilian ovarian cycles. At the end of the summer breeding season the lizard *Anolis carolinensis* becomes refractory to the effects of photoperiodic stimulation and this can be related to the presence of corpora atretica in the ovary. In addition these lizards, at this time, respond poorly to injected gonadotrophins but the response to these hormones could be increased five-fold if the corpora atretica were removed. The nature of the latter's inhibitory effect on the female reproductive system is unknown (Crews and Licht, 1974).

Progesterone, as we have seen, plays an important role in maintaining pregnancy in mammals but there is no conclusive evidence to indicate that this occurs in reptiles. Ovariectomy or hypophysectomy do not affect the course of pregnancy in a variety of viviparous and ovoviviparous species of reptiles (see Yaron, 1972). Progesterone, nevertheless, attains high concentrations during a reptilian pregnancy; so what is its function? Callard and Doolittle (1973) have suggested that its action in reptiles may be to inhibit gonadal growth during gestation and this may represent a 'more primitive' role than the regulation of the uterine environment that is seen in mammals. Corpora lutea are also formed following ovulation in oviparous reptiles and it has been found that when this tissue is removed in gravid *Sceloporous undulatus* (an oviparous lizard) an earlier oviposition occurs, suggesting that progesterone may control the period of egg retention in such reptiles (Roth, Jones and Gerrard, 1973).

In reptiles, as also in birds, estrogen, possibly in conjunction with growth hormone, stimulates the production of phospholipoproteins by the liver and these are incorporated into the egg. Prolactin, when injected, has been shown to exert an antigonadal effect but again, as in birds, the significance of this inhibition is unknown *in vivo*. It is nevertheless interesting that these actions of estrogens and prolactin are shared with the birds.

Despite the valuable experiments that have already been performed much additional, and more precise, information is needed before an adequate account of the role of hormones in the reptilian ovarian cycle can be given.

Amphibians

Most amphibians are oviparous though there are a few species that have ovoviviparous and even viviparous habits. An excellent account of the ovarian cycle in amphibians is given by Redshaw (1972). Complete maturation of the oocytes usually takes several years while formation of the yolk, in species that live in temperate zones, commences in the summer preceding spawning. The development of the ova is controlled by the adenohypophysis and the hypothalamus. Hypophysectomy or transplantation of the pituitary, so that it is no longer in contact with the hypothalamus, interrupts oogenesis. The available evidence seems to indicate that only a single gonadotrophin is present in the amphibian adenohypophysis.[1] This hormone (or hormones?) regulates the development of the ovum and the secretion of ovarian steroid hormones.

Estrogens are produced by the ovarian follicles and these contribute to the development of other sexual characters, including the oviduct, as well as vitellogenesis.

Progesterone is apparently also produced by the ovary, though the precise site of its formation is uncertain as most amphibians (except for ovoviviparous and viviparous species) do not form a distinct corpus luteum after they ovulate. Whether or not either or both gonadal steroid hormones exert a negative-feedback inhibition on the release of gonadotrophin is unknown.

Ovulation can be readily promoted in amphibians by the injection of gonadotrophin. This hormone may be obtained from amphibian pituitaries, but exogenous hormones, from other species, are also effective. The latter hormonal effects are particularly well known as they are the basis for a convenient pregnancy test for women. Human chorionic gonadotrophin (HCG) that is secreted in the urine during pregnancy induces ovulation in frogs and toads. Mammalian LH is usually considered to be more effective than FSH in inducing ovulation in frogs and so may have a closer structural resemblance to the amphibian gonadotrophin. It has recently been shown (Licht, 1973) that the mammalian gonadotrophins differ in their ability to induce spermiation in male amphibians; thus, LH is more potent than FSH in *Rana pipiens* but FSH is about twice as potent as LH in *Hyla regilla* and *Eleutherodactylus coqui*. It therefore seems likely that mammalian gonadotrophins may show similar differences in their ability to induce ovulation in different species of female amphibians. Ovulation is often, though not always, induced at the time of sexual pairing. *Xenopus* seems to be on the verge of ovulation for prolonged periods of time while in *Rana temporaria* the eggs are stored in the oviduct, from which they are expelled when mating occurs.

The precise hormonal events that initiate ovulation and oviposition are not known. The injection of HCG into the toad *Bufo bufo* results in ovulaion in about 24 hours and this is preceded by a release of a progesterone-like material into the blood (Fig. 9.16). It has also been shown, *in vitro*, that progesterone can induce ovulation. In addition, the injection of progesterone hastens ovulation while estrogen retards it; a situation reminis-

---

[1] Two distinct gonadotrophins, one with LH and the other with FSH activity, have been isolated from the bullfrog, *Rana catesbeiana* pituitary (Licht and Papkoff, 1974*b*).

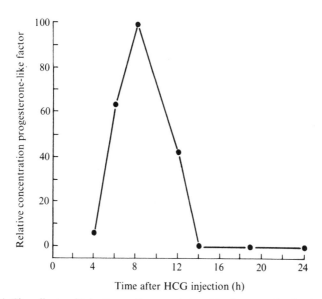

Fig. 9.16. The effects of injections of a gonadotrophin (human chorionic gonadotrophin, HCG) on the level of a progesterone-like factor in the plasma of female
'winter' toads, *Bufo bufo*.

The HCG results in ovulation in about 24 h and this is preceded by a rise and
then a decline in this 'factor'. Injections of progesterone can induce ovulation in
amphibians and it is possible that this is involved in the effect of the HCG (as well
as mediating release of 'jelly' from glands in the oviduct). (From Thornton, 1972.)

cent of that which occurs in birds. It is to be hoped that the use of radioimmunoassay procedures will allow the measurement of amphibian
gonadotrophin, progesterone and estrogens so that the natural events that
occur during the ovarian cycle in amphibians can be more directly observed.

The oviduct undergoes a distinct annual cycle in *Bufo bufo* and attains
its greatest size in the autumn (Jorgenson and Vijayakumar, 1970). A
decline in the weight of the oviduct takes place during spawning in April.
This decline is due to a loss of secretory contents that coat the eggs with a
'jelly'. The secretion of this jelly, like that of avidin in the fowl oviduct, is
controlled by progesterone.

A diagrammatic summary of the role of hormones in controlling ovarian
functions in amphibians is given in Fig. 9.17.

Some very interesting observations have been made on the effects of
progesterone on gestation in a viviparous frog *Nectophrynoides occidentalis*
(Xavier and Ozon, 1971; Zuber-Vogeli and Xavier, 1973; Xavier, 1974).
This frog lives in West Africa where it is subjected to periods of seasonal
drought during which it estivates in burrows. Following ovulation in
October the fertilized eggs are retained in the oviduct where development

PITUITARY GONADOTROPHIN

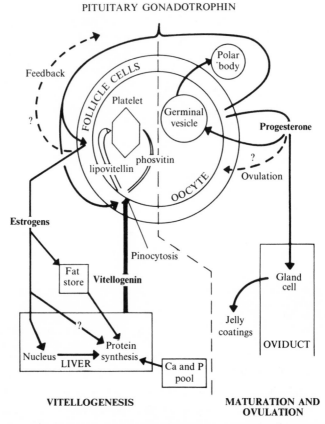

Fig. 9.17. The role of hormones in the production of eggs in anuran amphibians. The pituitary gonadotrophin controls the release estrogens and progesterone by the ovary.

Left. The control of the process of *vitellogenesis*. Estrogen, formed by the follicular cells, induces the formation of vitellogenin in the liver, which is incorporated into the yolk of the developing oocytes. The uptake process is stimulated by the gonadotrophin.

Right. The *maturation of the oocyte* and ovulation. These processes are controlled by the pituitary gonadotrophin which mediates the formation of a 'maturation agent' that seems to be progesterone. The progesterone also stimulates the oviductal glands to secrete the coating of 'jelly' with which the eggs are coated.

This summary is principally based on experiments on *Xenopus laevis* and *Bufo bufo*. (From Follett and Redshaw, 1974.)

proceeds until parturition the following June. In November, these pregnant frogs estivate and do not emerge until April (see Fig. 9.18). Corpora lutea are formed following ovulation, which apparently secrete progesterone.

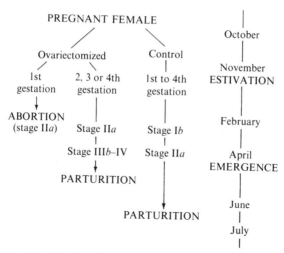

Fig. 9.18. Gestation in the viviparous frog, *Nectophrynoides occidentalis*, in relation to seasons.

Normally these frogs ovulate and become pregnant in October. The dry season commences in November when they estivate from which they emerge, with the onset of rain, in April. The young are born in June. If these frogs are ovariectomized early in pregnancy, they may either abort, if the animals are young (and this is their 1st pregnancy) or, if they are large and it is the second, third or fourth time of gestation, the development of the young is accelerated and they are born much earlier than usual. Ovarian progesterone is thought to delay the development of the young during the period of estivation. (From Zuber-Vogeli and Xavier, 1973.)

When the ovaries from pregnant frogs are incubated *in vitro* with the progesterone substrate, pregnenolone, they convert this steroid to progesterone. This ability to form progesterone declines as gestation progresses (Fig. 9.19). Following parturition, pregnenolone is converted to other steroids by the ovarian tissue and this process increases until ovulation again takes place. The preovulatory period is also the time when production of estrogens are thought to increase. If ovariectomy is performed early in gestation (see Fig. 9.18) of young frogs, during their first pregnancy, abortion occurs. In more mature frogs, however, development of the young is accelerated following ovariectomy and parturition takes place about three months earlier than usual. The implantation of progesterone into these frogs towards the end of gestation reduces the rate of growth of the embryos. Progesterone thus appears to slow the rate of development of the embryos during the prolonged period of estivation so that the young are delivered at a more appropriate and favorable time of the year. This is indeed a novel role for progesterone.

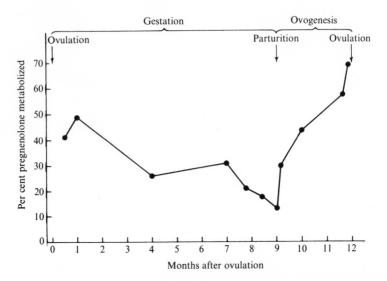

Fig. 9.19. The ability of the ovarian tissue of the viviparous frog *Nectophrynoïdes occidentalis* to metabolize pregnenolone (*in vitro*) at different stages of its ovarian cycle.

During gestation the principal steroid produced from the pregnenolone is progesterone but after parturition 17-hydroxyprogesterone, androstendione and testosterone are also formed. (From Xavier and Ozon, 1971.)

Fishes

Sporadically distributed information is available about the control of ovarian function in fishes. This is not unexpected considering the enormous numbers of piscine species, and the wide phyletic gaps that separate them. In addition, fish are usually somewhat diffident scientific collaborators.

Most fishes are oviparous but some species have evolved ovoviviparous and viviparous methods of reproduction. In teleosts the last two processes are usually rather different from the *in-utero* development common to many other vertebrates. The young teleosts thus may develop *in situ* in the follicle or be incubated in the hollow central cavity of the ovary. In some teleosts successive broods (superfetation) may develop in such follicles which are then delivered in waves. It would not be unexpected if such dramatic differences in the procedures for reproduction resulted in endocrine adaptations.

In teleosts the development of the oocytes and vitellogenesis are dependent on the pituitary. This gland appears to contain a single gonadotrophin which mediates these effects. The injection of homologous gonadotrophins can promote ovulation in teleosts. The roles of estrogens and progesterone in this process are, however, unknown. The endogenous release of gonado-trophin in teleosts is under the influence of the hypothalamus.

In cyclostome and chondrichthyean fishes, the pituitary, via the action of a gonadotrophin, also controls the development of the ova but this process does not appear to be under hypothalamic control. It is rather a mystery how cyclical control of reproduction is mediated in these verte-brates which are the only groups known to lack such a mechanism.

Little is known about the influence of gonadotrophin on gonadal steroid production in fishes though it is generally considered that estrogen secretion is controlled by such a hormone. The production of progesterone by the ovary and the possible role of corpora lutea is contentious. Many teleosts possess so-called 'pre-ovulatory corpora lutea' which are formed as a result of atresia of unovulated follices and are more aptly called corpora atretica. Post-ovulatory corpora lutea, as well as corpora atretica, are present in many chondrichthyeans and there is considerable speculation as to whether they contribute to successful gestation in ovoviviparous and viviparous species (see Chieffi, 1967; Dodd, 1972a). Hypophysectomy does not interrupt pregnancy, at least for the first three months, in the viviparous shark, *Mustelus canis*, suggesting that an adenohypophysial control of progesterone secretion is not vital. It is, however, unknown whether progesterone contributes to gestation in ovoviviparous and viviparous fishes. Hisaw in 1959 stated that 'the elimination of yolk during follicular atresia and material from ruptured follicles at ovulation is a primitive function of corpora lutea and endocrine functions such as luteinization of the granulosa by pituitary luteinizing hormone and secretion of pro-gesterone in response to pituitary luteotrophic hormone as seen in mam-mals, are more recent adaptations'. This interesting idea has, however, not been unquestionably accepted.

In many teleost and chondrichthyean fishes, the thyroid gland displays an increased activity during the breeding season. The latter is associated with many physiological and environmental changes so that it is difficult to be certain whether the endocrine events are primarily related to repro-duction. Sage (1973) considers that it is likely that the thyroid is involved in the reproductive process in fishes as it is necessary for gonadal maturation in some species. Such a role for thyroxine could reflect a primeval endocrine use of this hormone.

Oviposition and parturition in non-mammals – a role for the neurohypophysial hormones?

The passage of eggs or young along the female gonaducts and their exit into the outside world may be assisted by rhythmical contractions of the muscles that surround these ducts. Such muscles are usually non-striated (smooth) muscles (though striated muscle may also be present in some fishes), which have an inherent ability to contract even in the absence of nerves or hormones. The rate and pattern of contractility of smooth muscles can, nevertheless, be modified by such stimuli. As we have seen in mammals, oxytocin can promote contractions and aid in the process of parturition. It should be remembered, however, that smooth muscle readily reacts to local stimuli, and this may include the presence of an egg or fetus, so that oviposition or parturition can occur even in the absence of neural or hormonal stimuli.

The neurohypophysial hormones have a special ability to contract mammalian uterine smooth muscle and this response has also been shown to occur (mainly *in vitro*) in many other species of birds, reptiles, amphibians and fishes (Heller, 1972). Such effects can usually be elicited by low con-

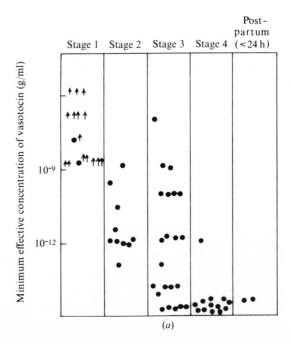

Fig. 9.20. Periodic changes in the responsiveness of the ability of the ovoviviparous ovary of a teleost fish *Poecilia*, and the oviduct of a urodele amphibian *Necturus maculosus*, to contract (*in vitro*) to vasotocin.

(*a*) *Poecilia* (the guppy); minimum effective concentration of vasotocin required to produce a contraction of the ovary during successive stages of gestation. The arrows indicate a lack of response; it can be seen that the preparations were most sensitive just prior to parturition (stage 4). (From Heller, 1972.)

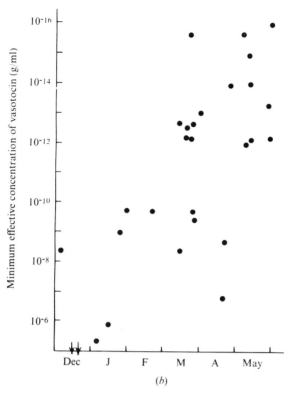

(b) *Necturus* (the mudpuppy); minimum effective concentration of vasotocin required to contract the oviduct at different seasons of the year. It can be seen that this was least in winter but progressively increased and was greatest in April and May. (From Heller, Ferreri and Leathers, 1970.)

centrations of such peptide hormones that represent only a small fraction of those stored in the neurohypophysis.

It is interesting that in poecilid teleosts that are ovoviviparous, the ovary contains smooth muscle and this is even more sensitive to the effects of vasotocin than the oviduct. The young in such fishes develop within the follicles so that their expulsion may conceivably be promoted by the contraction of these ovarian muscles.

Such contractile responses of the oviduct and teleost ovary to vasotocin can be seen to change at different stages of the sexual cycle and, like mammals, the sensitivity is greatest at the time of parturition or oviposition. This can be seen in the ovary of *Poecilia* (Fig. 9.20); an effect that can be mimicked by the implantation of gonadal steroids into these fish. The oviduct, *in vitro*, of the mudpuppy, *Necturus maculosus*, also shows seasonal changes in sensitivity to vasotocin (Fig. 9.20); the tissue may be as much as 100-fold more sensitive to this neurohypophysial hormone in May compared with its sensitivity in December and January.

Observations *in vitro*, however, must be interpreted with considerable caution as they may not take place *in vivo* and even if the latter does occur following injections of hormones it may not reflect a normal physiological response. In other words, we must be careful to distinguish the pharmacological from the physiological effects of these hormones on the oviduct and uterus. The problem of the evolution of the roles of the neurohypophysial hormones in vertebrates has intrigued endocrinologists for many years. As we have seen (Chapter 3), such hormones are present in all vertebrates. While the neurohypophysial hormones have a fairly well-characterized role in osmoregulation of tetrapods, their role in fishes is not at all clear. The neurohypophysial hormones can contract smooth muscle of the oviduct and uterus in representatives of all the main groups of vertebrates and it is tempting to speculate that these effects may be of physiological significance and even reflect a more ancient role for such hormones.

The evidence for this interesting endocrine hypothesis has been summarized by Heller (1972). H. Heller over a long scientific lifetime has made many fascinating contributions to the comparative endocrinology of the neurohypophysis of which the present interest is a more recent outcome. There have been sporadic reports, spread over many years, about the ability of injected neurohypophysial hormones to promote oviposition and parturition in different vertebrates. This effect has been observed in the domestic fowl, several species of lizards and the ovoviviparous fire-salamander. The injection of neurohypophysial hormones also stimulates spawning-like behavior in the killifish, *Fundulus heteroclitus*. It is interesting that stored isotocin disappears from the pituitaries of the female, but not the male, killifish during the breeding season in June (Sawyer and Pickford, 1963). This change probably reflects a release of this neurohypophysial hormone which could be involved in the reproductive process. The precise physiological significance of such observations, however, is uncertain.

Observations, *in vitro*, indicate that vasotocin is the most active of the neurohypophysial peptides in stimulating the contractility of the oviduct in non-mammals. An interesting exception is the oviduct of the dogfish, *Scyliorhinus canicula*, in which another neurohypophysial peptide present in these fish is the most active. It would thus seem that an evolution of the receptor sensitivity in the gonaducts can occur, that is related to the nature of the homologous hormones which are present. In mammals, oxytocin has a greater uterotonic effect than vasotocin, which is not an homologous hormone in this group of vertebrates. Such an evolution in the sensitivity of the gonaducts to the homologous neurohypophysial peptides also leads us to suspect that they may have a physiological role. We must not, however, carry this sort of thing too far!

If the neurohypophysial hormones do indeed contribute to the physiology of oviposition and parturition in non-mammals (and even ovulation in teleosts) it is probably just that – a contribution. The domestic hen can still lay an egg following neurohypophysectomy and ovoviviparous poecilid fish still produce young following hypophysectomy. As in mammals, such hormones could, however, facilitate parturition as well as oviposition.

In teleost fishes, it is possible that another neurosecretory hormone may be involved in eliciting contractions of the gonaducts. As described in Chapter 8, urotensin II, from the urophysis, exhibits an ability to contract smooth muscles from the urinary bladder and gonaducts of teleosts. This substance may have an hormonal role in the reproduction of these fishes, such as in assisting the process of spawning. Thus a decline in the stored urotensin II in the urophysis of the white sucker, *Catostomus commersoni*, has been observed following spawning in the male, but not the female, fish (Lederis, 1973).

The role of hormones in maturation and development of the young

Reproduction can only be considered complete when the young have attained independence from the parents and are themselves able to propagate the species. Growth, differentiation and maturation of the young, either in the egg or *in utero*, is a most remarkable process the intricacies of which were, it seems, at least until recently, more appreciated by biologists 50 years ago than in the succeeding period of time. The role of chemical substances in coordinating and directing such embryological processes is reminiscent of the action of hormones though such inductor substances are not usually classified as such and will not be dealt with here. There has, however, been speculation as to whether inductors that influence sexual differentiation are indeed identical to the gonadal steroid hormones.

It was appreciated more than 100 years ago that endocrine secretions could affect development (see Jost, 1971). The earliest observations were on the effects of congenital thyroid deficiency which is called cretinism in man. Such a thyroid deficiency results in inadequate development of the nervous system, skeleton and the reproductive organs. It seems that these hormones are, however, more important in early postnatal life than fetal development. Other hormones may be necessary earlier in embryonic development. In encephalectomized fetal rats and anencephalic human fetuses, the prenatal development of the adrenal cortex is retarded, apparently reflecting an absence of pituitary stimulation due to the lack of hypothalamic control. The parathyroids are also functional in the fetal rat and removal of these glands results in hypocalcemia. The stage of embryonic development when such hormones may become important varies a great deal. It should also be recalled that hormones such as thyroxine and the steroid hormones may cross the placenta in viviparous species so that the maternal endocrines may contribute to the development of the fetus. In extreme situations, such transfer of hormones across the placenta can result in fetal abnormalities such as masculinization of female human babies born to mothers to whom large doses of progesterone have been administered during pregnancy.

The effects of exogenous progesterone on sexual differentiation was to a considerable extent a predictable phenomenon. The observations of Lillie in 1917 on the free-martin effect in cattle led to considerable speculation as to the role of fetal sex hormones on sexual differentiation. Lillie observed that in some cows bearing twin fetuses of opposite sexes the sexual differentiation of the genetic female was changed so that it was born with testes and a male gonaduct system. The external genitalia remained female in character and the animal was infertile. This effect has been attributed to the passage, across the interdigitating fetal membranes of materials, possibly androgens, that effect the change. This interpretation as to the role of male sex hormones has not gone unchallenged and it has more recently been suggested that other factors may be involved.

These interesting observations, nevertheless, resulted in experimental testing of the ability of sex hormones to modify sexual differentiation in various other developing vertebrates. These experiments involve various techniques such as parabiotically joining, thus crossing, the circulation of embryos of the opposite sex, cross-grafting of fetal gonads, and the administration of gonadal steroid hormones. Such manipulations often resulted in considerable changes, even to the extent of reversing the recipient's predetermined genetic sex. A note of caution is, however, necessary as the results are usually not simple or predictable and the effects may depend on the age of the embryo, the dose of the steroidal hormones, the genetic sex of the recipient, and the particular species used.

Before describing some of these effects further, we should recall the patterns of sexual differentiation in embryonic mammals. The primordial gonad has the potential to develop into either a testis or an ovary and is divided into two distinct regions: a cortex, which is the definitive ovary, and a medulla that may become the testis. This segmentation of the primordial gonad applies to all vertebrates except cyclostomes and teleosts where a single structure is present. The ultimate predominance of one of these gonadal zones is normally determined genetically. Associated with the development of the gonads are their ducts (see Fig. 9.21): the Mullerian duct which may persist and develop into the oviduct or uterus, and the Wolffian duct which, in the male, becomes the vas deferens. The ultimate differentiation and persistence of the gonads, and their accessories, is dependent on the secretion of androgens or estrogens by the definitive

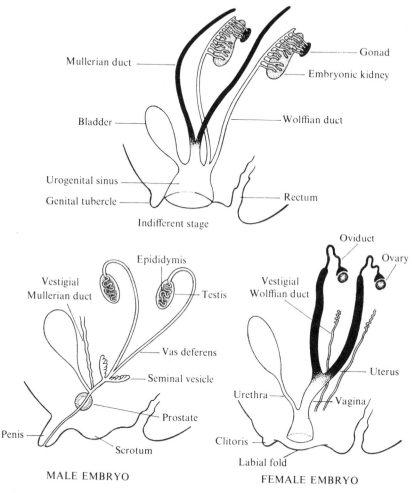

Fig. 9.21. Differentiation of the sexual organs in the embryonic mammal. The appropriate development of the genetic male is dependent on the presence of androgens and that of the genetic female on estrogens. This pattern can be changed by removal or antagonism of the natural hormones or by injecting hormones more typical of the opposite sex. (From Frye, 1967, by permission of Macmillan Publishing Co. Inc., New York. Copyright © 1967.)

gonad; thus, to some extent, one can manipulate their development by castration of the embryos and the injection of estrogens or androgens. It has recently become possible to distinguish such effects of hormones pharmacologically by administering a specific androgen antagonist, cyproterone. When this steroid is injected into male fetal rats (Elgar, Neumann and Berswordt-Wallrabe, 1971) differentiation of the Wolffian ducts and the vesicular and prostate glands is inhibited suggesting their normal dependence on androgens. The differentiation of the gonads and the regression of the embryonic Mullerian ducts is, however, not affected. The importance of androgens during the development of the male hypothalamic control mechanism of gonadotrophin release has already been described (Chapter 4).

The most extensive studies on experimental manipulation of embryonic sexual differentiation have been made in amphibians (see Dodd, 1960). Parabiotic union of male and female embryos of amphibians results in an inhibition of the growth or a masculinization of the ovary. This dominance of the testis over the ovary is typical in such experiments on amphibians. Parallel changes occur in the relative development of the Wolffian and Mullerian ducts. Normally in the male the latter become rudimentary but their growth can be stimulated by parabiosis with a female. Such effects were initially attributed to hypothetical inductor substances but can often be mimicked by administering gonadal steroid hormones. In some anurans (Ranidae), androgens have a masculinizing effect when they are administered to female embryos. Estrogens may have the opposite action and feminize male embryos but such effects of the female hormones are less predictable and pronounced. Other anurans and urodeles respond differently as the male embryos exhibit a stable feminizing response to administered estrogens; but androgens also have feminizing actions (paradoxal) and inhibit the medulla of the primordial gonad. These last two amphibian responses are similar to those observed in chondrichthyean fish (Chieffi, 1967). Estrogens and progesterone, as well as testosterone, have a feminizing effect on genetic males so that sex reversal is only possible in the female direction in these fish. In the domestic fowl, the injection of estrogens, in the early stages of incubation, into genetic males results in various degrees of feminization: they may become intersexual when the left testis becomes an ovo-testis, or even an ovary. The changes are accompanied by a persistence of the Mullerian ducts (see Parkes and Marshall, 1960). Such avian intersexes usually revert to their normal genetic sex later in life. Androgens may have the converse effects in birds but the results are rather variable and they may even have feminizing actions like those described above in amphibians and chondrichthyeans.

Sex-reversal (or sex-inversion) is a particularly interesting phenomenon in teleost fishes as it occurs normally during the life cycle of many species (see Reinbloth, 1970, 1972). This sex change may be from a male into a female (*protandry*) or more commonly a female into a male (*protogyny*). The stimuli that result in such sex inversions are not well understood but the removal of the male can initiate a protogynic sex change. The female fish then may become males that produce normal sperm. The injection of testosterone can mimic this transformation in wrasses (*Thalassoma bifasciatum*) but Reinbloth does not consider this to be proof that this is the normal mechanism mediating such a sex transformation. It is, however, an attractive hypothesis for the endocrinologist. Such a possibility is supported by the observation that in the protogynous symbranchid teleost *Monopterus albus*, the rice field eel, the gonads, *in vitro*, switch their steroid syntheses from a predominance of estrogens, in their female phase, to androgens in their male phase (Chan and Phillips, 1969).

True hermaphrodites, which can simultaneously produce both sperm and eggs and even be self-fertilizing have been described most commonly among the teleost fishes. Hermaphrodism is the normal situation in a fascinating oviparous teleost, *Rivulus marmoratus*, that lives in tidal-pluvial zones on the coasts of Florida and the West Indies (see Harrington, 1968). The differentiation of males can be promoted when embryonic development proceeds at temperatures below 20 °C, a situation that does not normally occur in this fish's native habitat. It is not known whether such sexual differentiation is mediated by endocrine changes under these conditions. The normal balance of sex hormones in hermaphroditic vertebrates constitutes rather an endocrine puzzle as it is difficult to conceive how different sex hormones can exist, be controlled, and act simultaneously within the same animal. This is an intriguing problem for the comparative endocrinologist.

The control of metamorphosis

Many fishes and amphibians exist in two or more distinct morphological and physiological forms during their life cycles. The transformation from one type to another is called metamorphosis. 'True' metamorphosis is considered to occur in preparation for life in a different habitat, such as fresh water (as opposed to the sea), or an aquatic compared to a terrestrial existence. These changes may involve a migration that is sometimes associated with sexual maturation and reproduction.

An interesting account of metamorphosis in fishes is given by Barrington (1968). Lampreys (Cyclostomata) undergo a prolonged period of larval development in fresh water which lasts for about $5\frac{1}{2}$ years. The metamorphosis of the ammocoete to the adult, which migrates to the sea, lasts for several weeks. Subsequently, lampreys again return to the rivers to breed and then undergo physiological changes, such as reflected by their inability to osmoregulate in sea-water. The environmental and physiological events that determine these metamorphic changes are unknown. Experimental manipulation of the lampreys' thyroid physiology, by placing them in solutions containing either thyroid hormones or anti-thyroid substances, have yielded inconclusive results.

Many teleost fish also undergo a metamorphosis. The best-known examples are seen in eels and salmon. The leptocephalus larval eel is transformed into the elver but little is known about the mechanism of this change that takes place in the sea and precedes migration into rivers. Salmon spawn in fresh water and the young parr, upon reaching a certain size, are transformed into smolt which migrate to the sea. Considerable endocrine changes occur at this time and these can be seen histologically in the pituitary and thyroid glands (Fontaine, 1954). The uptake of radio-iodine by the thyroid increases early in metamorphosis and the levels of 17-hydroxycorticosteroids in the plasma of the parr is five times greater than in the smolt. It is, however, not possible to decide whether such endocrine changes initiate metamorphosis or merely occur as a part of the general maturation process.

The possibility that changes in the endocrine glands, especially the thyroid, may initiate metamorphosis in fishes arose from the observation that thyroid hormones can initiate metamorphosis in most amphibians. The profound and dramatic morphological changes which accompany the metamorphosis of anuran tadpoles into adult frogs and toads have been a source of wonder to biologists for a long time. The transformation from a purely aquatic animal, with no limbs or lungs, into a terrestrial beast that breathes and hops about on four legs is also accompanied by many physiological and biochemical changes. The larval life and metamorphosis of amphibians may be relatively short, from several weeks in desert-dwelling species where water is available for only a short time to as long as three years in bullfrogs. The factors that determine the time of metamorphosis are not clear; they are partly genetic but they can also be modified by the environment. Bullfrog tadpoles from the tropical southern parts of the United States may metamorphose before the beginning of the first winter after hatching, while those in northern areas may endure three winters before this change occurs. Possible environmental factors that influence the time of metamorphosis include nutrition, temperature, the salinity and acidity of the water where they live, the relative proximity of other tadpoles ('crowding') and some experiments even suggest that light may stimulate this process (see Dent, 1968). One can foresee that such factors could exert their effects through the activation of endocrine glands, in this instance the thyroid through its hypothalamic and pituitary control mechanisms.

The feeding of thyroid gland extracts can produce metamorphosis in tadpoles far earlier than it would normally occur. Conversely the administration of anti-thyroid drugs prolongs, or even prevents, metamorphosis. Natural metamorphosis in tadpoles is accompanied by a sudden increase in thyroid gland activity as indicated by histological changes and an increase in the rate of uptake of radioactive iodine. There can be no doubt that the activity of the thyroid gland determines metamorphosis in tadpoles but this is only a part of the endocrine story. The thyroid in tadpoles, like that in other vertebrates, is under the control of TSH from the adenohypophysis. The injection of TSH into tadpoles thus also results in a premature metamorphosis. Hypophysectomized tadpoles do not metamorphose but grow, larger and larger, and attain 'giant' proportions. The hypothalamus and median eminence which are usually the next sites in the chain of thyroid control are also involved in metamorphosis for when the tadpole pituitary is transplanted to the tail, metamorphosis is prevented. Such tadpoles are dark in color due to an uncontrolled release of MSH, and they grow more rapidly than usual.

The latter effect on growth is an important clue which probably, at least partly, reflects a lack of the hypothalamic-inhibition of the release of prolactin. Injections of prolactin into bullfrog tadpoles have been shown to antagonize thyroxine-induced tail resorption and delay metamorphosis (Nicoll *et al.*, 1965; Etkin and Gona, 1967). The same effect has been achieved by grafts of tadpole pituitaries so that this effect probably also exists physiologically. Prolactin is thought to antagonize the peripheral actions of thyroxine and also exerts an inhibitory (goitrogenic) effect on the thyroid gland. Such effects are apparently confined to these amphibians and do not occur, for instance, in mammals.

Etkin (1970), after a careful assessment, has provided a description of how metamorphosis is normally regulated in tadpoles. This synthesis is summarized in Fig. 9.22 and is based on the observations already described which have been correlated with histological changes that occur in the hypothalamus, pituitary and thyroid gland. Metamorphosis is divided into three stages: (*a*) a period of rapid growth called premetamorphosis; (*b*) a time of reduced growth but increased differentiation called prometamorphosis and (*c*) metamorphic-climax when there are 'explosive changes'; the tail is resorbed and the frog emerges and takes up a terrestrial existence. Premetamorphosis is characterized endocrinologically as a time when thyroid hormone secretion is low, which reflects the anatomical immaturity of the hypothalamus–pituitary axis and a low rate of TSH secretion. This condition is stabilized further by the presence of large amounts of prolactin, possibly reflecting an immaturity of the inhibitory

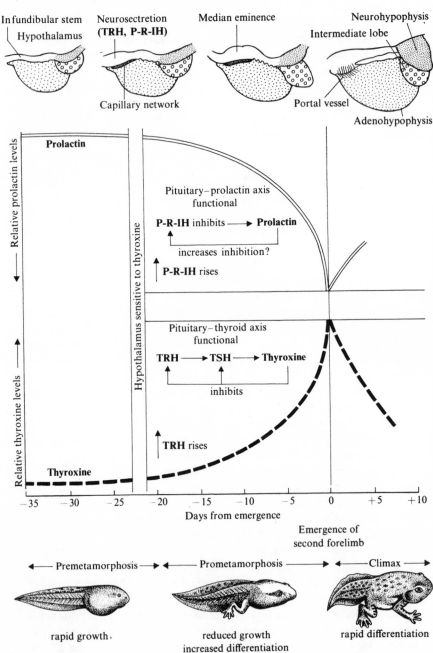

Fig. 9.22. The role of hormones in the growth and development of the tadpole of a frog.

influence of the hypothalamus (P-R-IH), which further inhibits any progress towards metamorphic change. With the progressive maturation of the hypothalamus–pituitary axis toward the beginning of prometamorphosis, TRH gradually increases and releases TSH. The thyroxine concentration thus rises progressively during prometamorphosis. The hypothalamus continues to mature under the influence of thyroxine (called a 'positive-feedback'), and when this is complete and the portal blood supply to the adenohypophysis is finally established, there is a massive stimulation of the thyroid, via TRH and TSH, and metamorphic-climax ensues. During this last period the levels of thyroxine are thought to be declining (due to the operation of the adult negative-feedback inhibition of TRH release).

The unequivocal proof of this theory of tadpole metamorphosis will depend on direct measurements of the hormone levels in tadpoles; a difficult, but with modern techniques, not an impossible task.

Newts and salamanders also undergo a metamorphic transformation but this is not as dramatic as in frogs and toads. Larval urodeles possess limbs and the most prominent morphological change is often the loss of the external gills. Other, less obvious changes nevertheless also occur. This metamorphosis is under pituitary–thyroid control and can be inhibited by prolactin (Gona and Etkin, 1970). A number of species of urodeles are neotenic; they retain their larval characters and never attain the normal adult morphology but can breed while in the larval form. Well-known examples of neoteny in urodeles are seen in axolotls (*Ambystoma mexicanum*) and mudpuppies (*Necturus maculosus*).

It was an event that caused some comment in the biological world when it was found that feeding thyroid glands to axolotls caused them to metamorphose into adult salamanders. The same effect can be elicited by TSH, indicating that the natural disability of these urodeles is due to a pituitary, rather than a thyroid, deficiency. These interesting experiments (see Dent, 1968) involved grafting of pituitaries from metamorphosing tiger salamanders (*Ambystoma tigrinum*) into axolotls which then metamorphosed themselves. The converse experiment, transplanting axolotl pituitaries into hypophysectomized tiger salamanders never resulted in metamorphosis of

---

Upper. The development of the pituitary, especially in relation to the establishment of its connections to the hypothalamus.

Lower. Changes in the activity and hormone concentrations of the pituitary–prolactin axis and the pituitary–thyroid axis. Major events are the acquisition by the hypothalamus of a sensitivity to thyroxine, which hastens its development, and the onset of the definitive roles of TRH and P-R-IH, so that TSH and thyroxine secretion increases while prolactin decreases. (Based on Etkin, 1970.)

the recipients which thereafter remained as larvae. Other neotenous urodeles, like the mudpuppy, however, do not metamorphose when given thyroid hormone; their tissues are not responsive to this hormone.

An interesting departure from the usual amphibian pattern is shown by the eastern spotted newt, *Notophthalmus* (*Diemictylus*) *viridescens*. This animal undergoes two metamorphoses during its life cycle. The first is from the aquatic larva to the adult, or red eft, that lives a terrestrial existence for one to three years, but undergoes a second metamorphosis when it returns to water to breed. This transformation is accompanied by morphological changes in the skin and the newt also loses its tongue. The second metamorphosis is thought to be due to a rise in prolactin levels which is called the 'eft water-drive effect' of this hormone (this has been referred to earlier). This change in the prolactin activity is accompanied by a decline in thyroid hormone (Gona, Pearlman and Etkin, 1970).

The dramatic effects of thyroxine on the development of amphibians has contributed considerably to our ideas about the role of the thyroid in mammalian development. It is, however, now generally considered that the two processes are not analogous and that the 'morphogenetic' effect of thyroid hormone in amphibians represents a special application of these hormones to their physiology. The aquatic and terrestrial phases in an amphibian's life cycle can be viewed as separate processes each of which has undergone evolutionary modification consistent with life in each environment. The transformation of one form to another has no parallel in other tetrapods. The use of the thyroid hormone, and probably also prolactin, to control such metamorphosis is an outstanding example of how adaptable the endocrine system is to the needs of evolutionary change.

Conclusions

Apart from nutrition, reproduction probably involves the most complex processes of humoral coordination that exists in vertebrates. It can also be considered as under 'multihormonal control'. Reproduction, and the preparations for this event, involve a variety of tissues, organs and several distinct, accurately timed, physiological events (for instance impregnation, ovulation, and parturition) and usually take a prolonged period of time to reach fruition. Hormones are especially suited to such needs for coordination. The nature of the reproductive process displays considerable morphological and physiological variation and certain types of mechanisms, such as viviparity in mammals and oviparity in birds, predominate in certain systematic groups, in which they are a feature, and involve special endocrine mechanisms. There are, however, many examples of what may be parallel evolution, such as viviparity in fishes and reptiles in which the special roles of hormones may have evolved independently but have a similar end result.

The diversity in reproductive processes does not appear to have involved many changes in the structure of the hormones themselves. The gonadal steroid hormones have the same structure in all vertebrates, though those from the pituitary show distinct differences which are probably of more consequence as regards limiting their action to a certain species than reflecting any functional predilection to coordinate novel processes. It is, however, of special note that pituitary gonadotrophin appears to exist as a single molecule in some vertebrates, especially the fishes, but in most tetrapods it now seems likely that two distinct hormones, one with LH and the other FSH activity, have emerged. This dichotomy presumably allows for the operation of a more specific and precise control mechanism but the physiological differences that, it would appear, must result are unknown. Another notable endocrine novelty has emerged in eutherian mammals and is the ability of the placenta to act as an additional site for steroid hormone synthesis and to form two pituitary-like hormones, chorionic gonadotrophin and somatomammotrophin. There is, however, considerable interspecific variability in the endocrine function of the eutherian placenta and it is suspected that this may have arisen on several separate evolutionary occasions and been perpetuated according to the requirements of the particular species involved.

Hormones contribute directly to the control of differentiation and growth of the embryo and behavioral and physiological processes concerned with the care of the young. The control of metamorphosis in amphibians is a unique and dramatic example of how endocrines can influence development but they also contribute more ubiquitously, especially to sexual differentiation, in all groups of vertebrates. The phylogenetically novel process of lactation in mammals is a clear example of the evolution of endocrine function, though hormones may also contribute to the process of parental care of the young in a variety of other vertebrates.

References

Abe, K., Robison, G. A., Liddle, G. W., Butcher, R. W., Nicholson, W. E. and Baird, C. E. (1969). Role of cyclic AMP in mediating the effects of MSH, norepinephrine and melatonin on frog skin color. *Endocrinology* **85**, 674–682.

Acher, R., Chauvet, J. and Chauvet, M. T. (1972). Phylogeny of the neuro-hypophysial hormones. Two new active peptides isolated from a cartilaginous fish, *Squalus acanthias*. *Eur. J. Biochem.* **29**, 12–19.

Agus, Z. S., Gardner, L. B., Beck, L. H. and Goldberg, M. (1973). Effects of parathyroid hormone on renal tubular reabsorption of calcium, sodium and phosphate. *Amer. J. Physiol.* **224**, 1143–1148.

Alvarado, R. H. and Johnson, S. R. (1966). The effects of neurohypophysial hormones on water and sodium balance in larval and adult bullfrogs (*Rana catesbeiana*). *Comp. Biochem. Physiol.* **18**, 549–561.

Amoroso, E. C. and Marshall, F. H. A. (1960). External factors in sexual periodicity. In *Marshalls Physiology of Reproduction* (edited by A. S. Parkes), vol. I (Pt 2) pp. 707–831. London: Longmans.

Anderson, L. L. (1973). Effects of hysterectomy and other factors on luteal function. In *Handbook of Physiology*, Section 7 *Endocrinology*, vol. II *Female reproductive system* (Pt. 2) pp. 69–86. Washington: American Physiological Society.

Anon (1970). Effects of sexual activity on beard growth in man. *Nature, Lond.* **226**, 869–870.

Ariëns Kappers, J. (1965). Survey of the innervation of the epiphysis cerebri and the accessory pineal organs of vertebrates. *Prog. Brain Res.* **10**, 87–151.

(1970). The pineal organ: An introduction. In *The Pineal Gland* (edited by G. E. W. Wolstenholme and J. Knight) pp. 3–25. Edinburgh and London: Churchill Livingstone.

Armstrong, D. T. and Kennedy, T. G. (1972). Role of luteinizing hormones in regulation of the rat estrous cycle. *Amer. Zool.* **12**, 245–255.

Augee, M. L. and McDonald, I. R. (1973). Role of the adrenal cortex in the adaptation of the monotreme *Tachyglossus aculeatus* to low environmental temperature. *J. Endocr.* **58**, 513–523.

Aurbach, G. D., Keitmann, H. T., Niall, H. D., Tregear, G. W., O'Riordan, J. L. H., Marcus, R., Marx, S. J. and Potts, J. T. (1972). Structure, synthesis, and mechanism of action of parathyroid hormone. *Rec. Prog. Hormone Res.* **28**, 353–392.

Axelrod, J. (1974). The pineal gland: a neurochemical transducer. *Science* **184**, 1341–1348.

Axelrod, J., Wurtman, R. J. and Snyder, S. H. (1965). Control of hydroxyindole *O*-methyltransferase activity in the rat pineal gland by environmental lighting. *J. biol. Chem.* **240**, 949–954.

Baber, E. C. (1876). Contributions to the minute anatomy of the thyroid gland of the dog. *Proc. Roy. Soc.* **24**, 240–241.

Babiker, M. M. and Rankin, J. C. (1973). Effects of neurohypophysial hormones on renal function in the freshwater- and sea-water-adapted eel (*Anguilla anguilla L.*) *J. Endocr.* **57**, xi–xii.

Bach, J-F., Dardenne, M., Papiernik, M., Barois, A., Levasseur, P. and Le Brigand, H. (1972). Evidence for a serum-factor secreted by the human thymus. *Lancet* **2**, 1056–1058.

Bagnara, J. T. (1969). Responses of pigment cells of amphibians to intermedin. *Colloques du C.N.R.S.* no. **177**, 153–158.

Bagnara, J. T. and Hadley, M. E. (1970). Endocrinology of the amphibian pineal. *Amer. Zool.* **10**, 201–216.

(1972). *Chromatophores and Color Change.* New Jersey: Prentice Hall.

Balazs, R., Cocks, W. A., Eayrs, J. T. and Kovacs, S. (1971). Biochemical effects of thyroid hormones on the developing brain. In *Hormones in Development* (edited by M. Hamburgh and E. J. W. Barrington), pp. 357–379. New York: Appleton–Century–Crofts.

Baldwin, R. L. (1969). Development of milk synthesis. *J. Dairy Sci.* **52**, 729–736.

Ball, J. N. and Baker, B. I. (1969). The pituitary gland: anatomy and histo-physiology. In *Fish Physiology* (edited by W. S. Hoar and D. J. Randall), vol II *The Endocrine System*, pp. 1–110. New York: Academic Press.

Ball, J. N., Chester Jones, I., Forster, M. E., Hargreaves, G., Hawkins, E. F. and Milne, K. P. (1971). Measurement of plasma cortisol levels in the eel *Anguilla anguilla* in relation to osmotic adjustments. *J. Endocr.* **50**, 75–96.

Ball, J. N. and Ensor, D. M. (1965). Effect of prolactin on plasma sodium in the teleost, *Poecilia latipinna. J. Endocr.* **32**, 269–270.

(1967). Specific action of prolactin on plasma sodium levels in hypophysecto-mized *Poecilia latipinna* (Teleostei). *Gen. comp. Endocr.* **8**, 432–440.

Ball, J. N. and Ingleton, P. M. (1973). Adaptive variations in prolactin secretion in relation to external salinity in the teleost *Poecilia latipinna. Gen. comp. Endocr.* **20**, 312–325.

Bargmann, W. (1943). Die Epiphysis cerebri. *Handbuch der mikroskopischen Anatomie des Menschen.* Vol. VI (4). Berlin: Springer Verlag.

Barrington, E. J. W. (1942). Blood sugar and the follicles of Langerhans in the ammocoete larva. *J. exp. Biol.* **19**, 45–55.

(1962). Hormones and vertebrate evolution. *Experimentia* **18**, 201–210.

(1968). Metamorphosis in lower chordates. In *Metamorphosis, A Problem in Developmental Biology* (edited by W. Etkin and L. I. Gilbert), pp. 223–270. New York: Appleton–Century–Crofts.

Barrington, E. J. W. and Dockray, G. J. (1970). The effect of intestinal extracts of lampreys (*Lampetra fluviatilis* and *Petromyzon marinus*) on pancreatic secretion in the rat. *Gen. comp. Endocr.* **14**, 170–177.

(1972). Cholecystokinin–pancreozymin-like activity in the eel *Anguilla anguilla. Gen. comp. Endocr.* **19**, 80–87.

Basu, S. L. (1969). Effects of hormones on the saliential spermatogenesis *in vivo* and *in vitro. Gen. comp. Endocr. Suppl.* **2**, 203–213.

Bates, R. W., Miller, R. A. and Garrison, M. M. (1962). Evidence in the hypo-physectomized pigeon of a synergism among prolactin, growth hormone, thyroxine, and prednisone upon weight of the body, digestive tract, kidney and fat stores. *Endocrinology* **71**, 345–360.

Bayliss, W. M. and Starling, E. H. (1902). The mechanism of pancreatic secretion. *J. Physiol., Lond.* **28**, 325–353.

(1903). On the uniformity of the pancreatic mechanism in vertebrata. *J. Physiol., Lond.* **29**, 174–180.

Bélanger, L. F., Dimond, M. T. and Copp, D. H. (1973). Histological observations on bone and cartilage of growing turtles treated with calcitonin, *Gen. comp. Endocr.* **20**, 297–304.

Bellamy, D. and Leonard, R. A. (1965). Effect of cortisol on the growth of chicks. *Gen. comp. Endocr.* **5**, 402–410.

Bentley, P. J. (1962). Studies on the permeability of the large intestine and urinary bladder of the tortoise (*Testudo graeca*) with special reference to the effects of neurohypophysial and adrenocortical hormones. *Gen. comp. Endocr.* **2**, 323–328.

(1966). Hyperglycaemic effect of neurohypophysial hormones in the chicken, *Gallus domesticus. J. Endocr.* **34**, 527–528.

(1969). Neurohypophysial function in Amphibia: hormone activity in the plasma. *J. Endocr.* **43**, 359–369.

(1971). *Endocrines and Osmoregulation. A Comparative Account of the Regulation of Water and Salt in Vertebrates.* New York, Berlin, Heidelberg: Springer Verlag.

(1972). Introductory remarks. Symposium on endocrinology and osmoregulation. *Fedn Proc.* **31**, 1583–1586.

(1973). Osmoregulation in the aquatic urodeles *Amphiuma means* (the congo eel) and *Siren lacertina* (the mud eel). Effects of vasotocin. *Gen. comp. Endocr.* **20**, 386–392.

Bentley, P. J. and Follett, B. K. (1963). Kidney function in a primitive vertebrate, the cyclostome *Lampetra fluviatilis. J. Physiol., Lond.* **169**, 902–918.

(1965). The effects of hormones on the carbohydrate metabolism of the lamprey *Lampetra fluviatilis. J. Endocr.* **31**, 127–137.

Berde, B. and Boissonnas, R. A. (1968). Basic pharmacological properties of synthetic analogues and homologues of the neurohypophysial hormones. In *Neurohypophysial Hormones and Similar Polypeptides* (edited by B. Berde), pp. 802–870. Berlin, Heidelberg, New York: Springer Verlag.

Berlind, A. (1972*a*). Teleost caudal neurosecretory system: release of urotensin II from isolated urophyses. *Gen. comp. Endocr.* **18**, 557–571.

Berlind, A. (1972*b*). Teleost caudal neurosecretory system: sperm duct contraction induced by urophysial material. *J. Endocr.* **52**, 567–574.

Berlind, A., Lacanilao, F. and Bern, H. A. (1972). Teleost caudal neurosecretory system: effects of osmotic stress on urophysial proteins and active factors. *Comp. Biochem. Physiol.* **42***A*, 345–352.

Bern, H. A. (1972). Comparative endocrinology – the state of the field and art. *Gen. comp. Endocr. Suppl.* **3**, 751–761.

Bern, H. A. and Nicoll, C. S. (1968). The comparative endocrinology of prolactin. *Rec. Prog. Hormone Res.* **24**, 681–713.

(1969). The zoological specificity of prolactin. *Colloques du C.N.R.S.* **177**, 193–202.

Bewley, T. A. and Li, C. H. (1970). Primary structures of human pituitary growth hormone and sheep pituitary lactogenic hormone compared. *Science* **168**, 1361–1362.

Binkley, S., MacBride, S. E., Klein, D. C. and Ralph, C. L. (1973). Pineal enzymes: regulation of avian melatonin synthesis. *Science* **181**, 273–275.

Blair-West, J. R., Coghlan, J. P., Denton, D. A., Nelson, J. F., Orchard, E., Scoggins, B. A., Wright, R. D., Myers, K. and Junqueira, C. L. (1968). Physiological, morphological and behavioural adaptation to a sodium deficient environment by native Australian and introduced species of animals. *Nature, Lond.* **217**, 922–928.

Blum, J. J. (1967). An adrenergic control system in *Tetrahymena*. *Proc. natn. Acad. Sci., USA* **58**, 81–88.

Boehlke, K. W., Church, R. L., Tiemeier, O. W. and Eleftheriou, B. E. (1966). Diurnal rhythm in plasma glucocorticoid levels in the channel catfish (*Ictalurus punctatus*). *Gen. comp. Endocr.* **7**, 18–21.

Boelkins, J. N. and Kenny, A. D. (1973). Plasma calcitonin levels in Japanese quail. *Endocrinology* **92**, 1754–1760.

Bourne, A. R. and Seamark, R. F. (1973). Seasonal changes in testicular function in the lizard *Tiliqua rugosa*. *J. Endocr.* **57**, x.

Bower, A. and Hadley, M. E. (1973). Catecholamine control of melanophore stimulating hormone (MSH) secretion *in vitro*. *Amer. Zool.* **13**, 1277.

Bower, A., Hadley, M. E. and Hruby, V. J. (1974). Biogenic amines and control of melanophore stimulating hormone release. *Science* **184**, 70–72

Bradshaw, S. D. (1972). The endocrine control of water and electrolyte metabolism in desert reptiles. *Gen. comp. Endocrin. Suppl.* **3**, 360–373.

Bradshaw, S. D., Shoemaker, V. H. and Nagy, K. A. (1972). The role of adrenal corticosteroids in the regulation of kidney function in the desert lizard *Dipsosaurus dorsalis*. *Comp. Biochem. Physiol.* **43A**, 621–635.

Bradshaw, S. D. and Waring, H. (1969). Comparative studies on the biological activity of melanin-dispersing hormone (MDH). *Colloques du C.N.R.S.* **177**, 135–151.

Braun, E. J. and Dantzler, W. H. (1972). Functions of mammalian-type and reptilian-type nephrons in kidney of desert quail. *Amer. J. Physiol.* **222**, 617–629.

(1974). Effects of ADH on single-nephron glomerular filtration rates in the avian kidney. *Amer. J. Physiol.* **226**, 1–28.

Bromer, W. W. (1972). Chemistry of glucagon and gastrin. *Handbook of Physiology*, Section 7, *Endocrinology*, vol I *Endocrine pancrease*, pp. 133–138. Washington: American Physiological Society.

Brooks, C. J. W., Brooks, R. V., Fotherby, K., Grant, J. K., Klopper, A. and Klyne, W. (1970). The identification of steroids. *J. Endocr.* **47**, 265–272.

Brown, P. S. and Brown, S. C. (1973). Prolactin and thyroid hormone interactions in salt and water balance in the newt *Notophththalmus viridescens*. *Gen. comp. Endocr.* **20**, 456–466.

Brown-Grant, K. and Ostberg, A. J. C. (1974). Lack of effect of pineal denervation on the responses of the female albino rat to exposure to constant light. *J. Endocr.* **62**, 45–50.

Browning, H. C. (1969). Role of prolactin in regulation of reproductive cycles. *Gen. comp. Endocr. Suppl.* **2**, 42–54.

Bullough, W. S. (1971). The actions of chalones. *Ag. Actions* **2**, 1–7.

Burgers, A. C. J. (1963). Melanophore-stimulating hormones in vertebrates. *Ann. N.Y. Acad. Sci.* **100**, 669–677.

Burzawa-Gerard, E. and Fontaine, Y. A. (1972). The gonadotropins of lower vertebrates. *Gen. comp. Endocr. Suppl.* **3**, 715–728.

Butler, D. G. (1966). Effect of hypophysectomy on osmoregulation in the European eel (*Anguilla anguilla* L.). *Comp. Biochem. Physiol.* **18**, 773–781.
(1969). Corpuscles of Stannius and renal physiology in the eel (*Anguilla rostrata*). *J. Fish. Res. Bd Can.* **26**, 639–654.

Butler, D. G. and Carmichael, F. J. (1972). (Na⁺–K⁺)-ATPase activity in eel (*Anguilla rostrata*) gills in relation to changes in environmental salinity: role of adrenocortical steroids. *Gen. comp. Endocr.* **19**, 421–427.

Butler, D. G., Clarke, W. C., Donaldson, E. M. and Langford, R. W. (1969). Surgical adrenalectomy of a teleost fish (*Anguilla rostrata* LESUEUR): effect on plasma cortisol and tissue electrolyte and carbohydrate concentrations. *Gen. comp. Endocr.* **12**, 503–514.

Cahill, G. F., Aoki, T. T. and Marliss, E. B. (1972). Insulin and muscle protein. *Handbook of Physiology*, Section 7 *Endocrinology*, vol. I *Endocrine pancreas*, pp. 563–577. Washington: American Physiological Society.

Callard, I. P. and Doolittle, J. P. (1973). The influence of intrahypothalamic injections of progesterone on ovarian growth and function in the ovoviviparous iguanid lizard *Sceloperus cyanogenys*. *Comp. Biochem. Physiol.* **44A**, 625–629.

Callard, I. P., Doolittle, J., Banks, W. L. and Chan, S. W. C. (1972). Recent studies on the control of the reptilian ovarian cycle. *Gen. comp. Endocr. Suppl.* **3**, 65–75.

Capelli, J. P., Wesson, L. G. and Aponte, G. E. (1970). A phylogenetic study of the renin–angiotensin system. *Amer. J. Physiol.* **218**, 1171–1178.

Celis, M. E., Hase, S. and Walter, R. (1972). Structure-activity studies of MSH-release-inhibiting hormone. *FEBS letters* **27**, 327–330.

Celis, M. E., Taleisnik, S. and Walter, R. (1971). Regulation of formation and proposed structure of the factor inhibiting the release of melanocyte-stimulating-hormone. *Proc. natn. Acad. Sci., USA* **68**, 1428–1433.

Chadwick, C. S. and Jackson, H. R. (1948). Acceleration of skin growth and molting in the red eft of *Triturus viridescens* by means of prolactin injections. *Anat. Rec.* **101**, 718.

Chan, D. K. O. (1972). Hormonal regulation of calcium balance in teleost fish. *Gen. comp. Endocr. Suppl.* **3**, 411–420.

Chan, D. K. O., Chester Jones, I., Henderson, I. W. and Rankin, J. C. (1967). Studies on the experimental alteration of water and electrolyte composition of the eel (*Anguilla anguilla* L.). *J. Endocr.* **37**, 297–317.

Chan, D. K. O., Phillips, J. G. and Chester Jones, I. (1967). Studies on electrolyte changes in the lip-shark, *Hemiscyllium plagiosum* (Bennett), with special reference to the hormonal influence on the rectal gland. *Comp. Biochem. Physiol.* **23**, 185–198.

Chan, S. T. H. and Phillips, J. G. (1969). The biosynthesis of steroids by the gonads of the ricefield eel, *Monopterus albus* at various stages during natural sex-reversal. *Gen. comp. Endocr.* **12**, 619–636.

Chan, S. W. C. and Phillips, J. G. (1971). Seasonal variations in production *in vitro* of corticosteroids by the frog (*Rana rugulosa*) adrenal. *J. Endocr.* **50**, 1–17.

Chan, S. W. C., Ziegel, S. and Callard, I. P. (1973). Plasma progesterone in snakes. *Comp. Biochem. Physiol.* **44A**, 631–637.

Channing, C. P., Licht, P., Papkoff, H. and Donaldson, E. M. (1974). Comparative activities of mammalian, reptilian and piscine gonadotropins in monkey granulosa cell cultures. *Gen. comp. Endocr.* **22**, 137–145.

Chavin, W., Kim, K. and Tchen, T. T. (1963). Endocrine control of pigmentation. *Ann. N. Y. Acad. Sci.* **100**, 678–685.

Chester Jones, I., Bellamy, D., Chan, D. K. O., Follett, B. K., Henderson, I. W., Phillips, J. G. and Snart, R. S. (1972). Biological actions of steroid hormones on nonmammalian vertebrates. In *Steroids in Nonmammalian Vertebrates* (edited by D. R. Idler), pp. 414–480. New York and London: Academic Press.

Chester Jones, I., Henderson, I. W., Chan, D. K. O., Rankin, J. C., Mosley, W., Brown, J. J., Lever, A. F., Robertson, J. I. S. and Tree, M. (1966). Pressor activity in extracts of the corpuscles of Stannius from the European eel (*Anguilla anguilla* L.). *J. Endocr.* **34**, 393–408.

Chieffi, G. (1967). The reproductive system of elasmobranchs: developmental and endocrinological aspects. In *Sharks, Skates and Rays* (edited by P. W. Gilbert, R. F. Mathewson and D. P. Rall), pp. 553–580. Baltimore: Johns Hopkins.

Chiu, K. W. and Lynn, W. G. (1972). Observations on thyroidal control of sloughing in the garter snake, *Thamnophis sirtalis, Copeia* 1972 (no. 1), 158–163.

Chiu, K. W. and Phillips, J. G. (1971a). The effect of hypophysectomy and of injections of thyrotrophin and corticotrophin into hypophysectomized animals on the sloughing cycle of the lizard *Gekko gecko* L. *J. Endocr.* **49**, 611–618.

(1971b). The role of prolactin in the sloughing cycle in the lizard *Gekko gecko* L. *J. Endocr.* **49**, 625–634.

Clark, N. B. (1967). Influence of estrogens upon serum calcium, phosphate and protein concentrations of fresh-water turtles. *Comp. Biochem. Physiol.* **20**, 823–834.

(1972). Calcium regulation in reptiles. *Gen. comp. Endocr. Suppl.* **3**, 430–440.

Clark, N. B. and Dantzler, W. H. (1972). Renal tubular transport of calcium and phosphate in snakes: role of parathyroid hormone. *Amer. J. Physiol.* **223**, 1455–1464.

Clarke, W. C., Bern, H. A., Li, C. H. and Cohen, D. C. (1973). Somatotropic and sodium-retaining effects of human growth hormone and placental lactogen in lower vertebrates. *Endocrinology* **93**, 960–964.

Cockburn, F., Hull, D. and Walton, I. (1968). The effect of lipolytic hormones and theophylline on heat production in brown adipose tissue in vivo. *Brit. J. Pharmacol.* **31**, 568–577.

Cofré, G. and Crabbé, J. (1965). Stimulation by aldosterone of active sodium transport by the isolated colon of the toad *Bufo marinus. Nature, Lond.* **207**, 1299–1300.

Collins, K. J. and Weiner, J. S. (1968). Endocrinological aspects of exposure to high environmental temperatures. *Physiol. Rev.* **48**, 785–839.

Colombo, L., Bern, H. A. and Pieprzyk, J. (1971). Steroid transformations by the corpuscles of Stannius and the body of *Salmo gairdnerii* (Teleostei). *Gen. comp. Endocr.* **16**, 74–84.

Cooper, W. E. and Ferguson, G. W. (1972). Steroids and color change during gravidity in the lizard *Crotaphytus collaris. Gen. comp. Endocr.* **18**, 69–72.

Coote, J. H., Johns, E. J., Macleod, V. H. and Singer, B. (1972). Effect of renal nerve stimulation, renal blood flow and adrenergic blockade on plasma renin activity in the cat. *J. Physiol., Lond.* **226**, 15–36.

Copp, D. H. (1969). The ultimobranchial glands and calcium regulation. In *Fish Physiology* (edited by W. S. Hoar and D. J. Randall), vol. II *The Endocrine System*, pp. 377–398. New York: Academic Press.

(1972). Calcium regulation in birds. *Gen. comp. Endocr. Suppl.* 3, 441–447.

Copp, D. H., Cameron, E. C., Cheney, B. A., Davidson, A. G. F. and Henze, K. G. (1962). Evidence for calcitonin – a new hormone from the parathyroid that lowers blood calcium. *Endocrinology* 70, 638–649.

Copp, D. H., Cockcroft, D. W. and Keuk, Y. (1967a). Calcitonin from ultimobranchial glands of dogfish and chickens. *Science* 158, 924–926.

(1967b). Ultimobranchial origin of calcitonin, hypocalcemic effect of extracts from chicken glands. *Can. J. Physiol. Pharmacol.* 45, 1095–1099.

Cortelyou, J. R. (1967). The effect of commercially prepared parathyroid extracts on plasma and urine calcium levels in *Rana pipiens*. *Gen. comp. Endocr.* 9, 234–240.

Cowie, A. T. (1972). Lactation and its hormonal control. In *Hormones in Reproduction* (edited by C. R. Austin and R. V. Short), pp. 106–143. London: Cambridge University Press.

Cowie, A. T. and Tindal, J. S. (1971). *The Physiology of Lactation*. Baltimore: Williams and Wilkins.

Crabbé, J. and De Weer, P. (1964). Action of aldosterone on the bladder and skin of the toad. *Nature, Lond.* 202, 278–279.

Crews, D. and Licht, P. (1974). Inhibition by corpora atretica of ovarian sensitivity and hormonal stimulation in the lizard, *Anolis carolinensis*. *Endocrinology* 95, 102–106.

Crighton, D. B., Hartley, B. M. and Lamming, G. E. (1973). Changes in the luteinizing hormone releasing activity of the hypothalamus, and in the pituitary gland and plasma luteinizing hormone during the oestrous cycle of the sheep. *J. Endocr.* 58, 377–385.

Crocker, A. D. and Holmes, W. N. (1971). Intestinal absorption in ducklings (*Anas platyrhynchos*) maintained on fresh water and hypertonic saline. *Comp. Biochem. Physiol.* 40A, 203–211.

Daughaday, W. H. (1971). Sulfation factor regulation of skeletal growth. *Amer. J. Med.* 50, 277–280.

Daughaday, W. H., Hall, K., Raben, M. S., Salmon, W. D., Van den Brande, J. L. and Van Wyke, J. J. (1972). Somatomedin: proposed designation for sulphation factor. *Nature, Lond.* 235, 107.

Davis, P. J., Gregerman, R. I. and Poole, W. E. (1969). Thyroxine-binding proteins in the serum of the grey kangaroo. *J. Endocr.* 45, 477–478.

Davis, P. J. and Jurgelski, W. (1973). Thyroid hormone-binding in opossum serum: evidence for polymorphism and relationship to haptoglobin polymorphism. *Endocrinology* 92, 822–832.

DeLuca, H. F. (1971). The role of vitamin D and its relationship to parathyroid hormone and calcitonin. *Rec. Prog. Hormone Res.* 27, 479–510.

DeLuca, H. F., Morii, H. and Melancon, M. J. (1968). The interaction of vitamin D, parathyroid hormone and thyrocalcitonin. In *Parathyroid Hormone and Thyrocalcitonin (Calcitonin)* (edited by R. V. Talmage and F. F. Belanger), pp. 448–454. Amsterdam: Excerpta Mecia Foundation.

DeLuise, M., Martin, T. J., Greenberg, P. B. and Michelangeli, V. (1972). Metabolism of porcine, human and salmon calcitonin in the rat. *J. Endocr.* 53, 475–482.

Dent, J. N. (1968). Survey of amphibian metamorphosis. In *Metamorphosis, a Problem in Developmental Biology* (edited by W. Etkin and L. I. Gilbert), pp. 271–311. New York: Appleton–Century–Crofts.

DeRoos, R. and DeRoos, C. C. (1972). Comparative effects of the pituitary-adrenocortical axis and catecholamines on carbohydrate metabolism in elasmobranch fish. *Gen. comp. Endocr. Suppl.* **3**, 192–197.

Desranleau, R., Gilardeau, C. and Chrétien, M. (1972). Radioimmunoassay of ovine beta-lipotropic hormone. *Endocrinology* **91**, 1004–1010.

Dharmamba, M., Mayer-Gostan, N., Maetz, J. and Bern, H. A. (1973). Effect of prolactin on sodium movement in *Tilapia mossambica* adapted to sea water. *Gen. com. Endocr.* **21**, 179–187.

Dicker, S. E. and Elliott, A. B. (1973). Neurohypophysial hormones and homeostasis in the crab-eating frog, *Rana cancrivora. Hormone Res.* **4**, 224–260.

Dodd, J. M. (1960). Gonadal and gonadotrophic hormones in lower vertebrates. In *Marshall's Physiology of Reproduction* (edited by A. S. Parkes), vol. I (Pt 2), pp. 417–582. London: Longmans.

—— (1972a). Ovarian control in cyclostomes and elasmobranchs. *Amer. Zool.* **12**, 325–339.

—— (1972b). The endocrine regulation of gametogenesis and gonad maturation in fishes. *Gen. comp. Endocr. Suppl.* **3**, 675–687.

Dodd, J. M. and Dodd, M. H. I. (1969). Phylogenetic specificity of thyroid stimulating hormone with special reference to the Amphibia. *Colloques du C.N.R.S.* **177**, 277–285.

Donaldson, E. M., Yamzaki, F., Dye, H. M. and Philleo, W. W. (1972). Preparation of gonadotropin from salmon (*Oncorhynchus tshawytscha*) pituitary glands. *Gen. comp. Endocr.* **18**, 469–481.

Donoso, A. O. and Segura, E. T. (1965). Seasonal variations of plasma adrenaline and noradrenaline in toads. *Gen. comp. Endocr.* **5**, 440–443.

Douglas, W. W. (1968). Stimulus–secretion coupling: the concept and clues from chromaffin and other cells. *Brit. J. Pharmacol.* **34**, 451–474.

Dousa, T., Hechter, O., Schwartz, I. L. and Walter, R. (1971). Neurohypophyseal hormone-responsive adenylate cyclase from mammalian kidney. *Proc. natn. Acad. Sci., USA* **68**, 1693–1697.

Dousa, T., Walter, R., Schwartz, I. L., Sands, H. and Hechter, O. (1972). Role of cyclic AMP in the action of neurohypophyseal hormones on kidney. *Advances in Cyclic Nucleotide Research* (edited by P. Greengard and G. A. Robison), **1**, 121–135. New York: Raven Press.

Ebling, F. J. and Hale, P. A. (1970). The control of the mammalian molt. *Mem. Soc. Endocr.* **18**, 215–235.

Eddy, J. M. P. and Strahan, R. (1968). The role of the pineal complex in the pigmentary effector system of lampreys, *Mordacia mordax* (Richardson) and *Geotria australis* Gray. *Gen. comp. Endocr.* **11**, 528–534.

Egami, N. and Ishii, S. (1962). Hypophysial control of reproductive functions in teleost fishes. *Gen. comp. Endocr. Suppl.* **1**, 248–253.

Elger, W., Neumann, F. and von Berswordt-Wallrabe, R. (1971). The influence of androgen antagonists and progestogens on the sex differentiation of different mammalian species. In *Hormones in Development* (edited by M. Hamburgh and E. J. W. Barrington), pp. 651–667. New York: Appleton–Century–Crofts.

Elliott, A. B. (1968). Effects of adrenaline on water uptake in *Bufo melanostictus*. *J. Physiol., Lond.* **197**, 87–88P.

Ensor, D. M. and Ball, J. N. (1972). Prolactin and osmoregulation in fishes. *Fedn Proc.* **31**, 1615–1623.

Ensor, D. M., Edmondson, M. R. and Phillips, J. G. (1972). Prolactin and dehydration in rats. *J. Endocr.* **53**, lix–lx.

Ensor, D. M., Simons, I. M. and Phillips, J. G. (1973). The effect of hypophysectomy and prolactin replacement therapy on salt and water metabolism in *Anas platyrhynchos*. *J. Endocr.* **57**, xi.

Epple, A. (1969). The endocrine pancreas. In *Fish Physiology* (edited by W. S. Hoar and D. J. Randall), vol. II *The Endocrine System*, pp. 275–319. New York: Academic Press.

Epstein, F. H., Cynamon, M. and McKay, W. (1971). Endocrine control of Na–K-ATPase and seawater adaptation in *Anguilla rostrata*. *Gen. comp. Endocr.* **16**, 323–328.

Epstein, F. H., Katz, A. I. and Pickford, G. E. (1967). Sodium- and potassium-activated adenosine triphosphatase of gills: role in adaptation of teleosts to salt water. *Science* **156**, 1245–1247.

Estler, C. J. and Ammon, H. P. T. (1969). The importance of the adrenergic beta-receptors for thermogenesis and survival of acutely cold-exposed mice. *Can. J. Physiol. Pharmacol.* **47**, 427–434.

Etkin, W. (1970). The endocrine mechanism of amphibian metamorphosis, an evolutionary achievement. *Mem. Soc. Endocri.* **18**, 137–153.

Etkin, W. and Gona, A. G. (1967). Antagonism between prolactin and thyroid hormone in amphibian development. *J. exp. Zool.* **165**, 249–258.

Fagerlund, U. H. M. (1967). Plasma cortisol concentration in relation to stress in adult sockeye salmon during the freshwater stage in their life cycle. *Gen. comp. Endocr.* **8**, 197–207.

Falkmer, S. and Patent, G. J. (1972). Comparative and embryological aspects of the pancreatic islets. In *Handbook of Physiology*, Section 7 *Endocrinology*, vol. I *Endocrine pancreas*, pp. 1–23. Washington: American Physiological Society.

Farer, L. S., Robbins, J., Blumberg, B. S. and Rall, J. E. (1962). Thyroxine-serum protein complexes in various animals. *Endocrinology* **70**, 686–696.

Fenwick, J. C. (1970). The pineal organ: photoperiod and reproductive cycles in the goldfish *Carassius auratus* L. *J. Endocr.* **46**, 101–111.

Fenwick, J. C. and Forster, M. E. (1972). Effects of Stanniectomy and hypophysectomy on total plasma cortisol levels in the eel (*Anguilla anguilla* L). *Gen. comp. Endocr.* **19**, 184–191.

Ferguson, D. R. and Heller, H. (1965). Distribution of neurohypophysial hormones in mammals. *J. Physiol., Lond.* **180**, 846–863.

Ferguson, G. W. and Chen, C. L. (1973). Steroid hormones, color change and ovarian cycling in free-living female collard lizards, *Crotaphytus collaris*. *Amer. Zool.* **13**, 1277.

Fernholm, B. (1972). Neurohypophysial–adenohypophysial relations in hagfish (Myxinoidea, Cyclostomata). *Gen. comp. Endocr. Suppl.* **3**, 1–10.

Fitzsimons, J. T. (1972). Thirst. *Physiol. Rev.* **52**, 468–561.

Fleming, W. R., Brehe, J. and Hanson, R. (1973). Some complicating factors in the study of calcium metabolism in teleosts. *Amer. Zool.* **13**, 793–797.

Fleming, W. R., Stanley, J. G. and Meier, A. H. (1964). Seasonal effects of external calcium, estradiol, and ACTH on the serum calcium and sodium levels of *Fundulus kansae*. *Gen. comp. Endocr.* **4**, 61–67.

Foà, P. P. (1972). The secretion of glucagon. *Handbook of Physiology*, Section 7 *Endocrinology*, vol. I, *Endocrine pancreas*, pp. 261–277. Washington: American Physiological Society.

Follett, B. K. (1963). Mole ratios of the neurohypophysial hormones in the vertebrate neural lobe. *Nature, Lond.* **198**, 693–694.

Follett, B. K. and Redshaw, M. R. (1974). The physiology of vitellogenesis. In *Physiology of the Amphibia* (edited by B. Lofts), vol. II, pp. 219–298. New York: Academic Press.

Follett, B. K. and Riley, J. (1967). Effect of the length of the daily photoperiod on thyroid activity in the female Japanese quail (*Coturnix coturnix japonica*). *J. Endocr.* **39**, 615–616.

Follett, B. K. and Sharp, P. J. (1969). Circadian rhythmicity in photoperiodically induced gonadotrophin release and gonadal growth in the quail. *Nature, Lond.* **223**, 968–971.

Fontaine, M. (1954). Du déterminisme physiologique des migrations. *Biol. Rev.* **29**, 390–418.

(1964). Corpuscules de Stannius et régulation ionique (Ca, K, Na) du milieu interiéur de l'Anguille (*Anguilla anguilla* L.). *C. R. Acad. Sci., Paris* **259**, 875–878.

Fontaine, M., Callamand, O. and Olivereau, M. (1949). Hypophyse et euryhalinité chez l'anguille. *C. R. Acad. Sci., Paris* **228**, 513–514.

Fontaine, Y-A. (1964). Characteristics of the zoological specificity of some protein hormones from the anterior pituitary. *Nature, Lond.* **202**, 1296–1298.

(1969a). La spécificité zoologique des protéines hypophysaires capables de stimuler la thyroide. *Acta Endocrin., Kobn, Suppl.* **136**, 1–154.

(1969b). La spécifité zoologique d'action des hormones thyréotropes. *Colloque du C.N.R.S.* **177**, 267–275.

Forrest, J. N., Cohen, A. D., Schon, D. A. and Epstein, F. H. (1973a). Na transport and Na–K-ATPase in gills during adaptation to seawater: effects of cortisol. *Amer. J. Physiol.* **224**, 709–713.

Forrest, J. N., MacKay, W. C., Gallagher, B. and Epstein, F. H. (1973b). Plasma cortisol response to saltwater adaptation in the American eel *Anguilla rostrata*. *Amer. J. Physiol.* **224**, 714–717.

Frantz, A. G., Kleinberg, D. L. and Noel, G. L. (1972). Studies on prolactin in man. *Rec. Prog. Hormone Res.* **28**, 527–573.

Fraps, R. M. (1955). Egg production and fertility in poultry. In *Progress in the Physiology of Farm Animals* (edited by J. Hammond) vol. II, pp. 661–740. London: Butterworths.

Freeman, H. C. and Idler, D. R. (1973). Effects of corticosteroids on liver transaminases in two salmonids, the rainbow trout (*Salmo gairdnerii*) and the brook trout (*Salvelinus fontinalis*). *Gen. comp. Endocr.* **20**, 69–75.

Fridberg, G. and Bern, H. A. (1968). The urophysis and the caudal neurosecretory system of fishes. *Biol. Rev.* **43**, 175–199.

Fritz, I. B. (1972). Insulin actions on carbohydrate and lipid metabolism. In *Biochemical Actions of Hormones* (edited by G. Litwack), Vol. II, pp. 165–214. New York and London: Academic Press.

Fritz, I. B. and Lee, L. P. K. (1972). Fat mobilization and ketogenesis. *Handbook of Physiology*, Section 7 *Endocrinology*, vol. I *Endocrine pancreas*, pp. 579–596. Washington: American Physiological Society.

Frye, B. E. (1967). *Hormonal Control in Vertebrates*, p. 104. New York: The Macmillan Co.

Frye, B. E., Brown, P. S. and Snyder, B. W. (1972). Effects of prolactin and somatotropin on growth and metamorphosis of amphibians. *Gen. comp. Endocr. Suppl.* **3**, 209–220.

Funder, J. W., Feldman, D. and Edelman, I. S. (1973). The roles of plasma binding and receptor specificity in the mineralocorticoid action of aldosterone. *Endocrinology* **92**, 994–1004.

Furr, B. J. A., Bonney, R. C., England, R. J. and Cunningham, F. J. (1973). Luteinizing hormone and progesterone in peripheral blood during the ovulatory cycle in the hen, *Gallus domesticus*. *J. Endocr.* **57**, 159–169.

Garabedian, M., Tanaka, Y., Holick, M. F. and DeLuca, H. F. (1974). Response of intestinal calcium transport and bone calcium mobilization to 1,25-dihydroxyvitamin D_3 in thyroparathyroidectomized rats. *Endocrinology* **94**, 1022–1027.

Geschwind, I. I. (1967). Growth hormone activity in the lungfish pituitary. *Gen. comp. Endocr.* **8**, 82–83.

(1969). The main lines of evolution of the pituitary hormones. *Colloques du C.N.R.S.* **177**, 385–400.

Geschwind, I. I., Huseby, R. A. and Nishioka, R. (1972). The effect of melanocyte-stimulating hormone on coat color in the mouse. *Rec. Prog. Hormone Res.* **28**, 91–129.

Gibbs, J., Young, R. C. and Smith, G. P. (1973). Cholecystokinin elicits satiety in rats with open gastric fistulas. *Nature, Lond.* **245**, 323–325.

Ginsburg, M. (1968). Production, release, transportation, and elimination of the neurohypophysial hormone. In *Neurohypophysial Hormones and similar Polypeptides* (edited by B. Berde), pp. 286–371. Berlin, Heidelberg and New York: Springer Verlag.

Godet, M. (1961). Le problème hydrique et son controle hypophysaire chez le Protptère. *Ann. Faculty Sciences de l'Université Dakar* **6**, 183–201.

Goldman, J. M. and Hadley, M. E. (1969). The beta adrenergic receptor and cyclic 3'-5'-adenosine monophosphate: possible roles in the regulation of melanophore responses of the spadefoot toad, *Scaphiopus couchi*. *Gen. comp. Endocr.* **13**, 151–163.

Goldstein, A. L., Hooper, J. A., Schulof, R. S., Cohen, G. H., Thurman, G. B., McDaniel, M. C., White, A. and Dardenne, M. (1974). Thymosin and the immunopathology of aging. *Fedn Proc.* **33**, 2053–2056.

Gona, A. G. and Etkin, W. (1970). Inhibition of metamorphosis in *Ambystoma tigrinum* by prolactin. *Gen. comp. Endocr.* **14**, 589–591.

Gona, O. and Gona, A. G. (1973). Action of human placental lactogen on second metamorphosis in the newt *Notophthalmus viridescens*. *Gen. comp. Endocr.* **21**, 377–380.

Gona, A. G., Pearlman, T. and Etkin, W. (1970). Prolactin–thyroid interaction in the newt, *Diemictlylus viridescens*. *J. Endocr.* **48**, 585–590.

Goodridge, A. G. (1964). The effect of insulin, glucagon and prolactin on lipid synthesis and related metabolic activity in migratory and non-migratory finches. *Comp. Biochem. Physiol.* **13**, 1–26.

Gorbman, A. (1940). Suitability of the common goldfish for assay of thyrotropic hormone. *Proc. Soc. exp. Biol. Med., New York* **45**, 772–773.

Gorbman, A. and Bern, H. A. (1962). *A Textbook of Comparative Endocrinology*, p. 220. New York: Wiley.

Gorbman, A. and Hyder, M. (1973). Failure of mammalian TRH to stimulate thyroid function in the lungfish. *Gen. comp. Endocr.* **20**, 588–589.

Gotshall, R. W., Davis, J. O., Shade, R. E., Spielman, W., Johnson, J. A. and Braverman, B. (1973). Effects of renal denervation on renin release in sodium-depleted dogs. *Amer. J. Physiol.* **225**, 344–349.

Goy, R. W. and Goldfoot, D. A. (1973). Hormonal influences on sexually dimorphic behavior. *Handbook of Physiology*, Section 7 *Endocrinology*, vol. II *Female reproductive system* (Pt 1), pp. 169–186. Washington: American Physiological Society.

Greenwood, A. W. and Blyth, J. S. S. (1935). Variation in plumage response of brown leghorn breast feather and its reaction to oestrone. *Proc. Zool. Soc. Lond. Ser. A* **109**, 247–288.

Greer, M. A. and Haibach, H. (1974). Thyroid secretion. *Handbook of Physiology*, Section 7 *Endocrinology*, vol. III *Thyroid*, pp. 135–146. Washington: American Physiology Society.

Gregory, R. A. (1962). *Secretory Mechanisms of the Gastrointestinal Tract*, p. 153. London: Edward Arnold.

Griffiths, M. (1965). Rate of growth and intake of milk in a suckling echidna. *Comp. Biochem. Physiol.* **16**, 383–392.

Guillemin, R. and Burgus, R. (1972). The hormones of the hypothalamus. *Scientific American* **227** (November) 24–33.

Habener, J. F., Singh, F. R., Deftos, L. J., Neer, R. M. and Potts, J. T. (1971). Explanation for unusual potency of salmon calcitonin. *Nature New Biol.* **232**, 91–92.

Hadley, M. E. (1972). Functional significance of vertebrate integumental pigmentation. *Amer. Zool.* **12**, 63–76.

Hall, P. F. (1969). Hormonal control of melanin synthesis in birds. *Gen. comp. Endocr. Suppl.* **2**, 451–458.

Hanaoka, T. (1953). Effect of melanophore hormone on regeneration of visual purple in solution. *Nature, Lond.* **172**, 866.

Handler, J. S., Bensinger, R. and Orloff, J. (1968). Effects of adrenergic agents on toad bladder response to ADH, 3',5'-AMP, and theophylline. *Amer. J. Physiol.* **215**, 1024–1031.

Hansel, W. and Echternkamp, S. E. (1972). Control of ovarian function in domestic animals. *Amer. Zool.* **12**, 225–243.

Hanstrom, B. (1966). Gross anatomy of the hypophysis in mammals. In *The Pituitary Gland* (edited by G. W. Harris and B. T. Donovan), vol. I, pp. 1–57. Berkeley and Los Angeles: University of California.

Harmeyer, J. and DeLuca, H. F. (1969). Calcium-binding protein and calcium absorption after vitamin D administration. *Arch. Biochem. Biophys.* **133**, 247–254.

Harper, A. A. and Raper, H. S. (1943). Pancreozymin, a stimulant of the secretion of pancreatic enzymes in extracts of the small intestine. *J. Physiol., Lond.* **102**, 115–125.

Harper, C. and Toverud, S. U. (1973). Ability of thyrocalcitonin to protect against hypercalcemia in adult rats. *Endocrinology* **93**, 1354–1359.

Harri, M. N. E. (1972). Effect of season and temperature acclimation on the tissue catecholamine level and utilization in the frog *Rana temporaria*. *Comp. gen. Pharmacol.* **3**, 101–112.

Harri, M. and Hedenstam, R. (1972). Calorigenic effect of adrenaline and nor-adrenaline in the frog, *Rana temporaria. Comp. Biochem. Physiol.* **41***A*, 409–419.

Harrington, R. W. (1968). Delimitation of the thermolabile phenocritical period of sex determination and differentiation in the ontogeny of the normally hermaphroditic fish *Rivulus marmoratus*, Poey. *Physiol. Zool.* **41**, 447–459.

Hartman, F. A. and Brownell, K. A. (1949). *The Adrenal Gland.* London: Henry Kimpton.

Hasan, S. H. and Heller, H. (1968). The clearance of neurohypophysial hormones from the circulation of non-mammalian vertebrates. *Brit. J. Pharmacol.* **33**, 523–530.

Hayashida, T. (1970). Immunological studies with rat pituitary growth hormone (RGH). II. Comparative immunochemical investigation of GH from representatives of various vertebrates classes with monkey antiserum to RGH. *Gen. comp. Endocr.* **15**, 432–452.

(1971). Biological and immunochemical studies with growth hormone in pituitary extracts of holostean and chondrostean fishes. *Gen. comp. Endocr.* **17**, 275–280.

(1973). Biological and immunochemical studies with growth hormone in pituitary extracts of elasmobranchs. *Gen. comp. Endocr.* **20**, 377–385.

Hayashida, T. and Lagios, M. D. (1969). Fish growth hormone: a biological, immunochemical, and ultrastructural study of sturgeon and paddlefish pituitaries. *Gen. comp. Endocr.* **13**, 403–411.

Hayashida, T., Licht, P. and Nicoll, C. S. (1973). Amphibian pituitary growth hormone and prolactin: immunochemical relatedness to rat growth hormone. *Science* **182**, 169–171.

Hazelwood, R. L. (1973). The avian endocrine pancreas. *Amer. Zool.* **13**, 699–709.

Hazelwood, R. L., Turner, S. D., Kimmel, J. R. and Pollock, H. G. (1973). Spectrum effects of a new polypeptide (third hormone?) isolated from chicken pancreas. *Gen. comp. Endocr.* **21**, 485–497.

Heap, R. B. (1972). Role of hormones in pregnancy. In *Hormones in Reproduction* (edited by C. R. Austin and R. V. Short), pp. 73–105. London: Cambridge University Press.

Heap, R. B., Perry, J. S. and Challis, J. R. G. (1973). Hormonal maintenance of pregnancy. *Handbook of Physiology*, Section 7 *Endocrinology*, vol. II *Female reproductive system* (Pt 2), pp. 217–260. Washington: American Physiological Society.

Heding, L. G. (1971). Radioimmunological determination of pancreatic and gut glucagon in plasma. *Diabetologia* **7**, 10–19.

Heins, J. N., Garland, J. T. and Daughaday, W. H. (1970). Incorporation of ^{35}S-sulfate into rat cartilage explants *in vitro*: effects of aging on responsiveness to stimulation by sulfation factor. *Endocrinology* **87**, 688–692.

Heller, H. (1941). Differentiation of an (amphibian) water balance principle from the antidiuretic principle of the posterior pituitary gland. *J. Physiol., Lond.* **100**, 125–141.

(1966). The hormone content of the vertebrate hypothalamo–neurohypophysial system. *Brit. Med. Bull.* **22**, 227–231.

(1972). The effect of neurohypophysial hormones on the female reproductive tract of lower vertebrates. *Gen. comp. Endocr. Suppl.* **3**, 703–714.

(1973). The effects of oxytocin and vasopressin during the oestrous cycle and pregnancy on the uterus of a marsupial species, the quokka (*Setonix brachyurus*). *J. Endocr.* **58**, 657–671.

(1974). Molecular aspects of comparative endocrinology. *Gen. comp. Endocr.* **22**, 315–332.

Heller, H., Ferreri, E. and Leathers, D. H. G. (1970). The effect of neurohypophysial hormones on the amphibian oviduct *in vitro*, with some remarks on the histology of this organ. *J. Endocr.* **47**, 495–509.

Heller, J. (1961). The physiology of the antidiuretic hormone VIII. The antidiuretic activity in the plasma of the mouse, guinea-pig, cat, rabbit and dog. *Physiol. Bohemoslov.* **10**, 167–172.

Heller, J. and Štulc, J. (1960). The physiology of the antidiuretic hormone. III. The antidiuretic activity of plasma in normal and dehydrated rats. *Physiol. Bohemslov.* **9**, 93–98.

Henderson, I. W. and Chester Jones, I. (1967). Endocrine influences on the net extrarenal fluxes of sodium and potassium in the European eel (*Anguilla anguilla* L.). *J. Endocr.* **37**, 319–325.

Henderson, I. W. and Wales, N. A. M. (1974). Renal diuresis and anti-diuresis after injections of arginine vasotocin in the fresh-water eel (*Anguilla anguilla* L.). *J. Endocr.* **41**, 487–500.

Henderson, J. R. (1969). Why are the Islets of Langerhans? *Lancet* ii, 469–470.

Hill, C. W. and Fromm, P. O. (1968). Response of the interrenal gland of rainbow trout (*Salmo gairdneri*) to stress. *Gen. comp. Endocr.* **11**, 69–77.

Hirano, T. (1969). Effects of hypophysectomy and salinity change on plasma cortisol concentration in the Japanese eel, *Anguilla japonica. Endocr. japon.* **16**, 557–560.

Hirsch, P. F. and Munson, P. L. (1969). Thyrocalcitonin. *Physiol. Rev.* **49**, 548–622.

Hisaw, F. L. (1959). The corpora lutea of elasmobranch fishes. *Anat. Rec.* **133**, 289.

Hogben, L. T. (1924). *The Pigmentary Effector System*, p. 67. Edinburgh: Oliver and Boyd.

(1942). Chromatic behaviour. *Proc. Roy. Soc. Lond. B* **131**, 111–136.

Hoffman, R. A. (1964). Terrestrial animals in cold: hibernators. *Handbook of Physiology*, Section 4 *Adaptation to the Environment*, pp. 379–403. Washington: American Physiological Society.

Holmes, R. L. and Ball, J. N. (1974). *The Pituitary Gland. A Comparative Account.* London: Cambridge University Press.

Holmes, W. N. (1972). Regulation of electrolyte balance in marine birds with special reference to the role of the pituitary–adrenal axis in the duck (*Anas platyrhynchos*) *Fedn Proc.* **31**, 1587–1598.

Holmes, W. N., Butler, D. G. and Phillips, J. G. (1961). Observations on the effects of maintaining glaucous-winged gulls (*Larus glaucescens*) on fresh water and sea water for long periods. *J. Endocr.* **23**, 53–61.

Horrobin, D. F., Burstyn, P. G., Lloyd, I. J., Durkin, N., Lipton, A. and Muiruri, K. L. (1971). Actions of prolactin on human renal function. *Lancet* ii, 352–354.

Howe, A. (1973). The mammalian pars intermedia: a review of its structure and function. *J. Endocr.* **59**, 385–409.

Humbel, R. E., Bosshard, H. R. and Zahn, H. (1972). Chemistry of insulin. *Handbook of Physiology*, Section 7 *Endocrinology*, vol. I *Endocrine pancreas*, pp. 111–132. Washington: American Physiological Society.

Idler, D. R. (editor). (1972). *Steroids in Nonmammalian Vertebrates.* New York and London: Academic Press.

Idler, D. R., Sangalang, G. B. and Truscott, B. (1972). Corticosteroids in the South American lungfish. *Gen. comp. Endocr. Suppl.* **3**, 238–244.

Idler, D. R. and Truscott, B. (1972). Corticosteroids in fish. In *Steroids in Non-mammalian Vertebrates* (edited by D. R. Idler), pp. 126–252. New York and London: Academic Press.

Ireland, M. P. (1969). Effect of urophysectomy in *Gasterosteus aculeatus* on survival in fresh water and sea-water. *J. Endocr.* **43**, 133–134.

——— (1973). Effects of arginine vasotocin on sodium and potassium metabolism in *Xenopus laevis* after skin gland stimulation and sympathetic blockade. *Comp. Biochem. Physiol* **44A**, 487–493.

Ivy, A. C. and Oldberg, E. (1928). A hormone mechanism for gall bladder contraction and evacuation. *Amer. J. Physiol.* **86**, 599–613.

Jackson, I. M. D. and Reichlin, S. (1974). Thyrotropin-releasing hormone distribution in hypothalamic and extrahypothalamic brain tissues of mammalian and submammalian chordates. *Endocrinology* **95**, 854–862.

Jackson, R. G. and Sage, M. (1973). A comparison of the effects of mammalian TSH on the thyroid glands of the teleost *Galeichthys felis* and the elasmobranch *Dasyatis sabina. Comp. Biochem. Physiol.* **44A**, 867–870.

Janský, L. (1973). Non-shivering thermogenesis and its thermoregulatory significance. *Biol. Rev.* **48**, 85–132.

Janssens, P. A. (1964). The metabolism of the aestivating African lungfish. *Comp. Biochem. Physiol.* **11**, 105–117.

——— (1967). Interference of metyrapone with the actions of cortisol in *Xenopus laevis* Daudin and the laboratory rat. *Gen. comp. Endocr.* **8**, 94–100.

Janssens, P. A., Vinson, G. P., Chester Jones, I. and Mosley, W. (1965). Amphibian characteristics of the adrenal cortex of the African lungfish (*Protopterus sp.*) *J. Endocr.* **32**, 373–382.

Jensen, E. V. and DeSombre, E. R. (1972). Estrogens and progestins. In *Biochemical Actions of Hormones* (edited by G. Litwack), vol. II pp. 215–255. New York and London: Academic Press.

——— (1973). Estrogen–receptor interaction. *Science* **182**, 126–134.

Joel, C. D. (1965). The physiological role of brown adipose tissue. *Handbook of Physiology*, Section 5 *Adipose Tissue*, pp. 59–85. Washington: American Physiological Society.

John, T. M. and George, J. C. (1973). Influence of glucagon and neurohypophysial hormones on plasma free fatty acid levels in the pigeon. *Comp. Biochem. Physiol.* **45A**, 541–547.

Johnston, C. I., Davis, J. O., Wright, F. S. and Howards, S. S. (1967). Effects of renin and ACTH on adrenal steroid secretion in the American bullfrog. *Amer. J. Physiol.* **213**, 393–399.

Jorgenson, C. B., Larsen, L. O. and Rosenkilde, P. (1965). Hormonal dependency of molting in amphibians: effect of radiothyroidectomy in the toad *Bufo bufo* L. *Gen. comp. Endocr.* **5**, 248–251.

Jorgenson, C. B. and Vijayakumar, S. (1970). Annual oviduct cycle and its control in the toad *Bufo bufo* L. *Gen. comp. Endocr.* **14**, 404–411.

Jorpes, E. and Mutt, V. (1966). Cholecystokinin and pancreozymin, one single molecule? *Acta physiol. scand.* **66**, 196–202.

Joss, J. M. P. (1973). Pineal–gonad relationships in the lamprey *Lampetra fluviatilus. Gen. comp. Endocr.* **21**, 118–122.

Jost, A. (1971). Hormones in development; past and present prospects. In *Hormones and Development* (edited by M. Hamburgh and E. J. W. Barrington), pp. 1–18. New York: Appleton–Century–Crofts.

Kamiya, M. (1972). Sodium–potassium-activated adenosinetriphatase in isolated chloride cells from eel gills. *Comp. Biochem. Physiol.* **43B**, 611–617.

Keller, N., Richardson, U. I. and Yates, F. E. (1969). Protein binding and the biological activity of corticosteroids: *in vivo* induction of hepatic and pancreatic alanine amino-transferases by corticosteroids in normal and estrogen-treated rats. *Endocrinology* **84**, 49–62.

Kenny, A. D. (1971). Determination of calcitonin in plasma by bioassay. *Endocrinology* **89**, 1005–1013.

Kenny, A. D. and Dacke, C. G. (1974). The hypercalcaemic response to parathyroid hormone in Japanese quail. *J. Endocr.* **62**, 15–23.

Kerkof, P. R., Boschwitz, D. and Gorbman, A. (1973). The response of hagfish thyroid tissue to thyroid inhibitors and to mammalian thyroid-stimulating hormone. *Gen. comp. Endocr.* **21**, 231–240.

King, J. R. and Farner, D. S. (1965). Studies of fat deposition in migratory birds. *Ann. N.Y. Acad. Sci.* **131**, 422–440.

Kirshner, N. and Viveros, O. H. (1972). The secretory cycle of the adrenal medulla. *Pharm. Revs.* **24**, 385–398.

Kleiber, M. (1961). *The Fire of Life. An Introduction to Animal Energetics*, pp. 312. New York: Wiley.

Klicka, J. and Mahmoud, I. Y. (1972). Conversion of pregenolone-4^{14}C to progesterone-4^{14}C by turtle corpus luteum. *Gen. comp. Endocr.* **19**, 367–369.

Krieger, D. T. (1971). The hypothalamus and neuroendocrinology. *Hospital Practice* September, 87–99.

(1972). Circadian corticosteroid periodicity: critical period for abolition by neonatal injection of corticosteroid. *Science* **178**, 1205–1207.

Krishna, G., Hynie, S. and Brodie, B. B. (1968). Effects of thyroid hormones on adenyl cyclase in adipose tissue and on free fatty acid mobilization. *Proc. natn. Acad. Sci., USA* **59**, 884–889.

Krishnamurthy, V. G. and Bern, H. A. (1969). Correlative histological study of the corpuscles of Stannius and the juxtaglomerular cells of teleost fishes. *Gen. comp. Endocr.* **13**, 313–335.

Kumar, M. A. and Sturtridge, W. C. (1973). The physiological role of calcitonin assessed through chronic calcitonin deficiency in rats. *J. Physiol., Lond.* **233**, 33–43.

Lam, T. J, (1972). Prolactin and hydromineral metabolism in fishes. *Gen. comp. Endocr. Suppl.* **3**, 328–338.

LaPointe, J. L. and Jacobson, E. R. (1974). Hyperglycemic effect of neuro-hypophysial hormones in the lizard, *Klauberina riversiana*. *Gen. comp. Endocrin.* **22**, 135–136.

Larsen, L. O. (1965). Effects of hypophysectomy in the cyclostome, *Lampetra fluviatilis* (L) Gray. *Gen. comp. Endocr.* **5**, 16–30.

(1969). Effects of gonadectomy in the cyclostome, *Lampetra fluviatilis*. *Gen. comp. Endocr.* **13**, 516–517.

(1973). Development in adult, freshwater river lampreys and its hormonal control, Thesis: University of Copenhagen.

Larsen, L. O. and Rosenkilde, P. (1971). Iodine metabolism in normal, hypophysectomized, and thyrotropin-treated river lampreys, *Lampetra fluviatilis* (Gray) L. (Cyclostomata). *Gen. comp. Endocr.* **17**, 94–104.

Larsson, A. L. (1973). Metabolic effects of epinephrine and norepinephrine in the eel *Anguilla anguilla* L. *Gen. comp. Endocr.* **20**, 155–167.

Larsson, A. and Lewander, K. (1972). Effects of glucagon administration to eels (*Anguilla anguilla* L.). *Comp. Biochem. Physiol.* **43A**, 831–836.

Lawson, D. E. M., Fraser, D. R., Kodicek, E., Morris, H. R. and Williams, D. H. (1971). Identification of 1,25-dihydroxycholecalciferol, a new kidney hormone controlling calcium metabolism. *Nature, Lond.* **230**, 228–230.

LeBrie, S. J. (1972). Endocrines and water and electrolyte balance in reptiles. *Fedn Proc.* **31**, 1599–1608.

Lederis, K. (1973). Current studies on urotensin. *Amer. Zool.* **13**, 771–773.

Lee, A. K. and Mercer, E. H. (1967). Cocoon surrounding desert-dwelling frogs. *Science* **157**, 87–88.

Lee, T. H., Lee, M. S. and Lu, M-Y. (1972). Effects of α-MSH on melanogenesis and tyrosinase in B-16 melanoma. *Endocrinology* **91**, 1180–1188.

LeFevre, M. D. (1973). Effects of aldosterone on the isolated substrate-depleted turtle bladder. *Amer. J. Physiol.* **225**, 1252–1256.

Leibson, L. and Plisetskaya, E. M. (1968). Effect of insulin on blood sugar level and glycogen content in organs of some cyclostomes and fish. *Gen. comp. Endocr.* **11**, 381–392.

Leloup, J. and Fontaine, M. (1960). Iodine metabolism in lower vertebrates. *Ann. N. Y. Acad. Sci.* **86**, 316–353.

Leloup-Hatey, J. (1970). Influence de l'ablation des corpuscules de Stannius sur le fonctionnment de l'interrenal de l'anguille (*Anguilla anguilla* L.). *Gen. comp. Endocr.* **15**, 388–397.

Lemon, M. (1972). Peripheral plasma progesterone during pregnancy and the oestrous cycle in the tammar wallaby, *Macropus eugenii*. *J. Endocr.* **55**, 63–71.

Lerner, A. B. and McGuire, J. S. (1961). Effect of alpha- and beta-melanocyte stimulating hormone on the skin colour of man. *Nature, Lond.* **189**, 176–179.

Li, C. H. (1969). Recent studies on the chemistry of human growth hormone. *Colloque du C.N.R.S.* **177**, 175–179.

—— (1972). Recent knowledge of the chemistry of lactogenic hormones. In *Lactogenic Hormones* (edited by G. E. W. Wolstenholme and J. Knight), pp. 7–22. Edinburgh and London: Churchill.

Li, C. H., Barnafi, L., Chrétien, M. and Chung, D. (1965). Isolation and amino-acid sequence of β-LPH from sheep pituitary glands. *Nature, Lond.* **208**, 1093–1094.

Licht, P. (1970). Effects of mammalian gonadotropins (ovine FSH and LH) in female lizards. *Gen. comp. Endocr.* **14**, 98–106.

—— (1972). Environmental physiology of reptilian breeding cycles: role of temperature. *Gen. comp. Endocr. Suppl.* **3**, 477–487.

—— (1973). Induction of spermiation in anurans by mammalian pituitary gonadotrophins and their subunits. *Gen. comp. Endocr.* **20**, 522–529.

Licht, P. and Hoyer, H. (1968). Somatotropic effects of exogenous prolactin and growth hormone in juvenile lizards (*Lacerta s. sicula*). *Gen. comp. Endocr.* **11**, 338–346.

Licht, P. and Papkoff, H. (1974*a*). Separation of two distinct gonadotropins from the pituitary gland of the snapping turtle (*Chelydra serpentina*). *Gen. comp. Endocr.* **22**, 218–237.

—— (1974*b*). Separation of two distinct gonadotropins from the pituitary gland of the bullfrog, *Rana catesbeiana*. *Endocrinology* **94**, 1587–1594.

Licht, P. and Stockell Hartree, A. (1971). Actions of mammalian, avian and piscine gonadotrophins in the lizard. *J. Endocr.* **49**, 113–124.

Lillie, F. R. (1917). The free-martin; a study of the actions of sex hormones in the foetal life of cattle. *J. exp. Biol.* **23**, 371–452.

Ling, J. K. (1972). Adaptive functions of vertebrate molting cycles. *Amer. Zool.* **12**, 77–93.

Lockett, M. F. and Nail, B. (1965). A comparative study of the renal actions of growth and lactogenic hormones in rats. *J. Physiol. Lond.* **180**, 147–156.

Lofts, B. (1964). Seasonal changes in the functional activity of the interstitial and spermatogenetic tissues of the green frog *Rana esculenta. Gen. comp. Endocr.* **4**, 550–562.

(1968). Patterns of testicular activity. In *Perspectives in Endocrinology. Hormones in the Lives of Lower Vertebrates* (edited by E. J. W. Barrington and C. B. Jorgenson), pp. 239–304. London and New York: Academic Press.

(1969). Seasonal cycles in reptilian testes. *Gen. comp. Endocr. Suppl.* **2**, 147–155.

Lofts, B. and Bern, H. A. (1972). The functional morphology of steroidogenic tissues. In *Steriods in Nonmammalian Vertebrates* (edited by D. R. Idler), pp. 37–125. New York and London: Academic Press.

Lofts, B., Follett, B. K. and Murton, R. K. (1970). Temporal changes in the pituitary gonadal axis. *Mem. Soc. Endocr.* **18**, 545–575.

Lofts, B., Murton, R. K. and Thearle, R. J. P. (1973). The effects of testosterone propionate and gonadotropins on the bill pigmentation and testes of the House sparrow (*Passer domesticus*). *Gen. comp. Endocr.* **21**, 202–209.

Lofts, B., Phillips, J. G. and Tam, W. H. (1971). Seasonal changes in the histology of the adrenal gland of the cobra, *Naja naja. Gen. comp. Endocr.* **16**, 121–131.

Lowry, P. J. and Chadwick, A. (1970). Purification and amino acid sequence of melanocyte stimulating hormone from the dogfish *Squalus acanthias. Biochem. J.* **118**, 713–718.

Luckey, T. D. (1973). Perspective of thymic hormones. In *Thymic Hormones* (edited by T. D. Luckey), pp. 275–314. Baltimore, London, Tokyo: University Park Press.

Lutherer, L. O., Fregly, M. J. and Anton, A. H. (1969). An interrelationship between theophylline and catecholamines in the hypothyroid rat acutely exposed to cold. *Fedn Proc.* **28**, 1238–1242.

Lyons, W. R. (1958). Hormonal synergism in mammary growth. *Proc. Roy. Soc. Lond. B.* **149**, 303–325.

MacLeod, R. M. and Lehmeyer, J. E. (1974). Studies on the mechanism of the dopamine-mediated inhibition of prolactin secretion. *Endocrinology* **94**, 1077–1085.

Maderson, P. F. A., Chiu, K. W. and Phillips, J. G. (1970). Endocrine–epidermal relationships in squamate reptiles. *Mem. Soc. Endocr.* **18**, 259–284.

Maderson, P. F. A. and Licht, P. (1967). Epidermal morphology and sloughing frequency in normal and prolactin treated *Anolis carolinensis* (Iquanidae, Lacertilia). *J. Morphol.* **123**, 157–172.

Maetz, J. (1969). Observations on the role of the pituitary–interrenal axis in the ion regulation of the eel and other teleosts. *Gen. comp. Endocr. Suppl.* **2**, 299–316.

(1971). Fish gills: mechanisms of salt transfer in fresh water and sea water. *Phil. Trans. Roy. Soc. Lond. B.* **262**, 209–249.

Maetz, J. and Rankin, J. C. (1969). Quelques aspects du rôle biologique des hormones neurohypophysaires chez les poissons. *Colloques du C.N.R.S.* **177**, 45–54.

Manning, M. and Sawyer, W. H. (1970). 4-Threonine-oxytocin: a more active and specific oxytocic agent than oxytocin. *Nature, Lond.* **227**, 715–716.

Manns, J. G., Boda, J. M. and Willes, R. F. (1967). Probable role of propionate and butyrate in control of insulin secretion in sheep. *Amer. J. Physiol.* **212**, 756–764.

Maher, M. J. (1965). The role of the thyroid gland in the oxygen consumption of lizards. *Gen. comp. Endocr.* **5**, 320–325.

Marshall, F. H. A. (1956). The breeding season. In *Marshall's Physiology of Reproduction* (edited by A. S. Parkes), vol. I (Pt I), pp. 1–42. London: Longmans.

Marx, S. J., Woodward, C. J. and Aurbach, G. D. (1972). Calcitonin receptors in kidney and bone. *Science* **178**, 999–1001.

Mayer, N., Maetz, J., Chan, D. K. O., Forster, M. and Chester Jones, I. (1967). Cortisol, a sodium excreting factor in the eel (*Anguilla anguilla* L.) adapted to sea water. *Nature, Lond.* **214**, 1118–1120.

McClanahan, L. (1967). Adaptations of the spadefoot toad *Scaphiopus couchi* to desert environments. *Comp. Biochem. Physiol.* **20**, 73–99.

McMillian, J. E. and Wilkinson, R. F. (1972). The effect of pancreatic hormones on blood glucose in *Ambystoma annulatum. Copeia* 1972 (no. 4), 664–668.

McNabb, R. A. (1969). The effects of thyroxine on glycogen stores and oxygen consumption in the leopard frog, *Rana pipiens. Gen. comp. Endocr.* **12**, 276–281.

McNatty, K. P., Cashmore, M. and Young, A. (1972). Diurnal variation in plasma cortisol levels in sheep. *J. Endocr.* **54**, 361–362.

Medica, P. A., Turner, F. B. and Smith, D. D. (1973). Hormonal induction of color change in female leopard lizards *Crotaphytus wislizenii. Copeia* 1973 (no. 4), 658–661.

Meier, A. H. (1970). Thyroxin phases the circadian fattening response to prolactin. *Proc. Soc. exp. Biol. Med., New York* **133**, 1113–1116.

Meier, A. H. and Farner, D. S. (1964). A possible endocrine basis for premigratory fattening in the white-crowned sparrow, *Zonotrichia leucophrys gambelii* (Nuttall). *Gen. comp. Endocr.* **4**, 584–595.

Meier, A. H., Trobec, T. N., Joseph, M. M. and John, T. M. (1971). Temporal synergism of prolactin and adrenal steroids in the regulation of fat stores. *Proc. Soc. exp. Biol. Med., New York* **137**, 408–415.

Meier, S. and Solursh, M. (1972). The comparative effects of several mammalian growth hormones on sulfate incorporation into acid mucopolysaccharides by cultured chick embryo chondrocytes. *Endocrinology* **90**, 1447–1451.

Minick, M. C. and Chavin, W. (1973). Effects of catecholamines upon serum FFA levels in normal and diabetic goldfish, *Carassius auratus* L. *Comp. Biochem. Physiol.* **44A**, 1003–1008.

Mirsky, I. A. (1965). Effect of biologically active peptides on adipose tissue. In *Handbook of Physiology*, Section 5 *Adipose Tissue*, pp. 407–415. Washington: American Physiology Society.

Mitnick, M. and Reichlin, S. (1972). Enzymatic synthesis of thyrotropin-releasing hormone (TRH) by hypothalamic 'TRH synthetase'. *Endocrinology* **91**, 1145–1153.

Monod, J. (1966). On the mechanism of molecular interactions in the control of cellular metabolism. *Endocrinology* **78**, 412–425.

Mueller, W. J., Brubaker, R. L., Gay, C. V. and Boelkins, J. N. (1973). Mechanisms of bone resorption in laying hens. *Fedn Proc.* **32**, 1951–1954.

Meuller, W. J., Hall, K. L., Maurer, C. A. and Joshua, I. G. (1973). Plasma calcium and inorganic phosphate response of laying hens to parathyroid hormone. *Endocrinology* **92**, 853–856.

Munday, K. A. (1957). The relation of endogenous melanophore-expanding hormone to hyperglycaemia in *Xenopus laevis. J. Endocr.* **15**, 190–198.

Myant, N. B. (1971). The role of thyroid hormone in the fetal and postnatal development of mammals. In *Hormones in Development* (edited by M. Hamburgh and E. J. Barrington), pp. 465–471. New York: Appleton–Century–Crofts.

Nalbandov, A. V. (1969). Specificity of action of gonadotrophic hormones. *Colloque du C.N.R.S.* **177**, 335–342.

—— (1973). Control of luteal function in mammals. In *Handbook of Physiology*, Section 7 *Endocrinology*, vol. II *Female reproductive system* (Pt 1), pp. 153–167. Washington: American Physiological Society.

Nature (1973). Reference to newt in Siberia, **242**, 369.

Niall, H. D., Hogan, M. L., Sauer, R., Rosenblum, I. Y. and Greenwood, F. C. (1971). Sequences of pituitary and placental lactogenic and growth hormones: evolution from a primordial peptide by gene reduplication. *Proc. natn. Acad. Sci., USA* **68**, 866–869.

Nicholls, T. J., Scanes, C. G. and Follett, B. K. (1973). Plasma pituitary luteinizing hormone in Japanese quail during photoperiodically induced gonadal growth and regression. *Gen. comp. Endocr.* **21**, 84–98.

Nichols, C. W. (1973). Somatotropic effects of prolactin and growth hormone in juvenile snapping turtles (*Chelydra serpentina*). *Gen. comp. Endocr.* **21**, 219–224.

Nicoll, C. S., Bern, H. A., Dunlop, D. and Strohman, R. C. (1965). Prolactin, growth hormone, thyroxine and growth in tadpoles of *Rana catesbeiana. Amer. Zool.* **5**, 738–739.

Nicoll, C. S. and Licht, P. (1971). Evolutionary biology of prolactins and somatotropins. ii. Electrophoretic comparison of tetrapod somatotropins. *Gen. comp. Endocr.* **17**, 490–507.

Nilsson, A. (1970). Gastrointestinal hormones in the holocephalian fish *Chimaera monstrosa* (L.). *Comp. Biochem. Physiol.* **32**, 387–390.

Nishimura, H., Ogawa, M. and Sawyer, W. H. (1973). Renin–angiotensin system in primitive bony fishes and a holocephalian. *Amer. J. Physiol.* **224**, 950–956.

Nishimura, H., Oguri, M., Ogawa, M., Sokabe, H. and Imai, M. (1970). Absence of renin in kidneys of elasmobranchs and cyclostomes. *Amer. J. Physiol.* **218**, 911–915.

Nolly, H. L. and Fasciola, J. C. (1973). The specificity of the renin–angiotensinogen reaction through the phylogenetic scale. *Comp. Biochem. Physiol.* **44A**, 639–645.

Novales, R. R. (1972). Recent studies of the melanin-dispersing effect of MSH on melanophores. *Gen. comp. Endocr. Suppl.* **3**, 125–135.

—— (1973). Discussion of 'Endocrine regulation of pigmentation' by Frank S. Abbott. *Amer. Zool.* **13**, 895–897.

O'Connor, J. M. (1972). Pituitary gonadotropin release patterns in pre-spawning brook trout, *Salvelinus fontinalis*, rainbow trout *Salmo gairdneri* and leopard frogs, *Rana pipiens. Comp. Biochem. Physiol.* **43A**, 739–746.

Odum, E. P. (1965). Adipose tissue in migratory birds. In *Handbook of Physiology*, Section 5 *Adipose Tissue*, pp. 37–43. Washington: American Physiology Society.

Ogawa, M., Oguri, M., Sokabe, H. and Nishimura, H. (1972). Juxtaglomerular apparatus in vertebrates. *Gen. comp. Endocr. Suppl.* **3**, 374–380.

Ogawa, M., Yagasaki, M. and Yamazaki, J. (1973). The effect of prolactin on water influx in isolated gills of the goldfish *Carassius auratus* L. *Comp. Biochem. Physiol.* **44A**, 1177–1183.

Oguro, C. (1973). Parathyroid gland and serum calcium concentration in the giant salamander, *Megalobatrachus davidianus*. *Gen. comp. Endocr.* **21**, 565–568.

Oguro, C. and Tomisawa, A. (1972). Effects of parathyroidectomy on serum calcium concentration of the turtle *Geoclemys reevesii*. *Gen. comp. Endocr.* **19**, 587–588.

Oishi, T. and Lauber, J. K. (1974). Pineal control of photo-endocrine responses in growing Japanese quail. *Endocrinology* **94**, 1731–1734.

Oksche, A. (1965). Survey of the development and comparative morphology of the pineal organ. *Prog. Brain Res.* **10**, 3–28.

Olivereau, M. (1967). Observations sur l'hypophyse de l'anguille femelle en particulier lors de la maturation sexuelle. *Z. Zellforsch. mikrosk. Anat.* **80**, 286–306.

O'Malley, B. W. and Means, A. R. (1974). Female steroid hormones and target cells nuclei. *Science* **183**, 610–620.

Oyer, P. E., Cho, S., Peterson, J. D. and Steiner, D. F. (1971). Studies on human proinsulin. *J. biol. Chem.* **246**, 1375–1386.

Ozon, R. (1972). Androgens in fishes, amphibians, reptiles and birds. In *Steroids in Nonmammalian Vertebrates* (edited by D. R. Idler), pp. 329–389. New York and London: Academic Press.

Pang, P. K. T. (1973). Endocrine control of calcium metabolism in teleosts. *Amer. Zool.* **13**, 775–792.

Pang, P. K. T., Pang, R. K. and Sawyer, W. H. (1973). Effects of environmental calcium and replacement therapy on the killifish, *Fundulus heteroclitus*, after the surgical removal of the corpuscles of Stannius. *Endocrinology* **93**, 705–710.

(1974). Environmental calcium and the sensitivity of the killifish (*Fundulus heteroclitus*) in bioassays for the hypocalcemic response to Stannius corpuscles from killifish and cod (*Gadus morhua*). *Endocrinology* **94**, 548–555.

Pang, P. K. T. and Sawyer, W. H. (1974). Effects of prolactin on hypophysectomized mud puppies *Necturus maculosus*. *Amer. J. Physiol.* **226**, 458–462.

Papkoff, H. (1972). Subunit interrelationships among the pituitary glycoprotein hormones. *Gen. comp. Endocr. Suppl.* **3**, 609–616.

Parkes, A. S. and Marshall, A. J. (1960). The reproductive hormones in birds. In *Marshall's Physiology of Reproduction* (edited by A. S. Parkes), vol. I (Pt 2), pp. 583–706. London: Longmans.

Patent, G. J. (1970). Comparison of some hormonal effects on carbohydrate metabolism in an elasmobranch (*Squalus acanthias*) and a holocephalan (*Hydrolagus colliei*). *Gen. comp. Endocr.* **14**, 215–242.

Pavel, S., Dorcescu, M., Petrescu-Holban, R. and Ghinea, E. (1973). Biosynthesis of a vasotocin-like peptide in cell cultures from pineal glands of human fetuses. *Science* **181**, 1252–1253.

Payan, P. and Maetz, J. (1971). Balance hydrique chez les elasmobranches: arguments en faveur d'un contrôle endocrinien. *Gen. comp. Endocr.* **16**, 535–554.

Peaker, M. (1971). Avian salt glands. *Phil. Trans. Roy. Soc. Lond.* B **262**, 289–300.

Penhos, J. C. and Ramey, E. (1973). Studies on the endocrine pancreas of amphibians and reptiles. *Amer. Zool.* **12**, 667–698.

Peter, R. E. (1971). Feedback effects of thyroxine on the hypothalamus and pituitary of the goldfish, *Carassius auratus. J. Endocr.* **51**, 31–39.

Phillips, J. G. and Ensor, D. M. (1972). The significance of environmental factors in the hormone mediated changes of nasal (salt) gland activity in birds. *Gen. comp. Endocr. Suppl.* **3**, 393–404.

Pic, P., Mayer-Gostan, N. and Maetz, J. (1973). Sea-water teleosts: presence of α- and β-adrenergic receptors in the gill regulating salt extrusion and water permeability. In *Comparative Physiology* (edited by L. Bolis, K. Schmidt-Nielsen and S. H. P. Maddrell), pp. 292–322. Amsterdam: North-Holland.

Pickering, A. D. (1972). Effects of hypophysectomy on the activity of the endostyle and thyroid gland in the larval and adult river lamprey, *Lampetra fluviatilis L. Gen. comp. Endocr.* **18**, 335–343.

Pickford, G. E. and Kosto, B. (1957). Hormonal induction of melanogenesis in hypophysectomized killifish (*Fundulus heteroclitus*). *Endocrinology* **61**, 177–196.

Pickford, G. E., Pang, P. K. T., Weinstein, E., Torretti, J., Hendler, E. and Epstein, F. H. (1970). The response of the hypophysectomized Cyprinodont, *Fundulus heteroclitus*, to replacement therapy with cortisol: effects on blood serum and sodium-potassium activated adenosine triphosphatase in the gills, kidney, and intestinal mucosa. *Gen. comp. Endocr.* **14**, 524–534.

Pickford, G. E. and Phillips, J. G. (1959). Prolactin, a factor in promoting survival of hypophysectomized killifish in fresh water. *Science* **130**, 454–455.

Pictet, R. and Rutter, W. J. (1972). Development of the embryonic endocrine pancreas. In *Handbook of Physiology*, Section 7 *Endocrinology*, vol. I *Endocrine pancreas*, pp. 25–66. Washington: American Physiological Society.

Pohorecky, L. A. and Wurtman, R. J. (1971). Adrenocortical control of epineephrine synthesis. *Pharmacol. Rev.* **23**, 1–35.

Potts, J. T., Keutmann, H. T., Niall, H. D., Habener, J. F. and Tregear, G. W. (1972). Comparative biochemistry of parathyroid hormone. *Gen. comp. Endocr. Suppl.* **3**, 405–410.

Prigge, W. F. and Grande, F. (1971). Effects of glucagon, epinephrine and insulin on *in vitro* lipolysis of adipose tissue from mammals and birds. *Comp. Biochem. Physiol.* **39B**, 69–82.

Quay, W. B. (1970). Endocrine effects on the mammalian pineal. *Amer. Zool.* **10**, 237–246.

——— (1972). Integument and the environment: glandular composition, function and evolution. *Amer. Zool.* **12**, 95–108.

Quevedo, W. C. (1972). Epidermal melanin units: melanocyte-keratinocyte interactions. *Amer. Zool.* **12**, 35–41.

Rall, J. E., Robbins, J. and Lewallen, C. G. (1964). The thyroid. In *The Hormones* (edited by G. Pincus, K. V. Thimann and E. B. Astwood), vol. V, pp. 159–439. New York: Academic Press.

Ralph, C. L. (1970). Structure and alleged functions of avian pineals. *Amer. Zool.* **10**, 217–235.

Ramsey, D. H. and Bern, H. A. (1972). Stimulation by ovine prolactin of fluid transfer in everted sacs of rat small intestine. *J. Endocr.* **53**, 453–459.

Randle, P. J. and Hales, C. N. (1972). Insulin release mechanisms. In *Handbook of Physiology*, Section 7 *Endocrinology*, vol. I *Endocrine pancreas*, pp. 219–235. Washington: American Physiological Society.

Rankin, J. C. and Maetz, J. (1971). A perfused teleostean gill preparation: vascular actions of neurohypophysial hormones and catecholamines. *J. Endocr.* **51**, 621–635.

Rasquin, P. and Rosenbloom, L. (1954). Endocrine imbalance and tissue hyperplasia in teleosts maintained in darkness. *Bull. Amer. Mus. nat. Hist.* **104**, 359–420.

Redshaw, M. R. (1972). The hormonal control of the amphibian ovary. *Amer. Zool.* **12**, 289–306.

Reinbloth, R. (1970). Intersexuality in fishes. *Mem. Soc. Endocr.* **18**, 515–541.

(1972). Hormonal control of the teleost ovary. *Amer. Zool.* **12**, 307–324.

Reiter, R. J. and Sorrentino, S. (1970). Reproductive effects of the mammalian pineal. *Amer. Zool.* **10**, 247–258.

Rippel, R. H., Johnson, E. S., White, W. F., Fujino, M., Yamazaki, I. and Nakayama, R. (1973). Ovulating and LH-releasing activity of a highly potent analog of synthetic gonadotropin-releasing hormone. *Endocrinology* **93**, 1449–1452.

Robertshaw, D., Taylor, C. R. and Mazzia, L. M. (1973). Sweating in primates: role of secretion of the adrenal medulla during exercise. *Amer. J. Physiol.* **224**, 678–681.

Robertson, D. R. (1968). The ultimobranchial gland in *Rana pipiens*. IV. Hypercalcemia and glandular hypertrophy. *Z. Zellforsch. mikroskop. Anat.* **85**, 441–542.

(1969a). The ultimobranchial body of *Rana pipiens*. VIII. Effects of extirpation upon calcium distribution and bone cell types. *Gen. comp. Endocr.* **12**, 479–490.

(1969b). The ultimobranchial body in *Rana pipiens*. IX. Effects of extirpation and transplantation on urinary calcium excretion. *Endocrinology* **84**, 1174–1178.

(1971). Cytological and physiological activity of ultimobranchial gland in the premetamorphic anuran *Rana catesbeiana*. *Gen. comp. Physiol.* **16**, 329–341.

Robertson, O. H., Krupp, M. A., Thomas, S. F., Favour, C. B., Hane, S. and Wexler, B. C. (1961). Hyperadrenocorticoidism in spawning migratory and nonmigratory rainbow trout (*Salmo gairdnerii*); comparison with Pacific salmon (Genus *Oncorhynchus*). *Gen. comp. Endocr.* **1**, 473–484.

Robertson, O. H. and Wexler, B. C. (1959). Histological changes in the organs and tissues of migrating and spawning Pacific salmon (Genus *Oncorhynchus*). *Endocrinology* **66**, 222–239.

Robinson, K. W. and MacFarlane, W. V. (1957). Plasma antidiuretic activity of marsupials during exposure to heat. *Endocrinology* **60**, 679–680.

Robison, G. A., Butcher, R. W. and Sutherland, E. W. (1971). *Cyclic AMP.* New York and London: Academic Press.

(1972). The catecholamines. In *Biochemical Actions of Hormones* (edited by G. Litwack), vol. II, pp. 81–111. New York and London: Academic Press.

Roth, J. J., Jones, R. E. and Gerrard, A. M. (1973). Corpora lutea and oviposition in the lizard *Sceloporus undulatus*. *Gen. comp. Endocr.* **21**, 569–572.

Rowan, W. (1925). Relation of light to bird migration and developmental changes. *Nature, Lond.* **115**, 494–495.

Rubin, B., Engel, S. L., Drungis, A. M., Dzelzkalns, M., Grigas, E. O., Waugh, M. H. and Yiacas, E. (1969). Cholecystokinin-like activities in guinea pigs and in dogs of the *C*-terminal octapeptide (SQ 19,884) of cholecystokinin. *J. Pharmac. Sci.* **58**, 955–959.

Rudinger, J. (1968). Synthetic analogues of oxytocin: an approach to problems of hormone action. *Proc. Roy. Soc., Lond. B* **170**, 17–26.

Rust, C. C. and Meyer, R. K. (1968). Effects of pituitary autografts on hair color in the short-tailed weasel. *Gen. comp. Endocr.* **11**, 548–551.

(1969). Hair color, molt, and testis size in male, short-tailed weasels treated with melatonin. *Science* **165**, 921–922.

Sage, M. (1973). The evolution of thyroidal function in fishes. *Amer. Zool.* **13**, 899–905.

Samols, E., Tyler, J., Megyesi, C. and Marks, V. (1966). Immunochemical glucagon in human pancreas, gut, and plasma. *Lancet* **ii**, 727–729.

Sandor, T. (1969). A comparative survey of steroids and steroidogenic pathways throughout the vertebrates. *Gen. comp. Endocr. Suppl.* **2**, 284–298.

Sandor, T. and Idler, D. R. (1972). Steroid methodology. In *Steroids in Non-mammalian Vertebrates* (edited by D. R. Idler), pp. 6–36. New York and London: Academic Press.

Sassin, J. F., Frantz, A. G., Weiztman, E. D. and Kapen, S. (1972). Human prolactin: 24-hour pattern with increased release during sleep. *Science* **177**, 1205–1207.

Sawyer, W. H. (1972a). Lungfishes and amphibians: endocrine adaptation and the transition from aquatic to terrestrial life. *Fedn Proc.* **31**, 1609–1614.

(1972b). Neurohypophysial hormones and water and sodium excretion in African lungfish. *Gen. comp. Endocr. Suppl.* **3**, 345–349.

Sawyer, W. H. and Pickford, G. E. (1963). Neurohypophyseal principles of *Fundulus heteroclitus*: characteristics and seasonal changes. *Gen. comp. Endocr.* **3**, 439–445.

Scanes, C. G., Dobson, S., Follett, B. K. and Dodd, J. M. (1972). Gonadotrophic activity in the pituitary gland of the dogfish (*Scyliorhinus canicula*). *J. Endocr.* **54**, 343–344.

Scanes, C. G., Follett, B. K. and Goos, H. J. Th. (1972). Cross-reaction in a chicken LH radioimmunoassay with plasma and pituitary extracts from various species. *Gen. comp. Endocr.* **19**, 596–600.

Schally, A. V., Arimura, A. and Kastin, A. J. (1973). Hypothalamic regulatory hormones. *Science* **179**, 341–350.

Schreibman, M. P. and Kallman, K. D. (1969). The effect of hypophysectomy on freshwater survival in teleosts of the order Antheriniformes. *Gen. comp. Endocr.* **13**, 27–38.

Schwartz, N. B. (1973). Mechanisms controlling ovulation in small mammals. In *Handbook of Physiology*, Section 7 *Endocrinology*, vol. II *Female reproductive system* (Pt 1), pp. 125–141. Washington: American Physiology Society.

Seal, U. S. and Doe, R. P. (1963). Corticosteroid-binding globulin: species distribution and small-scale purification. *Endocrinology* **73**, 371–376.

Sellers, E. A., Flattery, K. V. and Steiner, G. (1974). Cold acclimation in hypothyroid rats. *Amer. J. Physiol.* **226**, 290–294.

Senior, B. E. and Cunningham, F. J. (1974). Oestradiol and luteinizing hormone during the ovulatory cycle of the hen. *J. Endocr.* **60**, 201–202.

Shafrir, E. and Wertheimer, E. (1965). Comparative physiology of adipose tissue in different sites and in different species. In *Handbook of Physiology*, Section 5 *Adipose Tissue*, pp. 417–429. Washington: American Physiological Society.

Shapiro, M., Nicholson, W. E., Orth, D. N., Mitchel, W. M., Island, D. P. and Liddle, G. W. (1972). Preliminary characterization of the pituitary melanocyte stimulating hormones of several vertebrate species. *Endocrinology* **90**, 249–256.

Sharman, G. B. (1970). Reproductive physiology of marsupials. *Science* **167**, 1221–1228.

Shire, J. G. M. (1970). Genetic variation in adrenal structure: quantitative measurements on the cortex and medulla in hybrid mice. *J. Endocr.* **48**, 419–431.

Shoemaker, V. H., Nagy, K. A. and Bradshaw, S. D. (1972). Studies on the control of electrolyte excretion by the nasal gland of the lizard *Dipsosaurus dorsalis*. *Comp. Biochem. Physiol.* **42A**, 749–757.

Simpson, P. A. and Blair-West, J. R. (1972). Estimation of marsupial renin using marsupial renin-substrate. *J. Endocr.* **53**, 125–130.

Skadhauge, E. (1969). Activités biologiques des hormones neurohypophysaires chez les oiseaux et les reptiles. *Colloque du C.N.R.S.* **177**, 63–68.

Smith, H. W. (1930). Metabolism of the lungfish, *Protopterus aethiopicus*. *J. biol. Chem.* **88**, 97–130.

Smith, L. F. (1966). Species variation in the amino acid sequence of insulin. *Amer. J. Med.* **40**, 662–666.

Smith, P. M. and Follett, B. K. (1972). Luteinizing hormone releasing factor in the quail hypothalamus. *J. Endocr.* **53**, 131–138.

Sokabe, H. and Nakajima, T. (1972). Chemical structure and role of angiotensins in the vertebrates. *Gen. comp. Endocr. Suppl.* **3**, 382–392.

Sokabe, H., Nishimura, H., Ogawa, M. and Oguri, M. (1970). Determination of renin in the corpuscles of Stannius of the teleost. *Gen. comp. Endocr.* **14**, 510–516.

Sokabe, H., Nishimura, H., Kawabe, K., Tenmoku, S. and Arai, T. (1972). Plasma renin activity in varying hydrated states in the bullfrog. *Amer. J. Physiol.* **222**, 142–146.

Sokabe, H., Oide, H., Ogawa, M. and Utida, S. (1973). Plasma renin activity in Japanese eels (*Anguilla japonica*) adapted to sea-water or in dehydration. *Gen. comp. Endocr.* **21**, 160–167.

Spallanzani (1784). *Dissertations Relative to the Natural History of Animals and Vegetables* 2. Trans. from the Italian, London. Quoted by F. H. A. Marshall, 1956.

Speers, G. M., Perey, D. Y. E. and Brown, D. M. (1970). Effect of ultimobranchialectomy in the laying hen. *Endocrinology* **87**, 1292–1297.

Srivastava, A. K. and Meier, A. H. (1972). Daily variation in concentration of cortisol in plasma in intact and hypophysectomized gulf killifish. *Science* **177**, 185–187.

Stannius, H. (1839). Die Nebennieren bei Knochenfischen. *Arch. Anat. Physiol.* **97**, 97–101.

Steiner, D. F., Kemmler, W., Clark, J. L., Oyer, P. E. and Rubinstein, A. H. (1972). The biosynthesis of insulin. In *Handbook of Physiology*, Section 7

Endocrinology, vol. I *Endocrine pancreas*, pp. 175–198. Washington: American Physiological Society.

Sterling, K., Brenner, M. A. and Saldanha, V. F. (1973). Conversion of thyroxine to triiodothyronine by cultured human cells. *Science* **179**, 1000–1001.

Stetson, M. H. and Erickson, J. E. (1972). Hormonal control of photoperiodically induced fat deposition in white-crowned sparrows. *Gen. comp. Endocr.* **19**, 355–362.

Stewart, A. D. (1968). Genetic variation in the neurohypophysial hormones of the mouse. *J. Endocr.* **41**, xix–xx.

—— (1972). Genetic determination of the storage of vasopressin and oxytocin in neural lobes of mice. *J. Physiol., Lond.* **222**, 157P.

—— (1973). Sensitivity of mice to (8-arginine)- and (8-lysine)- vasopressins as antidiuretic hormones. *J. Endocr.* **59**, 195–196.

Stewart, J., Fraser, R., Papaioannou, V. and Tait, A. (1972). Aldosterone production and the zona glomerulosa: a genetic study. *Endocrinology* **90**, 968–972.

Strauss, J. S. and Ebling, F. J. (1970). Control and function of skin glands in mammals. *Mem. Soc. Endocr.* **18**, 341–368.

Sutherland, E. W. (1972). Studies on the mechanism of hormone action. *Science* **177**, 401–408.

Suzuki, S. and Kondo, Y. (1973). Thyroidal morphogenesis and biosynthesis of thyroglobulin before and after metamorphosis in the lamprey, *Lampetra reissneri*. *Gen. comp. Endocr.* **21**, 451–460.

Swift, D. R. and Pickford, G. E. (1965). Seasonal variations in the hormone content of the pituitary gland of the perch *Perca fluviatilis* L. *Gen. comp. Endocr.* **5**, 354–365.

Swaminathan, R., Bates, R. F. L. and Care, A. R. (1972). Fresh evidence for a physiological role for calcitonin in calcium homeostasis. *J. Endocr.* **54**, 525–526.

Swaminathan, R., Ker, J. and Care, A. D. (1974). Calcitonin and intestinal calcium absorption. *J. Endocr.* **61**, 83–94.

Takasugi, N. and Bern, H. A. (1962). Experimental studies on the caudal neurosecretory system in *Tilapia mossambica*. *Comp. Biochem. Physiol.* **6**, 289–303.

Tanabe, Y., Ishii, T. and Tamaki, Y. (1969). Comparison of thyroxine-binding plasma proteins of various vertebrates and their evolutionary aspects. *Gen. comp. Endocr.* **13**, 14–21.

Tanaka, Y., Frank, H. and DeLuca, H. F. (1973). Intestinal calcium transport: stimulation by low phosphorus diets. *Science* **181**, 564–566.

Tanner, J. M. (1972). Human growth hormone. *Nature, Lond.* **237**, 433–439.

Tashjian, A. H., Levine, L. and Wilhelmi, A. E. (1965). Immunochemical relatedness of porcine, bovine, ovine and primate growth hormones. *Endocrinology* **77**, 563–573.

Taylor, J. D. and Bagnara, J. T. (1972). Dermal chromatophores. *Amer. Zool.* **12**, 43–62.

Taylor, R. E., Tu, T. and Barker, S. B. (1967). Thyroxine-like actions of 3′-*iso*propyl-3′,5′-dibromo-L-thyronine, a potent iodine-free analog. *Endocrinology* **80**, 1143–1147.

Temple, S. A. (1974). Plasma testosterone titers during the annual reproductive cycle of starlings (*Sturnus vulgaris*). *Gen. comp. Endocr.* **22**, 470–479.

Tepperman, J. and Tepperman, H. M. (1970). Gluconeogenesis, lipogenesis and the Sherringtonian metaphor. *Fedn Proc.* **29**, 1284–1293.

Tewary, P. D. and Farner, D. S. (1973). Effect of castration and estrogen administration on the plumage pigment of the male House Finch (*Carpdacus mexicanus*). *Amer. Zool.* **13**, 1278.

Thody, A. J. and Plummer, N. A. (1973). A radioimmunoassay for β-melanocyte stimulating hormone in human plasma. *J. Endocr.* **58**, 263–273.

Thornton, V. F. (1972). A progesterone-like factor detected by bioassay in the blood of the toad (*Bufo bufo*) shortly before induced ovulation. *Gen. comp. Endocr.* **18**, 133–139.

Torresani, J., Gorbman, A., Lachiver, F. and Lissitzky, S. (1973). Immunological cross-reactivity between thyroglobulins of mammals and reptiles. *Gen. comp. Endocr.* **21**, 530–535.

Torrey, T. W. (1971). *Morphogenesis of the Vertebrates* (3rd edition), pp. 44–45. New York, London, Sydney, Toronto: John Wiley.

Tracy, H. J. and Gregory, R. A. (1964). Physiological properties of a series of synthetic peptides structurally related to gastrin 1. *Nature, Lond.* **204**, 935–938.

Tregear, G. W., Rietschoten, J. V., Greene, E., Keutmann, H. T., Niall, H. D., Reit, B., Parsons, J. A. and Potts, J. T. (1973). Bovine parathyroid hormone: minimum chain length of synthetic peptide required for biological activity. *Endocrinology* **93**, 1349–1353.

Turkington, R. W. (1972). Multiple hormonal interactions. The mammary gland. In *Biochemical Actions of Hormones* (edited by G. Litwack), vol. II, pp. 55–80. New York and London: Academic Press.

Tyndale-Biscoe, H. (1973). *Life of Marsupials.* New York: Elsevier.

Urasaki, H. (1972). Effects of restricted photoperiod and melatonin administration on gonadal weight in the Japanese killifish. *J. Endocr.* **55**, 619–620.

Urist, M. R. (1962). The bone-body fluid continuum: calcium and phosphorus in the skeleton and blood of extinct and living vertebrates. *Perspectus Biol. Med.* **6**, 75–115.

 (1963). The regulation of calcium and other ions in the serums of hagfish and lampreys. *Proc. N. Y. Acad. Sci.* **109**, 294–311.

 (1973). Testosterone-induced development of limb gills of the lungfish, *Lepidosiren paradoxa. Comp. Biochem. Physiol.* **44A**, 131–135.

Urist, M. R. and Scheide, A. O. (1961). Partition of calcium and proteins in the blood of oviparous vertebrates during estrus. *J. gen. Physiol.* **44**, 743–756.

Urist, M. R., Uyeno, S., King, E., Okada, M., and Applegate, S. (1972). Calcium and phosphorus in the skeleton and blood of the lungfish, *Lepidosiren paradoxa*, with comment on humoral factors in calcium homeostasis in the Osteichthyes. *Comp. Biochem. Physiol.* **42A**, 393–408.

Utida, S., Hirano, T., Oide, H., Ando, M., Johnson, D. W. and Bern, H. A. (1972). Hormonal control of the intestine and urinary bladder in teleost osmoregulation. *Gen. comp. Endocr. Suppl.* **3**, 317–327.

Uttenthal, L. O. and Hope, D. B. (1972). Neurophysins and posterior pituitary hormones in the Suiformes. *Proc. Roy. Soc. Lond. B* **182**, 73–87.

van Tienhoven, A. and Planck, R. J. (1973). The effect of light on avian reproductive activity. In *Handbook of Physiology*, Section 7 *Endocrinology*, vol. II, *Female reproductive tract* (Pt 1), pp. 79–107. Washington: American Physiological Society.

van Tienhoven, A. and Schally, A. V. (1972). Mammalian luteinizing hormone-releasing hormone induces ovulation in the domestic fowl. *Gen. comp. Endocr.* **19**, 594–595.

Valtin, H., Sawyer, W. H. and Sokol, H. W. (1965). Neurohypophysial principles in rats homozygous and heterozygous for hypothalamic diabetes insipidus (Brattleboro strain). *Endocrinology* **77**, 701–706.

Vander, A. J. (1967). Control of renin release. *Physiol. Rev.* **47**, 359–382.

Vizsolyi, E. and Perks, A. M. (1969). New neurohypophysial principle in foetal mammals. *Nature, Lond.* **223**, 1169–1171.

Waldo, C. M. and Wislocki, G. B. (1951). Observations on the shedding of the antlers of the virginia deer (*Odocoileus virginianus borealis*). *Amer. J. Anat.* **88**, 351–395.

Waring, H. (1936). Colour changes in the dogfish (*Scyllium canicula*). *Proc. Liverpool Biol. Soc.* **49**, 17–64.

(1938). Chromatic behaviour of elasmobranchs. *Proc. Roy. Soc. Lond. B* **125**, 264–282.

(1942). The co-ordination of vertebrate melanophore responses. *Biol. Rev.* **17**, 120–150.

(1963). *Color Change Mechanisms in Cold-blooded Vertebrates.* London and New York: Academic Press.

Weinstein, B. (1968). On the relationship between glucagon and secretin. *Experientia* **24**, 406–408.

Weisbart, M. and Idler, D. R. (1970). Re-examination of the presence of cortico-steroids in two cyclostomes, the Atlantic hagfish (*Myxine glutinosa* L.) and the sea lamprey (*Petromyzon marinus* L.). *J. Endocr.* **46**, 29–43.

Weiss, M. and McDonald, I. R. (1965). Corticosteroid secretion in the monotreme *Tachyglossus aculeatus*. *J. Endocr.* **33**, 203–210.

Wenberg, G. M. and Holland, J. C. (1973). The circannual variations of some of the hormones of the woodchuck (*Marmota monax*). *Comp. Biochem. Physiol.* **46A**, 523–535.

West, G. B. (1955). The comparative pharmacology of the suprarenal medulla. *Quart. Rev. Biol.* **30**, 116–137.

Wilson, J. F. and Dodd, J. M. (1973a). The role of the pineal complex and lateral eyes in the colour change response of the dogfish, *Scyliorhinus canicula* L. *J. Endocr.* **58**, 591–598.

(1973b). The role of melonophore-stimulating hormone in melanogenesis in the dogfish, *Scyliorhinus canicula* L. *J. Endocr.* **58**, 685–686.

Wingstrand, K. G. (1951). *The Structure and Development of the Avian Pituitary.* C. W. K. Gleerup: Lund.

(1966). Comparative anatomy and evolution of the hypophysis. In *The Pituitary Gland* (edited by G. W. Harris and B. T. Donovan), vol. I, pp. 58–146. Berkeley and Los Angeles: University of California.

Wise, J. K., Hendler, R. and Felig, P. (1972). Obesity: evidence of decreased secretion of glucagon. *Science* **178**, 513–514.

Woodhead, P. M. J. (1969). Effect of oestradiol and thyroxine upon the plasma calcium content of a shark, *Scyliorhinus canicula*. *Gen. comp. Endocr.* **13**, 310–312.

Woolley, P. (1957). Colour change in a chelonian. *Nature, Lond.* **179**, 1255–1256.

Wright, A., Chester Jones, I. and Phillips, J. G. (1957). The histology of the adrenal gland of prototheria. *J. Endocr.* **15**, 100–107.

Wurtman, R. J. and Axelrod, J. (1966). A 24-hr rhythm in the content of nore-pinephrine in the pineal and salivary glands of the rat. *Life Sciences* **5**, 665–669.

Wurtman, R. J., Axelrod, J. and Kelly, D. E. (1968). *The Pineal*. New York and London: Academic Press.

Xavier, F. (1974). La pseudogestation chez *Nectophyrnoïdes occidentalis* ANGEL. *Gen. comp. Endocr.* **22**, 98–115.

Xavier, F. and Ozon, R. (1971). Recherches sur l'activité endocrine de l'ovaire de *Nectophrynoïdes occidentalis* ANGEL (amphibien anoure vivipare). ii. Synthèse *in vitro* de stéroids. *Gen. comp. Endocr.* **16**, 30–40.

Yagil, R., Etzion, Z. and Berlyne, G. M. (1973). The effect of *d*-aldosterone and spironolactone on the concentration of sodium and potassium in the milk of rats. *J. Endocr.* **59**, 633–636.

Yaron, Z. (1972). Endocrine aspects of gestation in viviparous snakes. *Gen. comp. Endocr. Suppl.* **3**, 663–673.

Young, J. Z. (1935). The photoreceptors of lampreys. ii. The function of the pineal complex. *J. exp. Biol.* **12**, 254–270.

Zelnik, P. R. and Lederis, K. (1973). Chromatographic separation of urotensins. *Gen. comp. Endocr.* **20**, 392–400.

Zimmerman, E. A., Carmel, P. W., Husain, M. K., Ferin, M., Tannenbaum, M., Frantz, A. G. and Robison, A. G. (1973). Vasopressin and neurophysin: high concentrations in monkey hypophyseal portal blood. *Science* **182**, 925–927.

Zinder, O., Hamosh, M., Fleck, T. R. C. and Scow, R. O. (1974). Effect of pro-lactin on lipoprotein lipase in mammary gland and adipose tissue of rats. *Amer. J. Physiol.* **226**, 744–748.

Zipser, R. D., Licht, P. and Bern, H. A. (1969). Comparative effects of mammalian prolactin and growth hormone on growth in the toads *Bufo boreas* and *Bufo marinus*. *Gen. comp. Endocr.* **13**, 382–391.

Zuber-Vogeli, M. and Xavier, F. (1973). Les modifications cytologique de l'hypophyse distale des femelles de *Nectophrynoïdes occidentalis* Angel après ovariectomie. *Gen. comp. Endocr.* **20**, 199–213.

Index

(All hormones have been listed under their full names)

406   *Index*